Quantum Waveguide in Microcircuits

Second Edition

Quantum Waveguide in Microcircuits

Second Edition

Jian-Bai Xia
Duan-Yang Liu
Wei-Dong Sheng

JENNY STANFORD
PUBLISHING

Published by

Jenny Stanford Publishing Pte. Ltd.
101 Thomson Road
#06-01, United Square
Singapore 307591

Email: editorial@jennystanford.com
Web: www.jennystanford.com

British Library Cataloguing-in-Publication Data
A catalogue record for this book is available from the British Library.

Quantum Waveguide in Microcircuits (Second Edition)

ISBN 978-981-4968-30-0 (Hardback)
ISBN 978-1-003-65146-8 (eBook)

Contents

Introduction

Last century, the development of semiconductor microelectronic technology changed the whole world. The world entered the information society from the industrial society. The productive forces rose greatly, which promoted the development of human material and the spirit of civilization. Just because of the importance of semiconductor microelectronic technology, many governments and international companies invested heavily in developing the technology, hoping to make a break through and occupy an advanced position in the development of the whole information technology.

Integrated circuits were invented in 1958, and in subsequent years, development and progress in the degree of integration have largely followed Moore's law. Moore's law is a rule that combines technology development and economics to predict the degree of advancement in microelectronic circuit integration within a specified period. It predicts that the degree of microprocessor integration would double every 18 months in DRAM. Moore's law is still proving accurate today. However, as the sizes of circuit elements approach their physical limits, the optical method used in manufacturing 16-nm-node chips is also approaching a limit. Although the scaling of microelectronic circuit elements still follows Moore's law, the unit density of power consumption will become unacceptable. Therefore, on the one hand, people continuously develop microelectronic technology, while on the other hand, they consider the developing road after Moore's law is broken, that is, more Moore's law or more than Moore's law.

Physically, when the scale of the circuit element decreases to 10 nm or even less, the quantum effect will appear and play an increasingly important role. The electron transport becomes non-classical and non-linear, and even the electron motion likes the waveguide motion. This book consists of two parts: (i) non-classical, non-linear transport, and (ii) quantum waveguide theory.

The first part discusses the quantum correction effect in ultrasmall devices, including strong field transport and transport related to space (Chapter 2). The quantum mechanics effect is most obvious in the longitudinal transport of superlattices because the longitudinal length of the superlattice is about 10 nm, smaller than the electron mean free path. Quantum transport includes resonant tunneling (Chapter 3) and longitudinal transport of a superlattice (Chapter 4), which were observed early in the last century eighties. Due to the development of electron beam lithography in the last century nineties, people can fabricate an ultrathin metallic wire on a two-dimensional electron gas (2DEG). Applying a bias voltage on a metallic contact can form a small quantum dot in the 2DEG underneath the contact. In studying the transport of quantum dots and thin circuits, Landauer and Büttiker proposed their famous formulas named after them. This kind of transport is named mesoscopic transport (Chapter 5). People fabricated 3D quantum dots in the longitudinal direction of a quantum well by using lithography. The quantum dot is confined in the upper and lower directions by the barriers in the original quantum well, and its lateral direction is confined by vacuum due to the lithography. These kinds of quantum dots are similar to an artificial atom, in which the electrons are filled according to the shell. This characteristic is reflected in the quantum transport, for example, the Coulomb blockade (Chapter 6). Last, we introduce the applications of single-electron transport: single-electron transistor (Chapter 7) and single-electron memory (Chapter 8).

The second part studies quantum waveguide theory, mainly our own works. Since the Aharonov–Bohm effect (AB effect) was experimentally discovered by Webb et al., there have been many advances in the transport of mesoscopic systems. Electron transport in mesoscopic systems is not of the diffusing type but of the waveguide type because there are no electron collisions in such small systems. Transport of the waveguide type has many characteristics different from those of the diffusing type, and the theoretical research methods of these two types are also different. The former is based on quantum mechanics, while the latter is based on the classical statistical physics: Boltzmann equation. In application, mesoscopic systems, especially semiconductor mesoscopic systems, will be the basis of next-generation microelectronics.

This part summarizes the research results of our group in this field in the past 20 years. Chapter 9 covers the general concept of quantum transport. Chapter 10 discusses 1D quantum waveguide theory, which proposes two basic equations similar to Kirchhoff equations in electric circuits. Then the two basic equations are applied to many cases: AB rings, quantum interference devices, etc. Last, the theory is extended to the hole case, whose wave function has two components. Chapter 11 describes 2D quantum waveguide theory. When the width of the circuit is so large that the energy level spacing between the transverse modes in the circuit is comparable to the electron kinetic energy, we should consider the transport of multiple transverse modes, that is, 2D waveguide theory. In this chapter, the transfer matrix method, the scattering matrix method, and the theory of a waveguide with multiple terminals are developed. Chapter 12 discusses the 1D quantum waveguide theory of Rashba electrons. In recent years, much attention has been paid to the field of Rashba spin–orbit interaction (RSOI) in low-dimensional semiconductor structures because of its potential application in spintronic devices, which is based on the idea of the possible manipulation of electron spin by a magnetic or an electric field. Chapter 12 extends the 1D quantum waveguide theory of electrons without considering spin to the case of electrons with spin and RSOI, deriving the boundary conditions of the Rashba current. The theory is applied to study the transport of Rashba electrons in turning structures, spin-polarized devices, etc. Chapters 13 and 14 extend the 1D quantum waveguide theory of a Rashba electron in straight-line structures to curved-line structures. For this objective, the transfer matrix method is developed. With this method, the Rashba electron transport in the AB circular ring and square ring and related spin polarization modulation are studied. In Chapter 15, the 1D quantum waveguide theory of a Rashba electron is extended to the 2D case and some basic results are obtained.

The second edition contains two new chapters as follows. Chapter 16 theoretically investigates, by using the transfer matrix method, Rashba electron's spin transport in a straight waveguide with a stub which has a smooth boundary. Chapter 17 studies the spin transport of electrons in an elliptical ring with the Rashba spin–orbit interaction. It focuses on the spin flip in two kinds of rings and expects to find a suitable structure and conditions for a spin inverter.

In summary, the transport theories and experiments beyond classical transport quantum waveguide are introduced, which are prepared for future semiconductor micro- and nanoelectronics. They will be the basis of next-generation semiconductor electronics and industry. We believe that these theories will have more and more applications, popularization, and developments.

Jian-Bai Xia
Duan-Yang Liu
Wei-Dong Sheng

Part I

Non-Classical, Non-Linear Transport

Chapter 1

Properties of Quantum Transport

1.1 Characteristic Length [1]

In traditional transistors, whose length is larger than the average free path of electrons, the electron's movement is classical, which can be described by the Boltzmann equation. But when the scale of a device is so small that its scale is about or smaller than some characteristic lengths, there will be distinct quantum effects of current.

Electrons in two-dimensional electron gas (2DEG) are degenerate at low temperature, so their Fermi wave vector can be written as

$$k_{\mathrm{F}} = \sqrt{2\pi n_{\mathrm{s}}}, \tag{1.1}$$

where n_{s} is the area density of 2DEG. The electrons' Fermi wavelength is

$$\lambda_{\mathrm{F}} = \frac{2\pi}{k_{\mathrm{F}}} = \sqrt{\frac{2\pi}{n_{\mathrm{s}}}}. \tag{1.2}$$

For electrons with area density $n_{\mathrm{s}} = 5 \times 10^{11}\ \mathrm{cm}^{-2}$, the Fermi wavelength is about 35 nm, so it is obviously comparable to the scale of mesoscopic devices. At low temperature, electric current

Quantum Waveguide in Microcircuits (Second Edition)
Jian-Bai Xia, Duan-Yang Liu, and Wei-Dong Sheng

ISBN 978-981-4968-30-0 (Hardcover), 978-1-003-65146-8 (eBook)
www.jennystanford.com

is contributed mostly by electrons near the Fermi energy, and other electrons with energy less than the Fermi energy have no contribution to the conductance. Therefore, the Fermi wavelength is associated with the quantum effect of current.

Electrons which move in a semiconductor are scattered by impurities, defects, or phonons, and their momentum relaxation time τ_m and scattering time τ_s have the following relationship:

$$\frac{1}{\tau_m} = \frac{1}{\tau_s}\alpha_m, \tag{1.3}$$

where α_m is a constant between 0 and 1, and it denotes the effectiveness of different scatterings for the momentum relaxation.

The mean free path L_m is defined as the average distance travelled by an electron before it loses its momentum, so it can be written as

$$L_m = v_F\tau_m, \tag{1.4}$$

where v_F is the Fermi velocity, and for electrons with area density $n_s = 5 \times 10^{11}$ cm^{-2}. On the basis of Eq. 1.1, we obtain

$$v_F = \frac{\hbar k_F}{m^*} = \frac{\hbar}{m^*}\sqrt{2\pi n_s} = 3 \times 10^7 \text{ cm/s}, \tag{1.5}$$

where m^* is the electron's effective mass, m_0 is the static mass of an electron. If $m^* = 0.067m_0$, $\tau_m \cong 100$ ps, then it can be obtained from Eq. 1.4 that $L_m = 30$ μm.

The phase relaxation length is defined as the average distance travelled by an electron before the electron wave loses its phase. Similar to the momentum relaxation time τ_m, the phase relaxation time τ_φ has the following relationship with the scattering time τ_s:

$$\frac{1}{\tau_\varphi} = \frac{1}{\tau_c}\alpha_\varphi, \tag{1.6}$$

where α_φ is a constant between 0 and 1, and it denotes the effectiveness of different scatterings for the phase destruction.

To understand the concept of phase destruction, we imagine an experiment of AB ring. The incident electron beam splits in two at one end of the ring, then moves along up and down the two arms, then meets at the other end of the ring. If the lengths of each arm are the same, the phases of two waves at the meeting point are equal, and the amplitude increases. Assume that there is an impurity or defect (named scatterer) in one arm. Then it will scatter the electron wave passed. Because the scattering is elastic, it does not change the electron's energy, just its phase. When the two electron waves meet, their phases are unequal, so interference makes the amplitude reduced. Because the scatterer is fixed, the phase relationship between the two paths is definite.

If we apply a magnetic field perpendicular to the ring, the electron wave moving up or down the arm is added or reduced a phase factor, which is in proportion to the magnetic flux Φ through the ring (magnetic field strength multiplied by the area of the ring). Therefore the amplitude of the output electron wave oscillates with Φ, and it is the AB effect. In case there are scatterings in the arms, although the amplitude reduces when $\Phi = 0$ (the magnetic field is zero), when the strength of the field increases, the amplitude's oscillation with Φ doesn't change. When Φ reaches a certain value, the amplitude reaches its maximum. We can say that the loss of the phase at the scatterings has been compensated by the magnetic field. In this condition, we can consider

$$\alpha_{\varphi} = 0, \; \tau_{\varphi} \to \infty. \tag{1.7}$$

That is, the elastic scattering of static scatterers does not affect the phase relaxation time. This fact has been proved by the experiment. Experiments found that the length of the two arms of the AB ring is much larger than the mean free path L_{m} in general—i.e., an electron has experienced repeatedly the momentum elastic scattering, but experiments still observe the AB oscillation.

A major factor which affects the phase relaxation is the inelastic scattering of the electron-phonon. A phonon is a quantum mode of lattice vibrations, and it is not a fixed point as an impurity or a defect, so its scattering on electrons is random. As it is inelastic scattering, in every scattering the electron's energy will increase or decrease the energy of a phonon. When two electron waves scattered by phonons meet, the relationship of their phases is irregular, so the average amplitude of the electron wave reduces. In other words, fixed scatterers do not contribute to phase relaxation; only fluctuating scatterers do.

Assume that after time τ_{φ}, the mean square energy that the electron gains during phonon scattering is equal to the square of energy that the electron gains each time multiplied by the number of scatterings:

$$(\Delta\varepsilon)^2 = (\hbar\omega)^2 \left(\tau_\phi/\tau_\mathrm{S}\right), \tag{1.8}$$

where $\hbar\omega$ is the energy of a phonon. The phase relaxation time is defined as that after time τ_φ when the mean square value of an electron's phase change is of orders one:

$$\Delta\varphi \approx (\Delta\varepsilon)\,\tau_\varphi/\hbar \approx 1. \tag{1.9}$$

From Eqs. 1.8 and 1.9 we have

$$\tau_\varphi = \left(\frac{\tau_\mathrm{S}}{\omega^2}\right)^{1/3}. \tag{1.10}$$

Therefore the influence of a low-frequency phonon (acoustic phonon) on the phase relaxation is small. A major factor of phase relaxation is the optical phonon.

At low temperature, a major factor of the phase relaxation is electron–electron scattering. The frequency of electron–electron scattering is determined by the difference in electron energy E and Fermi energy E_F: $\Delta = E - E_\mathrm{F}$. Because Δ is small, the states which can be scattered are very few according to the Pauli exclusion principle. So the probability of scattering goes to zero. In 2DEG, it has been proved that

$$\frac{\hbar}{\tau_\varphi} \approx \frac{\Delta^2}{E_\mathrm{F}}\left[\ln\left(\frac{E_\mathrm{F}}{\Delta}\right) + \mathrm{const}\right]. \tag{1.11}$$

Because the hot electron's average energy $\Delta \approx k_\mathrm{B}T$, the relationship between τ_φ and T is the formula that Δ is replaced by $k_\mathrm{B}T$ in Eq. 1.11.

In high-mobility semiconductors, $\tau_\varphi \le \tau_\mathrm{S}$ in general. In low-mobility semiconductors, τ_φ changes less while τ_S decreases greatly, so $\tau_\varphi \gg \tau_\mathrm{S}$.

1.2 Non-equilibrium Transport

When the scale of a device is smaller than 100 nm, a few volts in the applied voltage can lead to very high electric fields of the order of 10,000 V/cm. These high fields lead to values of the carrier drift velocity of the order of 10^7 cm/s. Because there is not enough time for these electrons to exchange energy with the surroundings, their average energy is higher than the thermal average value $k_B T$, so that they are called "hot electrons."

When the electric field strength is small, the drift velocity of electrons is proportional to the electron field strength,

$$v_d = \mu E, \tag{1.12}$$

where the proportional coefficient μ is called mobility. The regime where Eq. 1.12 is valid is called the linear-response regime. In this regime, the Einstein relation expresses the diffusion in terms of the mobility and thermal equilibrium temperature as

$$D = \frac{\mu k_B T}{e}, \tag{1.13}$$

where e is the electron charge taken with its sign. The Nyquist relation expresses the available noise power P_{av} of a two-terminal network for unit bandwidth of frequency Δf in terms of the thermal equilibrium temperature as

$$\frac{P_{av}}{\Delta f} = k_B T. \tag{1.14}$$

Therefore, under the thermal equilibrium we have

$$\frac{P_{av}}{\Delta f} = k_B T = \frac{eD}{\mu}. \tag{1.15}$$

Equation 1.15 is just a macroscopic expression of the fluctuation-dissipation theorem.

Under hot-electron conditions, the Einstein and Nyquist relations no longer hold in general. Figure 1.1 schematically shows the drift velocity, diffuse coefficient, and white noise as functions of the electric field in the linear-response and hot-electron regimes [2].

The experimental evidences of the main effects related to the hot-electron conditions are as follows [2], as shown in Fig. 1.1:

(1) Deviation from Ohm's law through a nonlinear dependence of the drift velocity upon electric field strength
(2) Deviation from the Einstein relation of the diffusion coefficient through a diffusivity which depends on the electric field strength

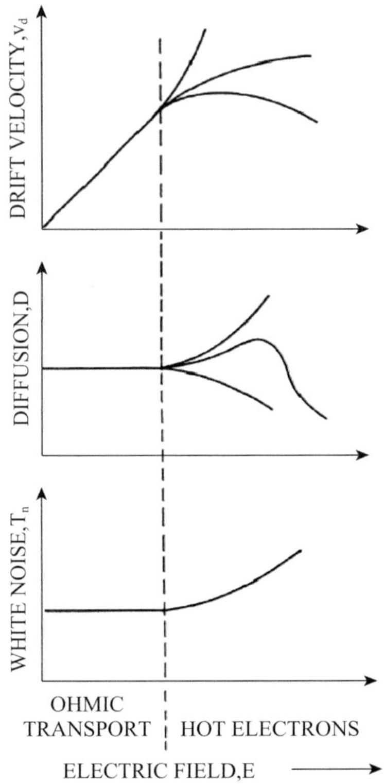

Figure 1.1 Schematic illustration of the drift velocity, diffusion coefficient, and white noise as functions of electric field strength.

(3) Deviation from the Nyquist relation of the noise power per unit bandwidth through an increase of the white-noise temperature T_n

(4) Anisotropy with respect to the direction of the applied electric field of the kinetic coefficients: differential mobility μ', diffusion coefficient D, white-noise temperature T_n

(5) Negative differential mobility for electric field strengths above a threshold value

In the linear-response regime, the transport behavior can be described by the Boltzmann equation, and it can be solved by the perturbation method. In the hot-electron regime, the transport is nonstatic, the distribution function deviates far from the balance distribution function, and its variation with the time is large. It is believed that if the concepts of the Boltzmann equation and the distribution function are still valid, three main assumptions will justify such an approach: (i) Carrier density should be sufficiently low so that only binary collisions occur. (ii) The time between successive collisions, τ, should be long enough when compared with the duration of a collision, τ_{coll}—i.e., $\tau_{\text{coll}}/\tau \ll 1$. (iii) Density gradients should be small over the range of the interparticle potential. In this case the perturbation method cannot be used to solve the Boltzmann equation, but the Monte Carlo method can be.

1.3 Quantum Effect

The hot-electron effect is a nonlinear transport behavior in ultrasmall devices, but the electron movement is still classical. If the scale of the devices becomes smaller and smaller, a series of quantum effects will appears [3].

1.3.1 Statistical Thermodynamics

Generally, modeling of quantum phenomena is more complicated than modeling of classical and/or semiclassical phenomena. For instance, one should consider the full nonlocal nature of the potential interactions in the dynamical variables. Consider a simple potential barrier system, where the potential distribution is

$$\begin{aligned} V(x) &= V_0 u(-x), \\ u(x) &= \begin{cases} 1, & x \geq 0 \\ 0, & x < 0 \end{cases} \end{aligned} \tag{1.16}$$

where $u(x)$ is the Heavyside step function. For the case $V_0 \to \infty$, the Wigner distribution as a function of space coordinate x and momentum k obtained quantum mechanically is shown in Fig. 1.2 [4]. It may be seen that far from the barrier, the distribution approaches the classical Maxwellian form, but near the barrier it differs greatly. The repulsion from the barrier is required by the vanishing of the wave function at the barrier, but the first peak of

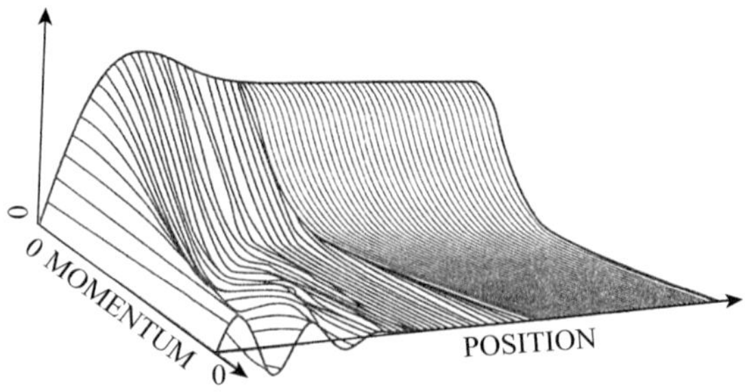

Figure 1.2 The Wigner distribution function as a function of space coordinate x and momentum k for a half infinite barrier.

the wave function, as one moves away from the barrier, occurs closer to the barrier for higher-momentum states. This variation exists for distances of the order of several thermal de Broglie wavelengths,

$$\lambda_D = \left(\frac{\hbar^2}{2m^* k_B T} \right)^{1/2} . \tag{1.17}$$

Thus, nonlocal variation can be expected over the range of 20–40 nm even at room temperature.

It is clear that the electron density no longer varies simply according to the Boltzmann distribution. The statistical mechanics should be modified with quantum mechanics, which is related to the electron momentum. One form introduces a quantum pressure term as a modification of the electron temperature,

$$\beta^{-1} = k_B T_e - \frac{\hbar^2}{8m^*} \nabla^2 \ln(n) . \tag{1.18}$$

Although the results obtained using this model are in agreement with the intuitive expectations, the correction term does not have the momentum dependence expected from Fig. 1.2.

People attempt to more properly treat quantum effects without such hydrodynamic approximations. A problem with the use of quantum approaches to device modeling is that most quantum discussions, especially those of quantum transport, tend to revolve around closed systems, whereas most devices are open systems. In treating such open systems quantum mechanically, it is quite difficult to properly define the reservoir (thermal equilibrium contact) regions, as well as the interface regions between the reservoirs and the active device region. Because of the nonlocal nature of the quantum system, errors in defining the contact region will propagate throughout the device, often leading to spurious results.

The double-barrier resonant tunneling diode (DBRTD) is a typical device of quantum mechanical effect. The Wigner distribution function is calculated from the density matrix through the Fourier transform, often called a Weyl transform. The Wigner distribution has been used to model the DBRTD. It is found that the Wigner function shows a depletion region in the cathode area, which arises from a contact potential drop and the tendency to form a bound state in this area. Such contact potential drops are typical of most open systems, whether classical or quantum. Generally, the cathode "barrier" will develop when there is a mismatch between the injection characteristics of the cathode reservoir and the dissipative nature of the active device region. It is largely eliminated if a lightly doped region is introduced adjacent to the barrier layers.

1.3.2 Phase-Coherent Effect

When electrons pass through a device, if the size of the device is smaller than electrons' coherent length (inelastic mean free path), different waves will interfere. The coherent effect will result in additional scattering, then reduce the conductivity. In addition, the coherent effect also will result in the Aharonov–Bohm effect, universal conductance fluctuation, etc. All these are named mesoscopic effects.

One of important problems is the conductance fluctuation caused by quantum effects. Assume that the action area of the gate is $0.1 \times 0.05\ \mu m^2$, and carrier concentration in the inversion layer is $2 \times 10^{12}\ cm^{-2}$, then there are only 100 electrons under the gate. The conductance change caused by the phase coherence is on the order of e^2/h, about 40 μS. If the normal conductivity is 1000 mS/mm, then for a device with a gate width of 0.1 μm, the total conductance is 100 μS. Therefore, the conductance fluctuation caused by the phase coherence is 40% of total conductance, so it will limit the performance of the device greatly.

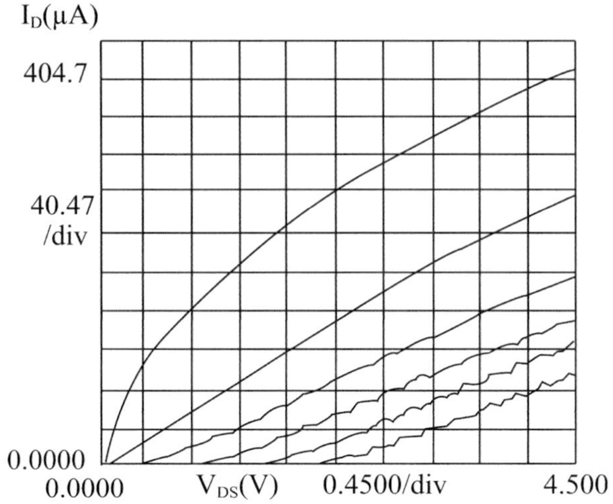

Figure 1.3 Current–voltage curves for a MESFET of 45 nm gate length and 20 μm gate width.

Figure 1.3 shows the current–voltage curves for a MESFET of 45 nm gate length and 20 μm gate width. Fluctuations, which are not time varying but d.c. fluctuations, are clearly evident in this figure, and these are estimated to have a rms (root-mean-square) amplitude of about 10 μS. In spite of the large gate width, it appears that the quantum fluctuations are clearly evident even at 300 K. Although the amplitudes of the conductance differ greatly, the fluctuations are of comparable rms amplitude. This strongly suggests a fundamental origin such as quantum interference.

A 0.1 μm gate width device is quite nearly a device formed on a quantum wire, which can be looked upon as a quantum wire device. The quantum wire has a series of the transverse bound states. When the gate potential increases, the number of the electron channels increases, resulting in a step-type jump in the conductance. The height of the step is $2e^2/h$. Figure 1.4 is the variation of the conductance of a wire with a rough boundary as the Fermi energy (gate potential) is varied. The upper curve is for $\Delta/\lambda_F = 0.1$, and the lower curve is for $\Delta/W = 0.1$, where Δ is the variation of the gate width, λ_F is the Fermi wavelength, W is the gate width, and L is the gate length. From Fig. 1.4 we see that the conductance curves

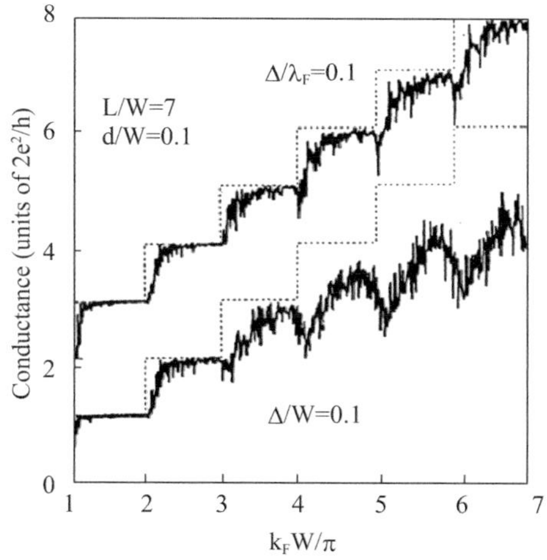

Figure 1.4 Variation of the conductance of a wire with the Fermi energy.

are basically of step form, and derivate from the step form due to the fluctuation of the gate width. Thus, the fluctuation of the gate width is also a reason behind the conductance fluctuation for narrow devices.

In addition, if the orbit of a ballistic electron surrounds an obstacle, such as an impurity atom, then interference between two orbits also will result in the phase-coherent effect. Especially when there is a perpendicular magnetic field in the area surrounded by two orbits, the Ahanorov–Bohm effect (AB effect) will occur. Here, the wave function in the closed loop can be expressed by the integral along the loop:

$$\Psi \approx \exp\left[\frac{e}{\hbar}\left(\int_1 \boldsymbol{A}\cdot d\boldsymbol{l} - \int_2 \boldsymbol{A}\cdot d\boldsymbol{l}\right)\right] = \exp\left[\frac{e}{\hbar}\oint \boldsymbol{B}\cdot d\boldsymbol{S}\right]$$
$$= \exp\left(\frac{\Phi}{\Phi_0}\right), \qquad (1.19)$$

where $\Phi_0 = h/e$ is the magnetic flux quantum. The electric current passing through the closed loop will oscillate periodically with the magnetic flux. When a current of 10 mA passes through the loop, if there is a magnetic field of 0.02 T, then the AB effect can result in distinct fluctuation.

1.3.3 Coulomb Blockade Effect

Experiments found that for a small quantum dot, only when the electric voltage is larger than a certain value, the electron can pass through the dot. The reason is that the Coulomb interaction between the electron in the quantum dot and electron in the circuit will reject the entry of the second electron to the quantum dot. Only when the energy of the electron exceeds the Coulomb interaction energy, the second electron can enter the quantum dot and the current can occur. This appearance is called Coulomb blockade effect, and the relationship curve between voltage and current has the shape of a step, which is called Coulomb staircase. The Coulomb blockade effects have been observed in 2DEG systems in Si-MOSFET and GaAs heterojunction.

From the point of view of large-scale integrated circuits, all the above effects are unfavorable, and we should avoid these in design. On the other hand, these mesoscopic effects may be utilized in the new generation of electronic devices. For example, in a device which has a size smaller than the inelastic scattering mean free path, electrons follow the motion law of quantum mechanics, so we can utilize properties of quantum interference and quantum waveguide to design a device and to control the movement of electrons. And we can design a single-electron transistor by using Coulomb blockade effects, which can greatly reduce the number of electrons stored in one bit, and can reduce power dissipation of memory devices. Because the spin of electron has a much longer scattering time and distance than that of the electric charge, we can utilize spin as the information carrier to design spintronics devices and so on.

1.4 Landauer–Büttiker Formula

Datta first proposed the concept of quantum device in 1989 [5], which can be understood only by considering the electron's wave property. Concepts which can describe the electron's movement are no longer the scattering probability, relaxation time, mobility, and so on. For a two-terminal device, assume that a^+, a^- and b^+, b^- are amplitudes of the electron wave in the input circuit and output

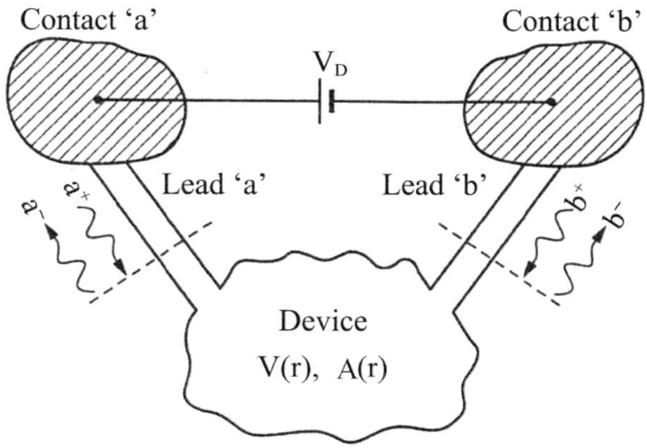

Figure 1.5 Two-terminal device, where a^+, a^- and b^+, b^- are amplitudes of the electron wave in the input circuit and output circuit, respectively.

circuit, respectively. Then, for a definite energy E, they can be related by a scattering matrix,

$$\begin{pmatrix} a^- \\ b^- \end{pmatrix} = \begin{pmatrix} r(E) & t'(E) \\ t(E) & r'(E) \end{pmatrix} \begin{pmatrix} a^+ \\ b^+ \end{pmatrix}, \tag{1.20}$$

where $r(E), r'(\mathrm{E})$ and $t(E), t'(E)$ denote reflection and transmission coefficients, respectively, which are determined by the internal potential and structure.

The circuit has a certain width, and there are transverse modes with different energy. In general, the transport in the longitudinal direction is of multimode, so the amplitudes a^+, a^- are column matrices with size $(M_a \times 1)$, b^+, b^- are column matrices with size $(M_b \times 1)$. Similarly, r is a matrix with size $(M_a \times M_b)$, t is a matrix with size $(M_b \times M_a)$, and so on. M_a and M_b are the number of transverse modes in the input circuit and output circuit, respectively.

In the Landauer–Büttiker formula, the conductance is expressed by the scattering matrix, so this theory is suitable for solving the conductance of a system which is made of different materials or conductors of different shapes that are put in contact with each other. This theory is the basic theory in the research on mescoscopic transport.

Suppose that an electron moves in an ideal conductor along the z axis, and neglect the positive ions in lattice points and scattering of positive electric charge on the electron. In the absence of a magnetic field (so take no account of spin), the Hamiltonian of the electron is

$$H = \frac{p^2}{2m} + V(x, y), \tag{1.21}$$

where $V(x)$ is the lateral confinement potential in the wire. On the basis of the Hamiltonian (1.21), we can obtain the system's eigenstate:

$$\Psi_{\alpha k}(x, y, z) = \frac{1}{\sqrt{2\pi}} e^{ikz} \phi_\alpha(x, y), \tag{1.22}$$

where k is the wave vector along the z direction of the wire, α denotes the α-th transverse eigenstate, and the eigenenergy corresponding to the α-th eigenstate is $E_{\alpha k} = E_\alpha + \hbar^2 k^2/2m$. The first term and the second term are lateral and longitudinal energy, respectively, and the lateral energy E_α is determined by the lateral confinement potential $V(x)$. We can use a square potential well to express the lateral confinement potential of an ideal wire. Suppose that the well in one direction is very narrow, and the width in another direction is W, then the lateral energy $E_\alpha = (n\hbar\pi)^2/2mW^2$ has discrete energy levels. Therefore, the energy spectrum of the system is made up of a series of parabolic and discrete energy subbands, and the minimum value of the corresponding curve is E_α.

When $T = 0$ K, the current in the α-th channel of the wire can be expressed as

$$\begin{aligned} I_\alpha = \frac{2}{2\pi} \int_0^{k_{\max}} e v_{\alpha k} dk &= \frac{2}{2\pi} \int_0^{k_{\max}} e v_{\alpha k} \frac{dk}{dE_{\alpha k}} dE_{\alpha k} \\ &= \frac{2e}{h} (\mu - E_\alpha), \end{aligned} \tag{1.23}$$

where $v_{\alpha k} = 1/\hbar \times dE_{\alpha k}/dk$ is the electron's group velocity, and μ is the chemical potential (Fermi energy level).

If the width of the ideal conductor is small enough, then the energy gap between two energy subbands will be large enough so that the electron can just occupy the lowest energy band, and then this energy band will become the one and only allowed channel. At present we can calculate the net current in the wires with the chemical potentials μ_1 and μ_2, respectively, connected to each other:

$$I = I_1 - I_2 = \left(\frac{2e}{h}\right)\mu_1 - \left(\frac{2e}{h}\right)\mu_2 = \left(\frac{2e^2}{h}\right)(V_1 - V_2). \tag{1.24}$$

Therefore, the conductance of the two-terminal and single-channel device is

$$G = \frac{I}{V_1 - V_2} = \frac{2e^2}{h}. \tag{1.25}$$

This is the Landauer formula of the two-terminal and single-channel system; $2e^2/h$ is called the quantum conductance.

If the wire has finite width, the electron may occupy several subbands. Assume that N subbands have been filled in, and $f(E)$ denotes the probability that the subbands have been occupied by the electron. At present the current can be written as

$$I = \frac{2e}{h}\int_0^{\infty} dE \left[f_1(E) \sum_i T_{\mathrm{i}}(E) - f_2(E) \sum_j T_{\mathrm{j}}(E) \right], \tag{1.26}$$

where $T_{\mathrm{i}} = \sum_j T_{ij} = \sum_j |t_{ij}|^2$ denotes the total transmission probability that an electron transport from all channels of wire 1 to the i-th channel of wire 2. $f_1(E)$ and $f_2(E)$ are the Fermi–Dirac distribution function in the wire 1 and wire 2, respectively. Therefore the current of the system is

$$I = \frac{2e}{h}\int dE\,[f(E) - f(E + V_{\mathrm{D}})]T(E), \tag{1.27}$$

where $f(E)$ is the Fermi–Dirac distribution function, $V_{\mathrm{D}} = V_1 - V_2$ is the bias voltage on the system, $T(E) = \Sigma_{\mathrm{i}} T_{\mathrm{i}}(E)$ is the total transmission probability. For a small bias, Eq. 1.27 can be simplified as

$$G = \frac{I}{V_{\mathrm{D}}} = \frac{2e^2}{h}\int dE \left(-\frac{\partial f}{\partial E}\right) T(E). \tag{1.28}$$

At low temperature, $-\partial f/\partial E = \delta(E - E_{\mathrm{F}})$, then in the end we obtain

$$G = \frac{2e^2}{h} T(E_{\mathrm{F}}) = \frac{2e^2}{h}\sum_{i=1}^{M_a}\sum_{j=1}^{M_b} \left| t_{ij}(E_{\mathrm{F}}) \right|^2. \tag{1.29}$$

where $t_{ij}(E_{\mathrm{F}})$ is the transmission coefficient shown in Eq. 1.20, and Eq. 1.29 is the Landauer–Büttiker formula for the two-terminal and multichannel device. It follows that the key point of utilizing the Landauer–Büttiker formula is solving the transmission coefficient matrix t. Most research studies on mesoscopic transport focus on this problem.

The two-terminal Landaure formula can be extended to the multi-terminal device. Considering a three-terminal device, we can write its scattering matrix functions similar to Eq. 1.20:

$$\begin{pmatrix} a^- \\ b^- \\ c^- \end{pmatrix} = \begin{pmatrix} r_{aa} & t_{ab} & t_{ac} \\ t_{ba} & r_{bb} & t_{bc} \\ t_{ca} & t_{cb} & r_{cc} \end{pmatrix} \begin{pmatrix} a^+ \\ b^+ \\ c^+ \end{pmatrix}. \tag{1.30}$$

Büttiker has proved that current I_i $(i = a, b, c)$ has the following relationship with the chemical potential of every circuit μ_i:

$$I_i = \frac{2e^2}{h} \sum_{j=a,b,c} \left(T_{ij}\mu_j - T_{ji}\mu_i\right), \tag{1.31}$$

where

$$T_{ij} = \sum_{m,n=1}^{M} \left|\left(t_{ij}\right)_{mn}\right|^2. \tag{1.32}$$

When there is no external magnetic field, $T_{ij} = T_{ji}$, Eq. 1.31 can be written as

$$I_i = \frac{2e^2}{h} \sum_{j=a,b,c} T_{ij}\left(\mu_j - \mu_i\right), \tag{1.33}$$

where $\mu_j - \mu_i$ corresponds to the bias voltage between two circuits. The conductance between each pair of circuits can be defined as

$$G_{ij} = \tfrac{2e^2}{h} T_{ij},\ i,\ j = a,\ b,\ c\ (i \neq j). \tag{1.34}$$

Equations 1.29 and 1.31 are the fundamental formulas of quantum conductance. They form the theoretical basis of future quantum devices. For a certain device, we should obtain scattering matrices $r_{aa}(E)$, $t_{ab}(E)$, ... It is shown in Eq. 1.31 that the quantum device is nonlocal, and change of any terminal's chemical potential will influence currents in all circuits.

1.5 Quantum Interference Transistor

By using the interference effect of the electron wave, Datta [5] proposed the concept of "quantum interference transistor," as shown in Fig. 1.6a. The structure consists of a conducting channel with a barrier in the middle. The length L is small enough so that electrons

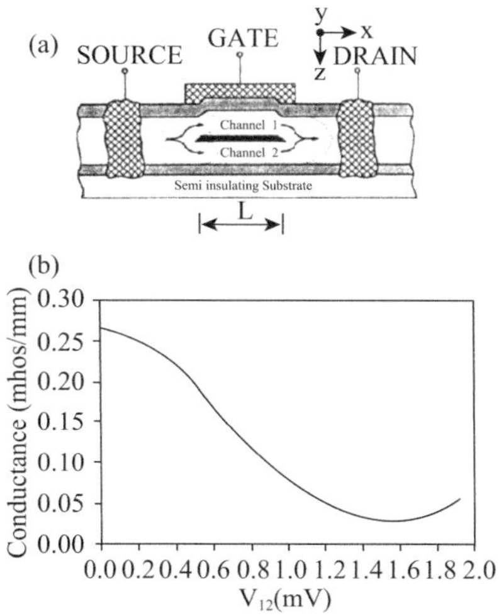

Figure 1.6 (a) A proposed quantum interference transistor, which consists of a conducting channel with a barrier in the middle. (b) Calculated conductance versus potential difference between the channels.

travel ballistically across it. The channels are narrow enough to be single-mode in the z direction. Without a gate voltage, the two channels have been assumed to be perfectly symmetric, so that the electrons passing channels 1 and 2 have the same phases when they meet at the drain of the device, the amplitude of the electron wave—i.e., the conductance will be maximum. But an applied gate voltage changes the average potential in channel 1 with respect to that in channel 2, resulting in the phase difference of electron waves traveling in channel 1 and channel 2, and the decrease in the conductance. Figure 1.6b shows the calculated conductance versus bias gate voltage, i.e., potential difference between the channels.

Figure 1.7a is another quantum interference transistor proposed by Datta [5]. It differs from the ordinary transistor in that the gate is not at the position between the source and the drain, but at the position beside the source and the drain. Thus it provides two primary paths from the source to the drain, with the gate controlling their interference, much like the Ahanonov–Bohm device. The length L can be controlled by the gate voltage, as shown in Fig. 1.7a. Figure 1.7b is the calculated conductance as a function of the

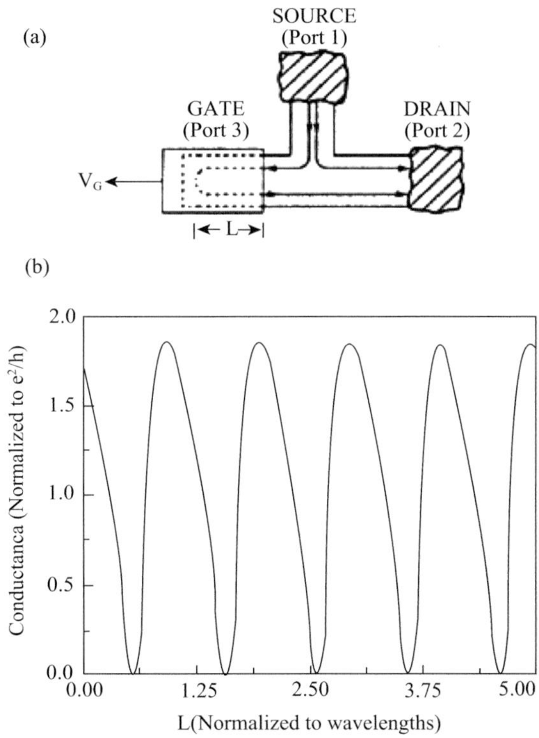

Figure 1.7 (a) Another proposed quantum interference transistor, which has the gate beside the source and the drain. (b) Calculated conductance versus the length *L*.

effective length *L*. From the figure we see that the conductance is modulated by 100% as the length *L* is changed. Therefore, the quantum interference transistor controls current not by changing the number of electrons, but by changing the phase.

In order to calculate the scattering matrix the authors proposed a one-dimensional quantum waveguide theory [6], which can be used to explain intuitively the above A-B devices and the quantum interference transistors, and which is also suitable for any shape and structure of one-dimensional quantum waveguide circuits. We also consider the quantum waveguide circuits with definite width. For details, see Chapter 9 and hereafter.

1.6 Spintronics Devices

Electron spin, or intrinsic electronic angular momentum, is another intrinsic property of electron besides electric charge. The electron spin has two discrete values $\pm\hbar/2$, which are generally named spin-up and spin-down states respectively. This is very similar to the high and low levels or switch off and switch on, which are used to realize logic 1 and 0 states. Thus, the 1/2 spin system is an ideal two-level system, and spin-up and spin-down states are an ideal choice to realize the quantum bit in quantum calculation. Recently, in traditional electronic industry, device integration has become higher and higher, and the size of devices smaller and smaller, which results in distinct quantum effects and too much energy consumption. Therefore, naturally, people think of utilizing spin-free degree, which was formerly always neglected, to replace or to combine the charge-free degree. The research on how to control spin-free degree effectively has attracted wide attention from scientists, and was named *spintronics* by S. A. Wolf et al. formally in 2001 [7]. Spintronics researches a particle's spin in a solid and applies it in devices, and it contains spin polarization's produce, control, transport, and exploitation. It is an interdiscipline field which involves electronics, magnetism, optics, and micro- and nanoprocessing technology. According to the materials researched, it can be grouped into two areas: metal spintronics and semiconductor spintronics.

Ideal spintronics devices realize functions by controlling spin orientation and spin, so they may not be restricted by the number of electric charges and energy dissipation. They have lower power dissipation and faster running speed. Moreover, electronic storage devices based on spin have the advantage of nonvolatility. These devices clearly show that spintronics will have wide prospects of development and application in areas of quantum calculation, communication, molecules, chemistry, and so on. Therefore, people expect spintronics to construct the physical foundation of a new generation of nanoelectronics, and spintronic devices to replace the current microelectronic devices.

The symbol of spintronics' birth is the discovery of the giant magnetoresistance (GMR) of the metallic multilayer structure in

1988 [8, 9]. The proposal of spin valves and room temperature magnetic sensor utilizing anisotropic magnetoresistance make GMRs a great success in commerce. Nowadays a new generation of the magnetic head of high-density disk is using the GMR principle. Thanks to the high sensitivity to the magnetic field of GMR, each GMR magnetic head has be reduced to 30–50 nm, and more than 4 T volume in desktops and more than 1 T volume in laptops have been realized.

Now spintronics devices that are practical are mostly made of metal magnetic materials, which belong to the first category of spintronics, i.e., magnetoelectronics. In 1995 scientists found a new phenomenon, tunneling magnetoresistance (TMR) [10]. When the moments of the above and underneath ferromagnetic layers change from parallel to antiparallel, the tunneling resistance changes 20–30%. The magnetic tunnel junction (MTJ) utilizing the TMR technology has been created, and new magnetoresistive random access memory (MRAM) is expected to replace nonvolatile memory based on CMOS. This new generation of memory has advantages such as no loss on data when power fails, rapid reading speed comparable to static random access memory (SRAM), and large memory space comparable to dynamic random access memory (DRAM), so it has wide application prospects.

The second category of spintronics mainly studies semiconductors, and it hopes to create an imbalance of spin number to realize spin transistor and spin valve, and then replace traditional electronic devices. These new spintronic devices have characteristics such as low energy consumption and rapid switching speed. Because ready-made mature semiconductor technology and equipment can be used to reduce developing cost, these devices have attracted wide attention. To utilize spin-free degree in semiconductors, we must produce, keep, control, and detect the spin polarization of carriers.

There are several methods which can produce spin current: ohmic injection, tunnel injection, ballistic electron injection, utilizing the giant Zeeman splitting of a dilute magnetic semiconductor (DMS) in a magnetic field, utilizing a ferromagnetic semicondutor (FMS) as spin calibrator, the optical method, and so on. Ohmic injection uses a ferromagnetic metal (FM) as electrode to inject spin-polarized electrons. If the ohmic contact is created in the junction between FM and semiconductor, then it can be expected that the spin-polarized current in the FM can be injected into the semiconductor. However, the ohmic contact by heavy doping produces scattering of spin flip, which reduces spin polarization. At $T < 10$ K, from FM-InAs ohmic contact, spin-polarized electrons of 4.5% have been obtained [11]. At room temperature, from the Fe-GaAs contact, spin-polarized electrons of 2% have been obtained [12].

As for how to control the spin polarization of carriers in a semiconductor, various semiconductor spintronic devices have been proposed. The spin field-effect transistor (SFET) proposed by Datta and Das in 1989 [13] is one such original device, and also the most famous one. Other important ones include the graphene spin field-effect transistor, metal-oxide-silicon spin field-effect transistor, Johnson spin switch, unipolar spin transistor, and so on. Despite making great efforts, people have not succeeded even with the original Datta–Das transistor yet, which shows that we have a long way to go, and there is much work to do.

The spin transistor first proposed by Datta and Das [13] is based on controlling the spin of an electron. The principle it works on is shown in Fig. 1.8a.

The principle of the electro-optic modulator shown in Fig. 1.8a should be introduced first. A polarizer at the input polarizes the light at 45° to the y axis (in the $y - z$ plane), which can be represented as combination of the z- and y-polarized light:

$$\begin{pmatrix} 1 \\ 1 \end{pmatrix}_{(45^\circ \text{ pol})} = \begin{pmatrix} 1 \\ 0 \end{pmatrix}_{(z\,\text{pol})} + \begin{pmatrix} 0 \\ 1 \end{pmatrix}_{(y\,\text{pol})}. \tag{1.35}$$

Because of the electro-optic material the dielectric constant ε_{zz} is slightly different from ε_{yy}. As this light passes through the electro-optic material, the two polarizations suffer different phase shifts $k_1 L$ and $k_2 L$, where L is the length of the crystal. The light emerging from the electro-optic material is represented as $((e^{ik_1^L})/(e^{ik_2^L}))$. The analyzer at the output lets the light with polarization (1/1) to pass through, so the output power P_0 is given by

$$P_0 \propto \left| \begin{pmatrix} 1 & 1 \end{pmatrix} \begin{pmatrix} e^{ik_1 L} \\ e^{ik_2 L} \end{pmatrix} \right|^2 = 4\cos^2 \frac{(k_1 - k_2)\,L}{2}. \tag{1.36}$$

The light output is modulated by a gate voltage that controls the difference between the phase shifts $\Delta\theta = (k_1 - k_2)L$.

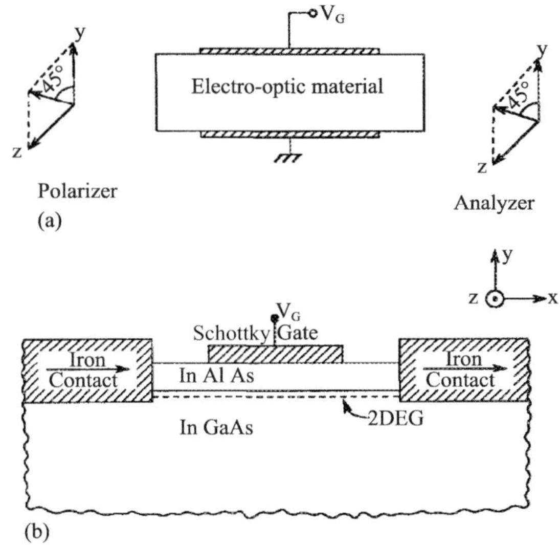

Figure 1.8 (a) Electro-optic modulator. (b) Spin transistor similar to electro-optic modulator.

The theorem of the spin transistor is based on the interaction of the electron orbit and spin, i.e., the Rashba interaction [14],

$$H_R = \frac{\alpha}{\hbar}(\sigma \times p)_y = i\alpha\left(\sigma_x \frac{\partial}{\partial z} - \sigma_z \frac{\partial}{\partial x}\right), \tag{1.37}$$

where α is named the Rashba coefficient, generally being $1 - 10 \times 10^{-10}$ eV · cm, and σ_x and σ_z are the Pauli matrices. In the case of Fig. 1.8b the gate voltage produces a perpendicular electric field to the two-dimensional electron gas (2DES); i.e., it produces a nonsymmetric potential with respect to the $x - z$ plane, which produces a Rashba interaction between the orbit and spin of electrons, as shown in Eq. 1.37. The Rashba coefficient α is proportional to the electric field.

In the spin transistor shown in Fig. 1.8b, both ends are FM iron contacts, which are magnetized in the x direction and correspond to the polarizer and the analyzer in the electro-optic modulator of Fig. 1.8a. Electrons injected through the source electrode are spin-polarized along the x direction, which can be represented as a linear combination of positive z-polarized and negative z-polarized electrons:

$$\begin{pmatrix} 1 \\ 1 \end{pmatrix}_{(+x \text{ pol})} = \begin{pmatrix} 1 \\ 0 \end{pmatrix}_{(+z \text{ pol})} + \begin{pmatrix} 0 \\ 1 \end{pmatrix}_{(-z \text{ pol})}. \tag{1.38}$$

Due to the Rashba term (1.30), in the absence of a magnetic field, the Rashba term causes the spin-up and spin-down states to split. If an electron is traveling in the x direction with $k_x \neq 0$ and $k_z = 0$, then considering the Rashba term, the energies of the electron can be written as

$$E\,(+z \text{ pol}) = \frac{\hbar^2 k_{x1}^2}{2m^*} + \alpha k_{x1},$$
$$E\,(-z \text{ pol}) = \frac{\hbar^2 k_{x2}^2}{2m^*} - \alpha k_{x2}. \tag{1.39}$$

Electrons with two kinds of spin polarization have the same energy, and it can be proved that they have different wave vectors. Their difference is

$$k_{x2} - k_{x1} = \frac{2m^*\alpha}{\hbar^2}. \tag{1.40}$$

Therefore there is a differential phase shift for spin-up and spin-down electrons after they pass through the junction area, and this shift is shown as

$$\Delta\theta = (k_{x2} - k_{x1})\,L = \frac{2m^*\alpha L}{\hbar^2}. \tag{1.41}$$

For InGaAs/InAlAs heterostructures, from the experimentally observed zero-field spin splitting, α was estimated to be $\sim 3.9^{-10}$ eV·cm. To make a phase difference of π, L is about 0.67 μm, which is less than the mean free path (≥ 1 μm) in high-mobility semiconductors at low temperatures. Therefore we can change the gate voltage, and then change the Rashba coefficient α to modulate the spin current just as the electric-optic modulator.

1.7 Carbon-Based Electronics [14]

In the last few decades, microelectronics has made dramatic advances, including the continuous miniaturization or "scaling" of electronic devices, particularly of the silicon-based large-scale integrated circuit (LSIC). However, this device scaling and performance enhancement cannot continue forever. A number of limitations in the fundamental scientific as well as technological nature of silicon devices place limits on their ultimate size and performance.

The approaches towards making a breakthrough in the limits are divided into two types. One is moving away from the traditional electron transport-based electronics: e.g., the development of spin-based devices. The other approach maintains the operating principles of the currently used devices, primarily those of the field-effect transistor, but replaces a key component of the device, the conducting channel, with carbon nanomaterials such as one-dimensional (1D) carbon nanotubes (CNTs) or two-dimensional (2D) graphene layers, which have superior electric properties. This book focuses mainly on the first kind of research and introduces only carbon-based devices in this section. The main problem with carbon-based devices is that it is difficult to integrate them into one large-scale circuit as silicon-based devices. Now the single carbon nanotube field-effect transistor (CNTFET) and graphene nano-ribbon field-effect transistor (GNRFET) have emerged.

Recently, graphene became the object of intense experimental study when it was realized that single layers, or a few layers, could be produced relatively easily by mechanical exfoliation of graphite [15], or by heating SiC [16]. Figure 1.9a shows the carbon atom structure in a honeycomb arrangement, and Fig. 1.9b shows the electronic band structure of this 2D material [14]. From Fig. 1.9b we see that the linear dispersion at low energies near the Γ point in the Brillouin

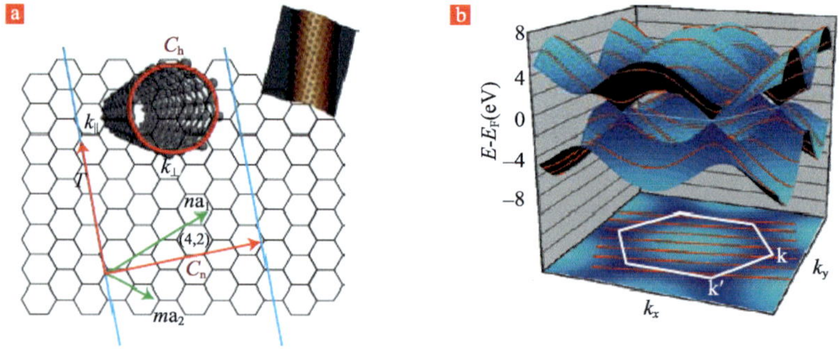

Figure 1.9 (a) Atomic structures of graphene and carbon nanotube. (b) Electronic structure of graphene.

zone makes the electrons and holes in graphene mimic relativistic particles that are described by the Dirac relativistic equation for particles with spin 1/2. Their dispersion,

$$E = \hbar v_{\mathrm{F}} \sqrt{k_x^2 + k_y^2}, \tag{1.42}$$

where $v_{\mathrm{F}} \approx 10^6$ m/s is the Fermi velocity. The linear dispersion means that quasi-particles in graphene display properties quite different from those of conventional semiconductor materials—e.g., the anomalous quantum Hall effect and half-integer quantization of the Hall conductivity. The quantum Hall effect in graphene can be observed even at room temperature.

1.7.1 Electronic Structure

As in Fig. 1.9a, the CNT is thought of as being formed by the rolling of a piece of a ribbon of graphene to form a seamless cylinder. The rolling process forming the nanotube and the resulting nanotube structure are specified by a pair of integers (n, m) defining the chiral vector,

$$\mathbf{C}_{\mathrm{h}} = n\mathbf{a}_1 + m\mathbf{a}_2, \tag{1.43}$$

where $\mathbf{a}_1$ and $\mathbf{a}_2$ are the unit vectors of the graphene honeycomb lattice. The length of $\mathbf{C}_{\mathrm{h}}$ describes the circumference of the nanotube $\mathbf{C}_{\mathrm{h}} = \pi d_{\mathrm{CNT}}$. The periodic boundary conditions around the circumference of a nanotube require that the component of the momentum along the circumference, $\mathbf{k}_{\perp}$, be quantized: $\mathbf{C}_{\mathrm{h}}\mathbf{k}_{\perp} = 2\pi\nu$, where ν is a non-zero integer. This quantization leads to the formation of metallic and semiconducting nanotubes.

Whereas the infinite 2D graphene is a semi-metal, if the graphene is a narrow ribbon, there will be a gap in the energy band due to the quantization of k_x or k_y. The bandgap is inversely proportional to the width W of the GNR, and is approximately given by

$$E_{\mathrm{g}} \approx 2\pi\hbar v_{\mathrm{F}}/(3W). \tag{1.44}$$

The electronic states in GNRs are not degenerate, whereas those of CNTs are doubly degenerate. This is due to the difference in the boundary conditions: in a GNR the wavefunction has to vanish at the edges, whereas in a CNT the wavefunction is periodic in the direction of the circumference.

1.7.2 Electric Properties

The 1D nature of the electronic states of CNTs leads to a new type of quantized resistance related to its contacts with three-dimensional (3D) macroscopic objects such as metal electrodes. The confinement of the electrons in the CNT around its circumference produces a small number of discrete states (modes) that overlap the continuous states of the metal electrodes. This mismatch of the number of states that can transport the current in the CNTs and the electrodes leads to a quantized contact resistance, R_Q. The resistance is determined by the number of modes, M, in the CNT that have energies lying between the Fermi levels of the electrodes:

$$R_Q = \frac{h}{2e^2 M}. \tag{1.45}$$

For a metal CNT, $M = 2$, so that $R_Q = h/4e^2 = 6.45\ \text{k}\Omega$.

Excluding this quantized contact resistance and other contact resistances such as those caused by the Schottky barriers, the transport in the CNT is ballistic; i.e., no carrier scattering or energy dissipation takes place in the body of the CNT. The length over which a CNT behaves as a ballistic conductor depends on its structure perfection, temperature, and the size of the driving electric field. In general, ballistic transport can be achieved over lengths ≤ 100 nm. Even in long CNTs, or at high bias, many scattering collisions can take place. The mobility in CNTs can still be very high, as much as 1000 times higher than in bulk silicon.

Due to the 1D nature of the CNT, small-angle scattering is forbidden. There will only be backscattering, so that the elastic scattering in CNTs is weak, and the inelastic scattering determines their transport properties. At low temperature and low bias, only low-energy acoustic phonons can scatter the electrons, which results in an inverse temperature dependence of the carrier mobility in semiconducting CNTs, while in bulk materials the acoustic phonon scattering leads typically to a $\sim 1/T^5$ temperature dependence of mobility due to the small-angle scattering. Thus the CNT has very high mobility even at room temperature.

In addition to the low-energy acoustic phonons, electron scattering by the optical phonon of the radial breathing mode (RBM) is important in the low-bias regime. The RBM phonon energy is inversely proportional to the tube diameter, and its energy is comparable to the thermal energy at room temperature for tubes in the diameter range of 1.5–2.0 nm. For electrons to emit an optical phonon, their energy must be larger than the optical phonon energy. This can be achieved only under a very high bias of a few V cm^{-1}. Such scattering processes were first observed in metallic tubes, in which the current was found to saturate at about 25 μA.

As the carrier energy increases further, other inelastic process can take place, in particular impact excitation. The high-energy electron can excite an electron–hole pair and lose its energy. The electron–hole interaction is very strong in CNTs, which leads to the formation of excitons with large binding energy of a few tenths of an electronvolt. The calculations suggest that the impact excitation processes in CNTs are much more efficient than in conventional bulk semiconductors—about four orders of magnitudes higher.

Figure 1.10 shows the calculated scattering probabilities, $1/\tau$, for a (19,0) nanotube as a function of the energy higher than that of the first subband [14], where Δ_1, Δ_2, Δ_3, and Δ_4 denote the energies of subsequent subbands, respectively. The peaks labeled by A-Ph are the longitudinal acoustic ones; RBM, the radial breathing mode; O-Ph, the optical phonon; and I-Exc, the impact excitation exciton scatterings. The various curves represent angular momentum bands with different circumferences.

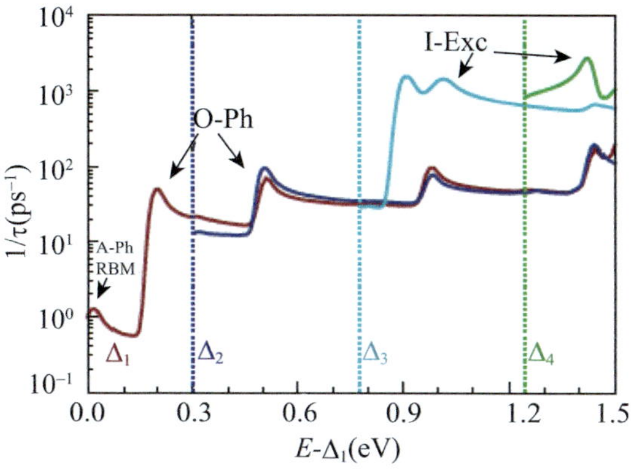

Figure 1.10 Calculated scattering probabilities, $1/\tau$, for a (19,0) nanotube as a function of the energy higher than that of the first subband.

1.7.3 Carbon Nanotube Field-Effect Transistor

The first carbon nanotube field-effect transistors (CNTFETs) were reported in 1998 [17, 18]. In a CNTFET one (or more) CNT plays the role of the channel. Figure 1.11 shows two different CNTFET structures: a top-gate CNTFET and an array of CNTFETs with wrap-around gates. The CNT as an FET channel has a number of advantages. Its small diameter allows optimum coupling between the gate and the channel. This strong coupling makes the CNT the ultimate thin-body semiconductor system and allows the device to be made shorter while avoiding the dreaded "short-channel effects." All bonds in the CNT are satisfied and the surface is smooth, so that the scattering by surface and roughness is absent. The key advantage is the low scattering in the CNT and the high mobility of the FET channel.

In MOSFETs, in addition to the channel, the source, drain, and gate are all made of heavily doped Si and the contacts are ohmic. This is not generally true for the CNT–metal contacts used in CNT

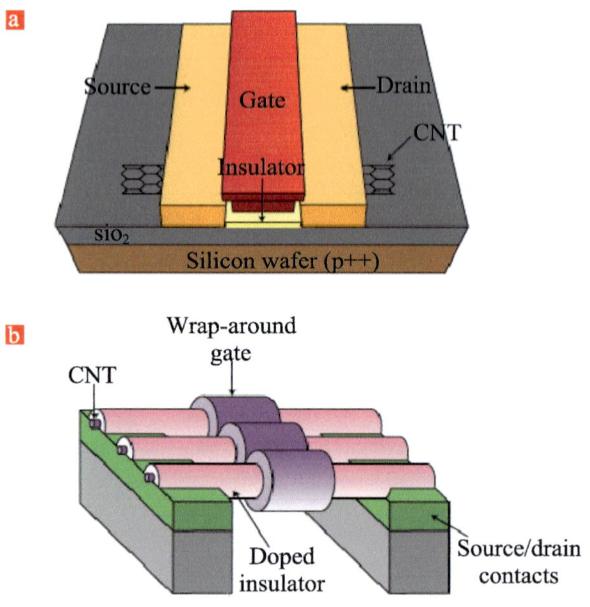

Figure 1.11 Schematic of two different CNTFET structures.

electronics. Metals like Au, Ti, Pd, and Al are used for the source and drain electrodes. The different work functions of the metal and the CNT lead to transfer of charge at their interface. The resulting interface dipole produces an energy barrier, the so-called Schottky barrier. The Schottky barrier height depends on their respective work function (Φ), the CNT bandgap, and the details of chemical bonding at the interface.

There are two Schottky barriers in an FET, one at the source and the other at the drain, as schematically shown in the insets of Fig. 1.12. Generally, it transports only one type of carrier: electrons or holes. From the left inset of Fig. 1.12 we see that if the electrode is a metal with high work function, such as Pd, and the valence band of the CNT is close to the metal Fermi level E_F, a nearly barrierless contact for holes will be formed. But electron injection at the other end would experience the maximum barrier of E_g, so that it is a unipolar device, a p-type CNTFET operation. If the electrode is a low work function metal such as Al, then it allows the electron transport and forbids the hole transport, as shown in the right inset of Fig. 1.12. Figure 1.12 shows the source–drain current as a function

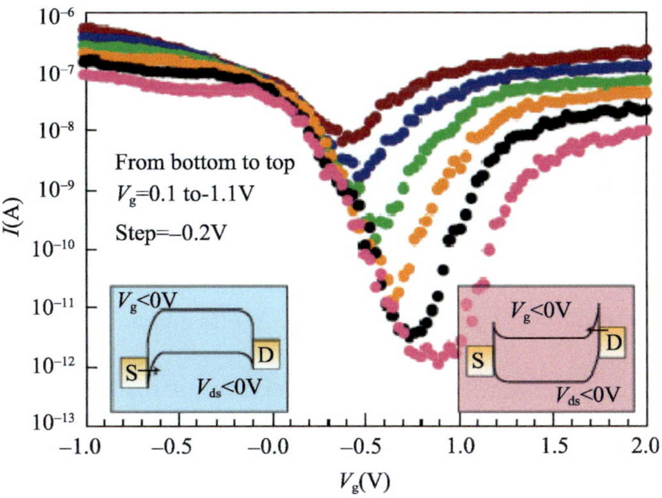

Figure 1.12 The source–drain current as a function of the gate voltage for different drain biases, increasing from 0.1 to −1.1 V in −0.2 V steps. The insets are the contact potential between the electrodes and the CNT.

of the gate voltage for different drain biases, increasing from 0.1 to 1.1 V in -0.2 V steps. From the figure we see that in the gate voltage variation range of 1 eV, the current can vary $5 \sim 6$ orders of magnitudes.

In general, the key advantage of CNTFETs over Si MOSFETs in logic applications is their much low capacitance of $\sim$10 aF (for a $d_{\text{CNT}} = 1$ nm, $L = 10$ nm, $t_{\text{ox}} = 5$ nm device) and their somewhat lower operating voltage. Thus, the CNTFETs have lower switching energy per logic transition. The dynamic switching energy of a device is given by $0.5(C_{\text{dev}} + C_{\text{wire}})V^2$, where C_{dev} and C_{wire} are the device and wire capacitance contributions, respectively. To minimize the switching energy, minimum-sized devices and interconnects should be used, as well as minimum supply voltage. The CNTFETs have the $I_{\text{on}}/I_{\text{out}}$ ratios in the range of $10^5 \sim 10^7$. Then the CNT can have a considerable advantage, of up to a factor of 6 over the convenient MOSFET.

More complex structures such as ring oscillators have been fabricated [19]. To build these circuits the complementary MOS architecture is preferred, which involves pairs of n- and p-type transistors. CNTs, in which the valence and conduction bands are mirror images of each other (equal effective masses for electrons and holes), are ideally suited for such application. The two types (p- and n-) of FETs can be made by doping the CNTs, but it is difficult to control doping in nanoscale devices. The ambipolar behavior of an undoped CNT can be successfully utilized to implement the CMOS architecture. For a given undoped CNTFET with a fixed energy gap and oxide thickness, tuning the work function of the gate metal is the only way. When the work functions are properly selected, the two characteristics can be relatively shifted toward each other, and this leads to a distinguishable on-state in one and an off-state in the other.

Figure 1.13 shows the scanning electron microscope image of a CNT ring oscillator circuit, and the frequency response in different supply voltages [14]. It is a ring oscillator using Pd gates for p-FETs and Al gates for n-FETs. The frequency response shows a strong dependence on the supply voltage. A 72 MHz frequency is measured for a supply voltage of about 1 V.

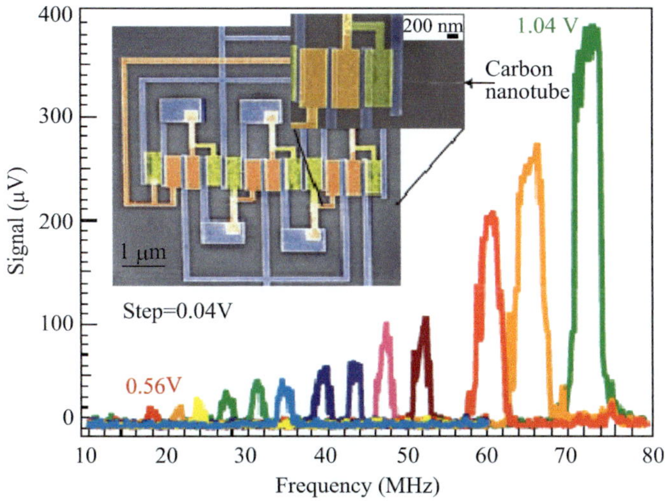

Figure 1.13 Scanning electron microscope image of a CNT ring oscillator circuit and the frequency response in different supply voltages.

1.7.4 Graphene Ribbon Transistor

The sub-10 nm width graphene ribbon field-effect transistor (GNRFET) has been fabricated [20]. The schematic structure of the GNRFET is shown in Fig. 1.14a, and the AFM images of two GNRs are shown in Fig. 1.14b,c. The height of the GNR is 1.5 nm, which is assigned as two-layer GNRs. The Ti-As is used as contact electrodes (source and drain) to minimize the Schottky barrier for holes in a p-type transistor. The GNRFET is placed on a 10 nm thick SiO_2 layer, and p^{++} Si is used as backgate. The widths of the GNRs in Figs. 1.13b and 1.14c are 5 and 60 nm, respectively.

Figure 1.15 shows the current–voltage characteristics for the $w \approx 2$ nm, $L = 236$ nm GNR device shown in Fig. 1.14b [20]. Figure 1.15a is current vs. gate voltage ($I_{ds} \sim V_{gs}$) under various V_{ds}, and the dashed lines are the turn-on and cut-off states of the current $I = 3.2$ μA and 10^{-12} A, respectively. Thus, at room temperature, when $V_{ds} = 0.5$ V, the ratio of the turn-on and cut-off currents $I_{on}/I_{off} > 10^6$. Figure 1.15b is ($I_{ds} \sim V_{ds}$) under various V_{gs}. From the figure, the subthreshold slope, ~210 mV/decade, and transconductance, ~1.8 μS, are obtained.

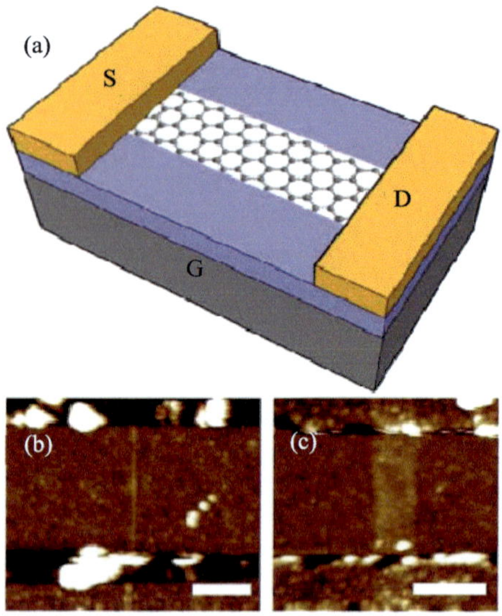

Figure 1.14 (a) Schematic structure of the GNRFET. (b, c) The AFM images of two GNRs.

Comparing experiments with theoretical modeling, it was found that the narrow device (Fig. 1.15) delivered about 21% of the ballistic current at $V_{ds} = 1$ V. The highest proportion that can be reached is 38%. The reason is that the optical phonon (OP) or zone

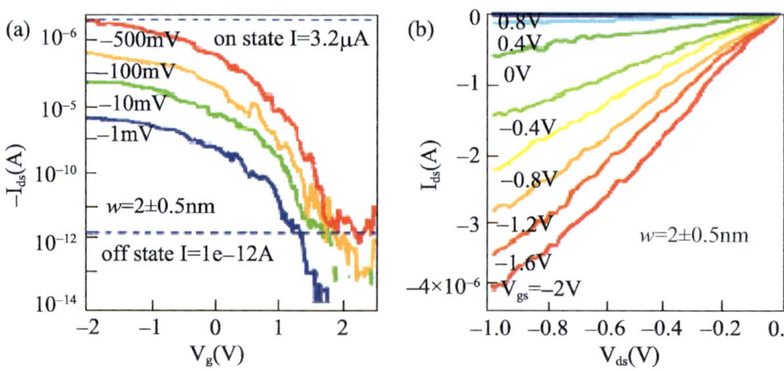

Figure 1.15 Transistor performance of the GNRFET shown in Fig. 1.14b. (a) ($I_{ds}\sim V_{gs}$) under various V_{ds}. (b) ($I_{ds}\sim V_{ds}$) under various V_{gs}.

boundary phonon (ZBP) emission needs higher energy (~0.2 eV). A carrier backscattered by emitting an OP/ZBP does not have enough energy to overcome the barrier near the source end of the channel, so that the OP/ZBP has a small effect on the current in GNRFETs. The wide GNRFETs (Fig. 1.14c) deliver higher current density of ~2000–3000 μA/μm.

1.7.5 Future of Carbon-Based Devices

Carbon nanotubes are ideal 1D model systems to study electrical and optical phenomena on the nanometre scale. Some new states of condensed matter such as the Luttinger–Tomonaga liquid can be studied. Graphene has exhibited a number of unique phenomena such as anomalous quantum Hall effect, Klein paradox, and so forth. New information on the physics of the nanoscale is expected to be found through the study of nanotubes and graphene. At the same time, the nanotechnology to handle and process nanomaterials has been developed. Besides electronic switching, emission/detection, and sensitive devices, nanotubes offer the potential of very fast (THz) transistors, ultimately scaled logic devices, and simpler and cheaper self-assembly-based fabrication. The excellent electrical conduction of metallic CNTs may eventually allow the development of electronic systems, where both active and interconnects are based on the same material: CNTs. Further integration to include optics could lead to a unified electronic-optoelectronic technology.

The large electronic and spin coherence lengths of both materials could lead to quantum interference and spintronic devices. Spin-polarized injection can be achieved using manganite electrodes. The weak spin-orbit coupling and short transit times of carriers in nanotubes tend to preserve the spin polarization. Spin transport and spin precession over micrometer-scale distances have been demonstrated in single graphene sheets at room temperature.

Single-wall carbon nanotubes (SWNTs) are potential candidates for application in electronic devices. The most severe bottleneck is to grow uniform, single type and scale of SWNTs, because the as-synthesized SWNTs are a mixture of semiconducting (SC) and metallic (M) tubes. Recently, the separation of SC and M tubes

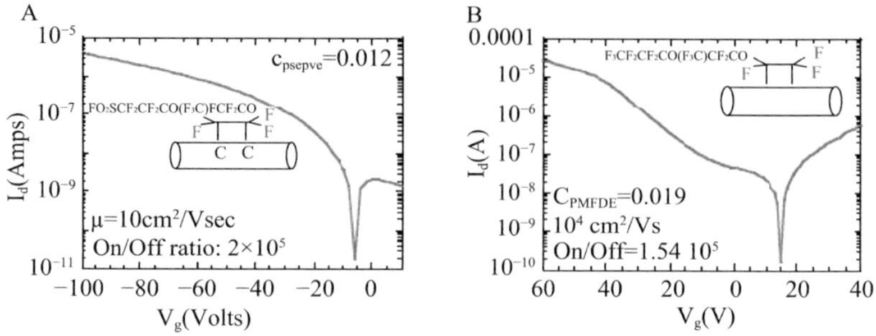

Figure 1.16 Source–drain current versus gate voltage for two functionalized SWNT FETs at $V_{ds} = -0.1$ V and -0.01 V, respectively.

has achieved some advances. The metallic M-SWNT cab be etched from the mixture through a controlled cycloaddition reaction with fluorinated polyolefins. The as-grown commercial nanotube mats are functionalized into a network of tubes that can then be dispersed in an organic solvent [21]. The resulting "denser" semiconducting inks, when coated as a percolating array, lead to high-mobility devices without requiring further nanotube separation.

Figure 1.16 shows the source–drain current versus gate voltage for two functionalized SWNT FETs [21]. The field-effect mobilities deduced from the linear regime are 10 cm^2/V·s and 104 cm^2/V·s with on/off ratio in excess of 10^5.

Along with the progressive improvement in the rapidly evolving technology of CNTs and graphene, the prospects of carbon-based devices are very bright.

References

1. Datta, S. (1995). *Electron Transport in Mesoscopic Systems* (Cambridge, London), p. 16.
2. Reggiani, L. (1985). *Hot-electron Transport in Semiconductors* (Spring-Verlag, Berlin), p. 7–11.
3. Ferry, D. K., Takagaki, Y., and Zhou, J. R. (1994). Future ultra-scale integration: transport physics in semiconductor nanostructures. *Jpn. J. Appl. Phys.* **33**, 873–878.

4. Kriman, A. M., Kluksdahl, N. C., and Ferry D. K. (1987). Scattering states and distribution functions for microstructures. *Phys. Rev. B* **36**, 5953–5959.
5. Datta, S. (1989). *Supperlatt. Microstruct.* **6**, 83.
6. Xia, J. B. (1992). *Phys. Rev. B* **45**, 3593.
7. Wolf, S. A., Awschalom, D. D., R. A. Buhrman, R. A., et al. (2001). *Science*, 294, pp. 1488.
8. Baibich, M. N., Broto, J. M., Fert, A., et al. (1988). *Phys. Rev. Lett.* **61**, 2472.
9. Binasch, G., Grunberg, P., Saurenbach, F., et al. (1989). *Phys. Rev. B* **39**, 4828.
10. Moodera, J. S., Kinder, L. R., Wong, T. M., et al. (1995). *Phys. Rev. Lett.* **74**, 3273.
11. Hu, C. M., Nitta, J., Jensen, A., et al. (2001). *Phys. Rev. Lett.* **63**, 125333.
12. Zhu, H. J., Ramsteiner, M., Kostial, H., et al. (2001). *Phys. Rev. Lett.* **87**, 016601.
13. S. Datta, S., and Das, B. (1990). Electron analog of the electro-optic modulator. *Appl. Phys. Lett.* **58**, 665.
14. Avouris, P., Chen, Z. H., and Perebeinos, V. (2007). Carbon-based electronics. *Nature Nanotechnol.* **2**, 605–615.
15. Novoselov, K. S., Geim, A. K., Morozov, S. V., et al. (2004). Electric field effect in atomically thin carbon films. *Science* **306**, 666–669.
16. Berger, C., Song, Z., Li, X., et al. (2006). Electronic confinement and coherence in patterned epitaxial graphene. *Science* **312**, 1191–1196.
17. Tans, S. J., Verscheuren, A. R. M., and Dekker, C. (1998). Room-temperature transistor based on a single carbon nanotube. *Nature* **393**, 49–52.
18. Martel, R., Schmidt, T., Shea, H. R., et al. (1998). Single- and multi-wall carbon nanotube field-effect transistors. *Appl. Phys. Lett.* **73**, 2447–2449.
19. Chen, Z., Appenzeller, J., et al. (2006). An integrated logic circuit assembled on a single carbon nanotube. *Science* **311**, 1735.
20. Wang, X. R., Ouyang, Y. J., Li, X. L., et al. (2008). Room-temperature all semiconducting sub-10-nm graphene nanoribbon field-effect transistors. *Phys. Rev. Lett.* **100**, 206803.
21. Kanungo, M., Lu, H., Malliars, G. G., and Blanchet, G. B. (2009). Suppression of metallic conductivity of single-walled carbon nanotubes by cycloaddition reactions. *Science* **323**, 234–237.

Chapter 2

Non-equilibrium Transport

In devices scaling larger than 100 nm the transport is described by the classical Boltzmann equation,

$$\frac{\partial f}{\partial t}+\mathbf{v}\cdot\frac{\partial f}{\partial \mathbf{r}}+\frac{e\mathbf{E}}{\hbar}\cdot\frac{\partial f}{\partial \mathbf{k}}=\int d\mathbf{k}'\left[f\left(\mathbf{k}'\right)W\left(\mathbf{k}',\mathbf{k}\right)-f\left(\mathbf{k}\right)W\left(\mathbf{k},\mathbf{k}'\right)\right], \tag{2.1}$$

where $f(\mathbf{k}, \mathbf{r}, t)$ is the single-particle distribution function at time t, momentum space $\mathbf{k}$, and real space $\mathbf{r}$. For semiconductors, the group velocity of the carrier is given by

$$\mathbf{v}=\frac{1}{\hbar}\frac{\partial\varepsilon\left(\mathbf{k}\right)}{\partial\mathbf{k}}. \tag{2.2}$$

When the scale of the devices becomes smaller, the electron transport will be nonlinear and nonstatic, as shown in Fig. 1.1. The distribution function is a derivate of the equilibrium distribution farther and largely varies with space and time. In some cases, it is recognized that the concepts of the Boltzmann equation and the distribution function are still effective, whose three main assumptions, that is, effective mass, energy band model, and electron collisions at space and time are instantaneous, are all tenable. But the Boltzmann equation cannot be solved by the ordinary perturbation method, while it can be by other methods, for example, the Monte Carlo method.

Quantum Waveguide in Microcircuits (Second Edition)
Jian-Bai Xia, Duan-Yang Liu, and Wei-Dong Sheng

ISBN 978-981-4968-30-0 (Hardcover), 978-1-003-65146-8 (eBook)
www.jennystanford.com

2.1 Monte Carlo Method

The Monte Carlo method utilizes the numerical method to simulate the movement of one or many electrons in the applied fields and scattering processes [1]. The macrophysical quantities, for instance, drift velocity, diffusion coefficient, and energy, are the assemble averages of corresponding physical quantities. If the system is a steady, space homogeneous distributed system, it is enough to simulate the movement of an electron. The statistical distribution of a given subsystem does not depend on the initial state of any other small part of the same system, since over a sufficiently long time the effect of this initial state will be entirely outweighed by the effect of the much larger remaining parts of the system. It is also independent of the initial state of the particular small part considered, since in time this part passes through all possible states, any of which can be taken as the initial state. It means that the average of a variable ϕ over the assemble equals the average of that variable over a sufficiently long time,

$$\bar{\phi} = \frac{\int \phi(\mathbf{k}) f(\mathbf{k}) d\mathbf{k}}{\int f(\mathbf{k}) d\mathbf{k}} = \frac{\int_0^T \phi[\mathbf{k}(t)] dt}{T} + O\left(T^{-1/2}\right), \tag{2.3}$$

where $O(T^{-1/2})$ represents the fluctuation proportional to $T^{-1/2}$. When T is long enough, the statistical error can be neglected and the simulation reaches convergence.

If the system is not steady, or inhomogeneous, then the movement of many particles (assemble) should be studied. At any time the transport properties of the assemble can be obtained by the average of the simulation of these particles. This Monte Carlo simulation of many particles is generally used to simulate the various instantaneous properties of a homogeneous system in the varied electric field. In the simulation of many particles the initial distribution of particles should be given and the instantaneous characteristics of the transport property actually depend on the initial distribution. This method is called a many-particle Monte Carlo simulation.

The free flight of electrons at intervals of two collisions satisfies Eq. 2.2 and

$$\hbar \frac{d\mathbf{k}}{dt} = -e\mathbf{E}. \tag{2.4}$$

Assume that the scattering probability of electrons is $\lambda(t)$, there are N_0 particles have the same initial state and the number of particles that have not collided at time t is denoted by $N(t)$. Then the number of first collisions between the time t and $t+dt$ is $-dN(t)$, and

$$-dN(t) = N(t)\lambda(t)dt. \tag{2.5}$$

Solving Eq. 2.5 we obtain

$$\frac{N(t)}{N_0} = \exp\left[-\int_0^t \lambda(t')\,dt'\right]. \tag{2.6}$$

Equation 2.6 is the probability of particles that have not collided at time t. Hence the probability of particles that have collided at time t is

$$\begin{aligned} P(t) &= 1 - \exp\left[-\int_0^t \lambda(t')\,dt'\right] \\ &= \int_0^t \lambda(t') \exp\left[-\int_0^{t'} \lambda(t'')\,dt''\right] dt' \\ &= \int_0^t f(t')\,dt', \end{aligned} \tag{2.7}$$

where

$$f(t) = \lambda(t)\exp\left[-\int_0^t \lambda(t')\,dt'\right], \tag{2.8}$$

representing the collision distribution function at time t.

The Monte Carlo method is semi-classical, which simulates the random movement of a single or many particles. On this basis, the needed properties are obtained through a proper statistical average. In the semi-classical picture, the microscopic movement of charged particles can be looked as a series of free flight and stochastic scatterings. The Monte Carlo method deals with the stochastic events related to scattering by a series of stochastic numbers between 0 and 1. On the other hand, for a given electric field, initial state, and free flight time, the whole flight process is completely determined.

From Eq. 2.7 the free flight time t_f can be obtained with the stochastic number r_t through the integration equation

$$r_t = 1 - \exp\left[-\int_0^{t_f} \lambda(t')\,dt'\right]. \tag{2.9}$$

From Eq. 2.9 we obtain

$$\int_0^{t_f} \lambda(t)\,dt = -\ln(1 - r_t)\,. \tag{2.10}$$

Actually the solution of the integration equation (Eq. 2.10) for such a scattering event is difficult. Rees [2] introduced the concept of self-scattering, making the problem simpler. Assume that the total scattering probability Γ, including the self-scattering, is a constant, independent of energy,

$$\Gamma = \lambda(E) + \lambda_s(E)\,, \tag{2.11}$$

where $\lambda_s(E)$ is the self-scattering probability. Inserting Γ into Eq. 2.10 instead of $\lambda(t)$, we obtain

$$t_f = -\frac{1}{\Gamma}\ln(1 - r_t)\,. \tag{2.12}$$

The flight time t_f can be easily calculated from the stochastic number r_t.

Suppose that the initial state of each free flight is $\mathbf{k}_a$. Then from the movement Eqs. 2.2 and 2.4 we can determine the momentum $\mathbf{k}_b$ and energy of the final state. With this method the scattering probability of the final state by various scattering mechanisms $\lambda_i(\mathbf{k}_b)$ can be calculated. Given a stochastic number r, if

$$\sum_{i=1}^{j} \lambda_i(\mathbf{k}_b) < r\Gamma < \sum_{i=1}^{j+1} \lambda_i(\mathbf{k}_b), \tag{2.13}$$

then we think that the collision of electrons is caused by the j-th scattering mechanism, whose scattering probability is $\lambda_j(\mathbf{k}_b)$. From the momentum $\mathbf{k}_b$ and energy of the final state and $\lambda_j(\mathbf{k}_b)$, we can determine the electron energy and momentum of the initial state of the next free flight.

If the Eq. 2.13 is not satisfied and

$$\sum_{i=1}^{N} \lambda_i(\mathbf{k}_b) < r\Gamma < \Gamma, \tag{2.14}$$

then we think that the electron undergoes a self-scattering and the electron energy and momentum of the next flight's initial state is the same as those of the above flight's final state. Repeating the above described process, we can obtain the movement course of a single electron in the long period of time or the situations at various times and positions of assemble electrons.

The physical quantity in the static state is given by Eq. 2.3, in the Monte Carlo simulation,

$$\bar{\phi} = \frac{1}{T}\sum_{i}\int_{0}^{t_i} \phi\left[k\left(t'\right)\right] dt', \tag{2.15}$$

where t_i is the flight time of each flight.

The distribution function can be calculated by a similar method. At the beginning of the calculation, we prepare a **k**-space mesh and in the process of simulation record the duration of the electron in each mesh lattice. If the time is long enough, the normalized duration time represents the distribution function of electrons.

2.2 Time-Related Transport Behaviors in Homogeneous Semiconductors

2.2.1 Drift Diffusion Model

The Monte Carlo simulation needs the knowledge of the relaxation mechanisms, of energy band shapes, and of the coupling constants and requires rather long computation times. Moreover, the transient response is scarcely given. Nougier et al. [3] developed a simple model based on the relaxation time approximation to study the transient response in semiconductors and compared the results with the numerically computed curves.

The dynamical equations, in the relaxation approximation, are balanced equations related to the drift velocity $\bar{v}(t)$ and to the average energy $\bar{\varepsilon}(t)$ of the carriers. One then defines the momentum and energy relaxation time τ_{p} and τ_{ε}, which are the functions of energy. The balanced equations are

$$\frac{m\partial\bar{v}\left(t\right)}{\partial t} = qE - \frac{m\bar{v}\left(t\right)}{\tau_{\mathrm{p}}\left(\bar{\varepsilon}\right)}$$

and

$$\frac{\partial\bar{\varepsilon}\left(t\right)}{\partial t} = qE\bar{v}\left(t\right) - \frac{\bar{\varepsilon}\left(t\right) - \varepsilon_0}{\tau_{\varepsilon}\left(\bar{\varepsilon}\right)}, \tag{2.16}$$

where E is the electric field, q is the electron charge, and ε_0 is the average energy in thermal equilibrium.

In the steady-state condition, equating $d(m\bar{v})/dt$ and $d\bar{\varepsilon}/dt$ to zero in Eq. 2.16 gives

$$\tau_{\mathrm{p}}(\varepsilon_{\mathrm{s}}) = \frac{m^* v_{\mathrm{s}}}{qE}, \quad \tau_{\varepsilon}(\varepsilon_{\mathrm{s}}) = \frac{\varepsilon_{\mathrm{s}} - \varepsilon_0}{qEv_{\mathrm{s}}} \tag{2.17}$$

where v_{s} and ε_{s} are the drift velocity and energy in the steady-state condition, which are all functions of the electric field or the energy $\bar{\varepsilon}$. Inserting Eq. 2.17 into Eq. 2.16 gives

$$\frac{d(m\bar{v})}{dt} = q\left[E - \frac{\bar{v}}{v_{\mathrm{s}}(\bar{\varepsilon})}E_{\mathrm{s}}(\bar{\varepsilon})\right] = q\left[E - \frac{\bar{v}}{\mu_{\mathrm{s}}(\bar{\varepsilon})}\right]$$

and

$$\frac{d\bar{\varepsilon}}{dt} = q\left[E\bar{v} - E_{\mathrm{s}}(\bar{\varepsilon})v_{\mathrm{s}}(\bar{\varepsilon})\right] = q\left[E\bar{v} - \mu_{\mathrm{s}}(\bar{\varepsilon})E_{\mathrm{s}}^2(\bar{\varepsilon})\right], \tag{2.18}$$

where $\mu_{\mathrm{s}}(\bar{\varepsilon}) = v_{\mathrm{s}}(\bar{\varepsilon})/E_{\mathrm{s}}(\bar{\varepsilon})$ is the steady-state mobility.

Figure 2.1 shows a comparison between the transient energy $\bar{\varepsilon}(t)$ in p-Ge deduced from the solution of the Boltzmann equation by an iterative method and from the relaxation time approximation. From the figure we see that the electron energies increase with time and the electric field, and approach saturation, but have no overshoot. The discrepancy between the results of the two methods is less than 10% until at least 20 kV/cm.

2.2.2 Transport in a Strong Electric Field

The calculated τ_{p} and τ_{ε} as functions of the difference between the average and thermal energies for the case of electrons in Si and GaAs at 293 K are shown in Figs. 2.2 and 2.3, respectively [4]. From the figure we see that in the hot-electron regime the relaxation times are not constants, which depend on the electron energy. For GaAs, τ_{p} and τ_{ε} increase first, then decrease with $\varepsilon - \varepsilon_0$ increasing, and for Si, τ_{p} decreases and τ_{ε} increases with $\varepsilon - \varepsilon_0$. The momentum relaxation times are larger than the energy relaxation times by 1 order of magnitude.

The balanced equations Eq. 2.18 have already been used successfully to determine the dependence on time of the drift velocity and the average energy; when a given time configuration of the electric field is applied to the semiconductor, a good agreement with more exact calculations has been observed. Figure 2.4 shows

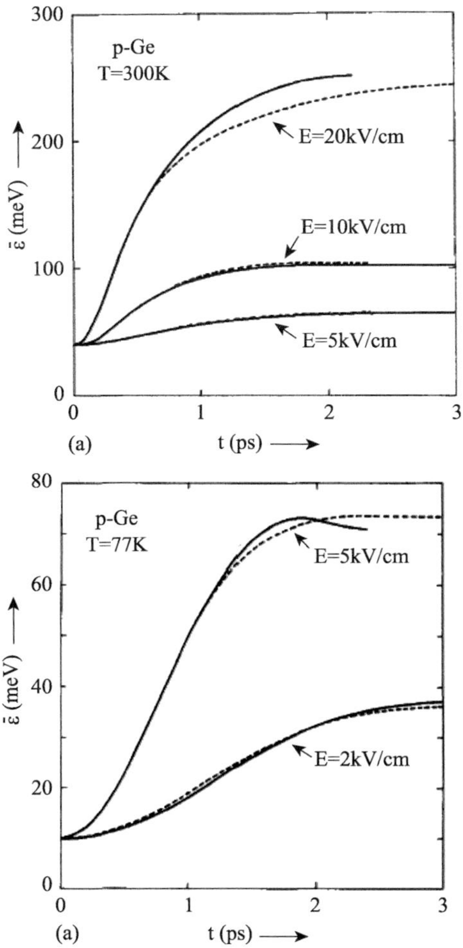

Figure 2.1 Comparison between the transient energy $\bar{\varepsilon}\,(t)$ in p-Ge deduced from the solution of the Boltzmann equation by an iterative method and from the relaxation time approximation: (a) at 300 K and (b) at 77 K.

a comparison of the draft velocities as functions of time when a step of the electric field is applied between the results obtained with the relaxation time description (dots) and Monte Carlo or the iterative method (curves) [4]. It can be seen that for the five semiconductors studied, the results obtained by both methods are in good agreement. Consequently, owing to its simplicity, the balance

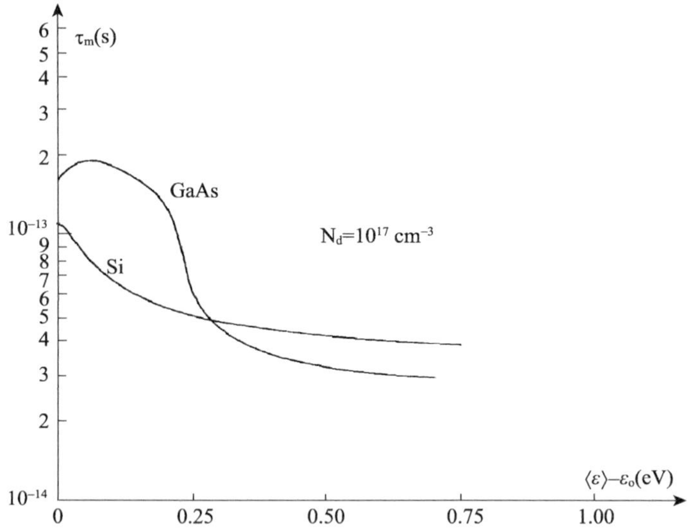

Figure 2.2 The momentum relaxation time as a function of the difference between the average and thermal energies for the case of electrons in Si and GaAs at 293 K.

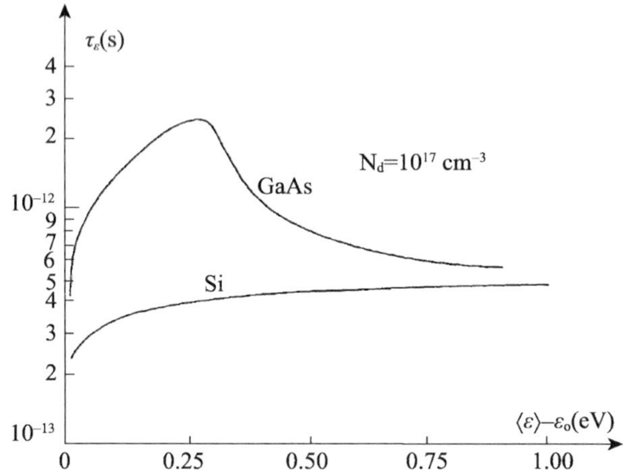

Figure 2.3 The energy relaxation time as a function of the difference between the average and thermal energies for the case of electrons in Si and GaAs at 293 K.

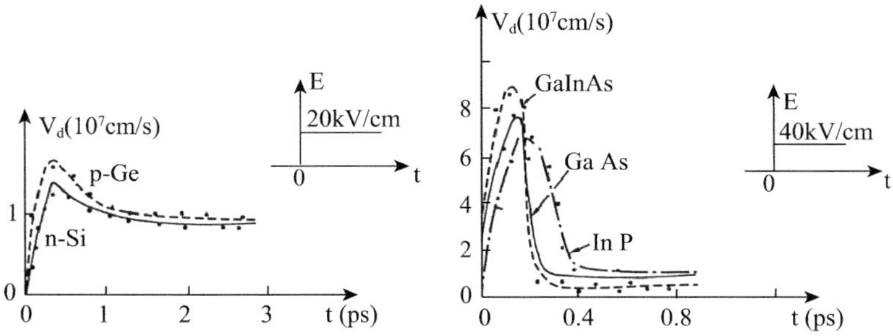

Figure 2.4 Comparison of the draft velocities as functions of time when a step of the electric field is applied between the results obtained with the relaxation time description (dots) and the Monte Carlo or the iterative method (curves).

equation method with the relaxation time approximation is very useful for the study of time-dependent phenomena.

From Fig. 2.4 we see that in a time of less than 1 ps the draft velocities rise rapidly, reaching $10^7 \approx 10^8$ cm/s. Then due to the momentum and energy relaxations, whose relaxation times are $0.1 \approx 1$ ps, the draft velocities decrease gradually and approach the steady-state values. The rapid increase of the draft velocity in a strong electric field is called a velocity overshoot. It is because at the initial stage of applying an electric field the electrons are accelerated suddenly by the strong electric field and do not have enough time to be scattered, so they can reach a very high draft velocity. Transport at this stage is called ballistic transport. The scattering plays a role with passage of time, which compensates the accelerating action of the electric field and makes the draft velocity approach a steady-state value.

In a transistor the size of the base region can be reduced to 100 nm or less without allowing the increase of the base resistance. In such a thin-base-region transistor the length of the electron-traveling region is very short, the transport of electrons could be near ballistic by a hot-electron injection structure, and then the working frequency will be very high. Hence such a transistor can be used to design a high-frequency transistor. Figure 2.5 shows a model of n-p-n heterojunction bipolar transistors of types (A), (B), and (C) [5]. The material in the n^+-emitter and p^+-base regions is

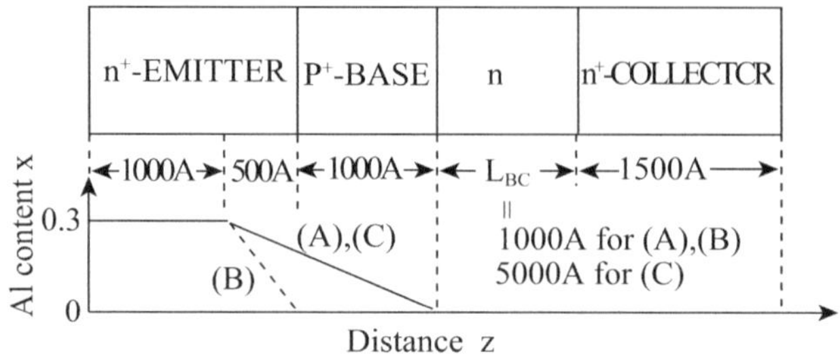

Figure 2.5 Model of n-p-n heterojunction bipolar transistors of types (A), (B), and (C).

$Al_xGa_{1-x}As$. The compositional grading x is linear and starts with $x = 0.3$ at 500 Å inside the emitter from the emitter-base junction and ends with $x = 0$ at the base-collector junction (types (A) and (C)) or at the emitter-base junction (type (B)). The grading for transistors of types (A) and (C) produces a quasi-electric field of about 20 kV/cm in the base region.

Figure 2.6 shows the profile of the mean electron velocity in a type (A) transistor (solid curve) and a type (B) transistor (dashed curve). The mean electron velocity in (A) reaches as high as 4.5×10^7 cm/s at base-collector junction. This is because electrons in the graded base region travel more or less ballistically as a result of the accelerating force of the quasi-electric field. The mean electron velocity in (B) is as low as 5×10^6 cm/s (notice that the ordinates in Fig. 2.6 are different for (A) and (B)) in the substantial part of the base region, suggesting that the electron motion in the nongraded base region is diffusion dominated.

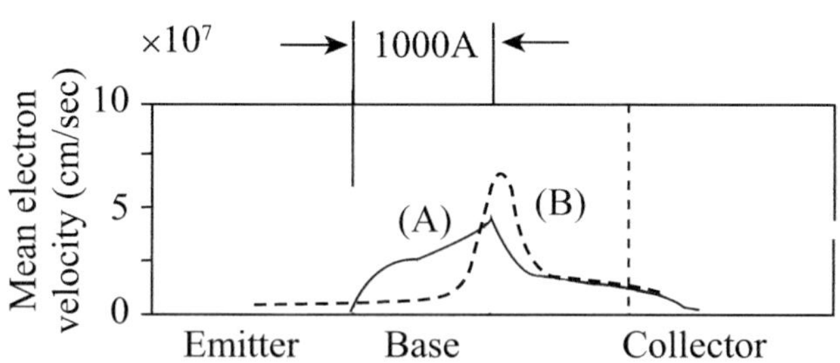

Figure 2.6 Profile of the mean electron velocity in a type (A) transistor (solid curve) and a type (B) transistor (dashed curve).

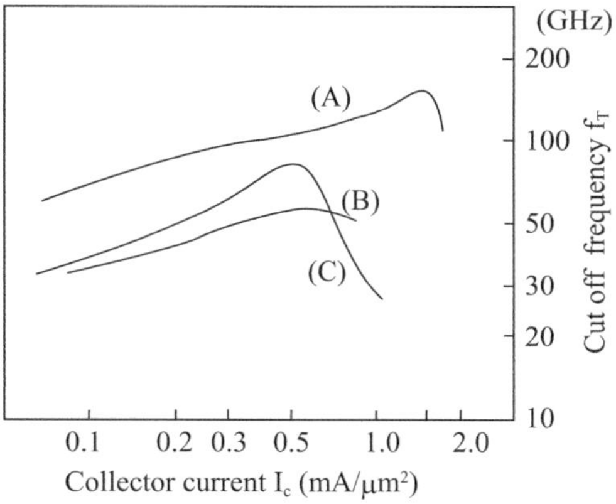

Figure 2.7 Current-gain cutoff frequencies f_T of transistors of types (A), (B), and (C).

Figure 2.7 shows the current-gain cutoff frequencies f_T of transistors of types (A), (B), and (C). It is seen in Fig. 2.7 that the f_T of a type (A) transistor shows an extremely high maximum value of about 150 GHz at $I_c = 1.4$ mA/μm^2. The maximum f_T of a type (B) transistor reaches only 58 GHz at $I_c = 0.6$ mA/μm^2. This is due to a large base capacitance caused by the diffusion-dominated electron transport in the base region. The maximum f_T of a type (C) transistor reaches only 80 GHz at $I_c = 0.5$ mA/μm^2, smaller than that of type (A). The decrease of f_T is due to the appearance of the base-push-out effect.

2.2.3 Application of a Balance Equation

From the above discussion we see that the results obtained by the relaxation time description (the balance equation) and the Monte Carlo or the iterative method are in good agreement. Due to its simplicity, the balance equation method with the relaxation time approximation is very useful for the study of time-dependent phenomena. The balance equation can be used to study the transport properties not only in the strong electric field case but also in the weak electric field case. In the weak electric field case, the momentum and energy relaxation times can be taken as constants; the calculation will be very simple. In the following we give an example: the terahertz (THz)-photocurrent resonances in miniband superlattices.

When a laser beam from the free electron laser (FEL) irradiates the GaAs/$Al_{0.3}Ga_{0.7}As$ superlattice in the differential conductance (DC) bias, giving pulses of intense THz radiation with a pulse length of several microseconds. The GaAs/$Al_{0.3}Ga_{0.7}As$ superlattice consists of 40 periods of an 8 nm wide GaAs well and 2 nm thick AlGaAs barrier. The width of the miniband is $\Delta = 22$ meV, and the electron density is 3×10^{15} cm^{-3}. The measured DC currents as functions of the DC bias for different laser powers are shown in Fig. 2.8 [6]. The FEL frequency was fixed to 0.6 THz (Fig. 2.8a) and 1.5 THz (Fig. 2.8b). The curves in the figure are shown for increasing FEL intensity from up to down. From Fig. 2.8 we see that at low intensities an additional peak emerges in the negative differential conductance (NDC) region. When the intensity is increased further, the first peak starts to decrease and a second peak at about twice the voltage of the first peak is observed due to a two-photon resonance.

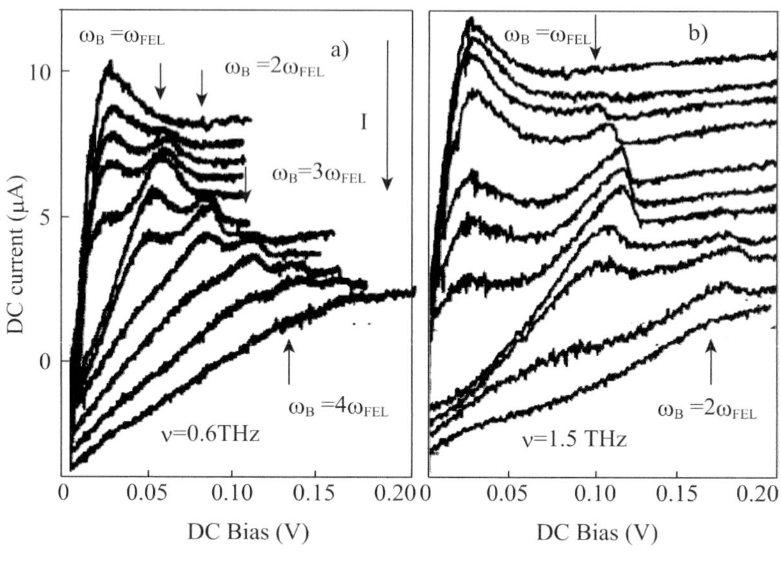

Figure 2.8 DC current-voltage for increasing the FEL intensity (the curves are shifted downward for an increasing laser intensity). The FEL frequency was fixed to 0.6 THz (a) and 1.5 THz (b).

The position of the peaks does not change with the intensity of the FEL. The authors attribute the first additional peak to a resonance of the external laser field where the Bloch frequency $\omega_B = \omega$. In Fig. 2.8b the laser frequency is 1.5 THz and the peaks are shifted to higher voltages and are much more pronounced. However, only the fundamental and the second harmonics are observed.

The Bloch oscillation is a special phenomenon in the longitudinal transport of the superlattice. In the intense electric field F the electron moves periodically in the one-dimensional (1D) Brillouin zone (BZ) of k space, $-\pi/L \leq k \leq \pi/L$, where L is the period of the superlattice. The oscillation is called the Bloch oscillation, whose oscillation frequency is $\omega_B = eFL/h$, proportional to the DC voltage and the superlattice period.

We use the balanced equations (Eq. 2.16) to study the miniband transport of superlattices under a direct bias and an alternating electric field and compare the theoretical results with the experimental results [7]. We take the relaxation time τ_v and τ_ε as constant, and the balanced equations are rewritten as

$$\frac{dv}{dt} = \frac{eF(t)}{m^*} - \frac{v}{\tau_v}$$

and

$$\frac{d\varepsilon}{dt} = eF(t)\,v - \frac{\varepsilon - \varepsilon_0}{\tau_\varepsilon}, \tag{2.19}$$

where m^* is the effective mass along the electron draft direction (longitudinal direction), which depends on the energy ε, and

$$m^* = \frac{m_0^*}{1 - 2\varepsilon/\Delta} \quad \text{and} \quad m_0^* = \frac{2\hbar^2}{\Delta d^2}, \tag{2.20}$$

where the origin of energy is at the conduction band bottom, Δ is the width of the miniband, and d is the period of the superlattice. In calculation we take the experimental parameters $\Delta = 23$ meV, $d = 10$ nm, $\varepsilon_0 = 0.05\,\Delta$, $\tau_v = 1.3$ ps, $\tau_\varepsilon = 6.5$ ps, and the frequency of the alternating field $\nu = 0.6$ THz. The electric field is applied to the superlattice from $t = 0$,

$$F(t) = F_0 + F_1 \cos \omega t. \tag{2.21}$$

The balanced equations (2.19) are solved by a numerical method, and the velocity $v(t)$ and energy $\varepsilon(t)$ are obtained as functions of t.

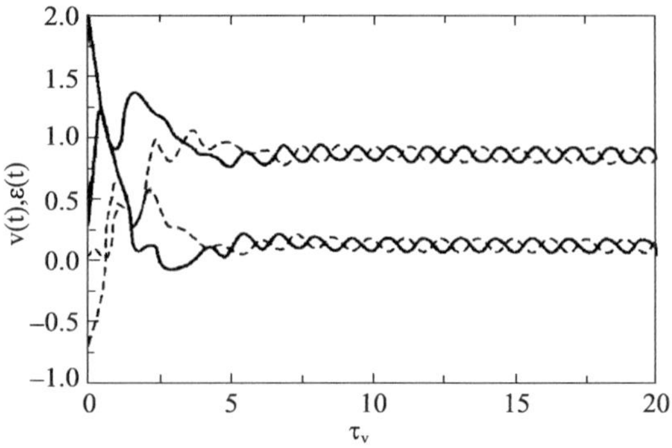

Figure 2.9 The $v(t)$ and $\varepsilon(t)$ for $F_0 = 0.5$ and $F_1 = 1$ (solid curve) and $F_0 = 0.5$ and $F_1 = -1$ (dashed curve).

Figure 2.9 shows the $v(t)$ and $\varepsilon(t)$ for $F_0 = 0.5$ and $F_1 = 1$ (solid curve) and $F_0 = 0.5$ and $F_1 = -1$ (dashed curve). The unit of the electric field is 1000 V/cm in the following. From Fig. 2.9 we see that after 10 periods of the external alternating field the $v(t)$ and $\varepsilon(t)$ reach steady states. At the initial stage the velocity and energy of electrons all reach very large values, which means there are "overshoot" and "overhot" behaviors, and the electrons do not have enough time to collide. The $v(t)$ and $\varepsilon(t)$ for $F_0 = 0.5$ and $F_1 = -1$ are similar to those for $F_0 = 0.5$ and $F_1 = 1$, only different from a phase π.

In the steady state the $v(t)$ is a periodic function with a period $T = 2\pi/\omega$, which consists of DC term and harmonic terms of ω,

$$v(t) = v_0 + \sum_{n=1} v_{\mathrm{n}} \cos(n\omega t + \varphi_{\mathrm{n}}). \tag{2.22}$$

Here for definition we assume that all v_{n} (except v_0) are positive. By Fourier transform, we obtained all v_{n} and $\cos\varphi_{\mathrm{n}}$. If $\cos\varphi_1 < 0$, that means that the first harmonic current is in the opposite phase with the driven alternating field and the energy will be transferred to the alternating field, that is, the alternating field will be amplified.

The DC current is given by $j_0 = ev_0 n$, where n is the electron concentration, which is taken as 10^{15} cm^{-3} in the following. The DC currents as functions of F_0 for different F_1 are shown in Fig. 2.10 [7]. From Fig. 2.10 we see that the global characteristic is in good agreement with the experimental results [7]: Fig. 2.8. The

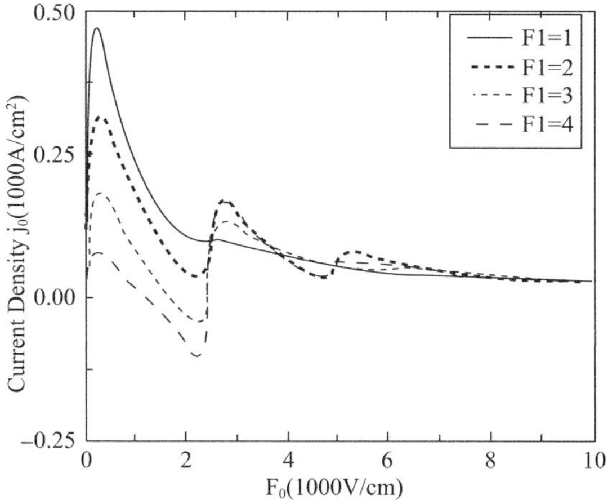

Figure 2.10 DC currents as functions of F_0 for different F_1.

DC currents show resonance peaks in the fields F_0 corresponding to the Bloch frequency $\omega_B = \omega, 2\omega, 3\omega, \ldots$. The resonance peaks in the DC current grow with increasing laser intensity (F_1), and simultaneously the current at the low bias side decreases.

The harmonic components of current j_0, j_1, j_2, and j_3 as functions of F_0 for $F_1 = 2$ are shown in Fig. 2.11 [7]. From Fig. 2.11

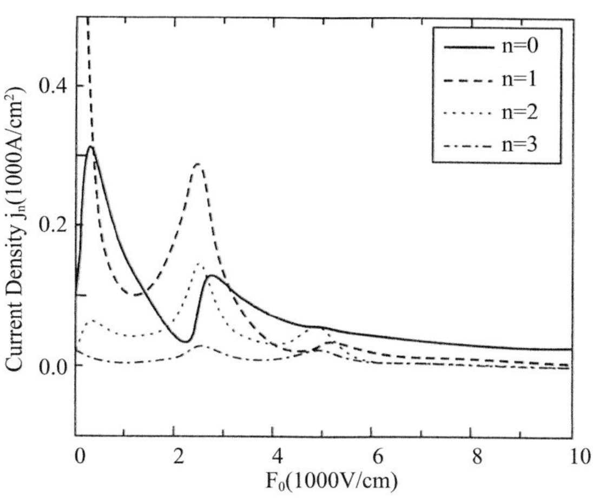

Figure 2.11 Harmonic components of current j_0, j_1, j_2, and j_3 as functions of F_0 for $F_1 = 2$.

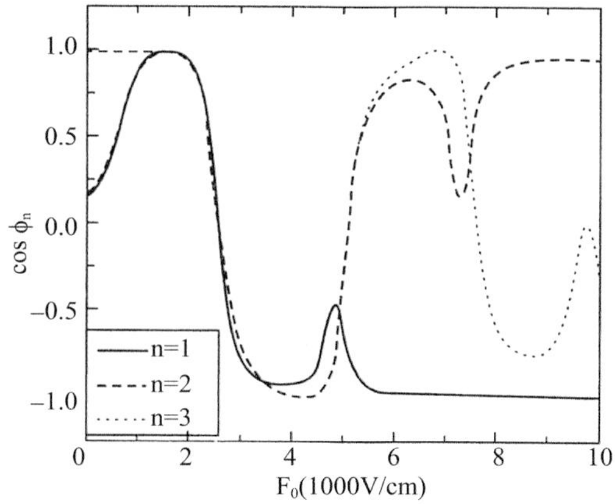

Figure 2.12 Cosines of the phase angle for each component $\cos\varphi_n$ as functions of F_0 for $F_1 = 2$.

we see that they also have resonance peaks at the resonance frequencies $\omega_B = \omega$, 2ω, and 3ω. At $\omega_B = \omega$, the values of j_1 and j_2 exceed that of j_0. The cosines of the phase angle for each component $\cos\phi_n$ as functions of F_0 for $F_1 = 2$ are shown in Fig. 2.12. From Fig. 2.12 we see that $\cos\phi_n$ oscillates with the DC field F_0 and becomes negative at the field range corresponding to the Bloch frequency ω_B between ω and 2ω. Cos ϕ_1 is negative for $\omega_B > \omega$, and with resonance peaks at $\omega_B = 2\omega, 3\omega, \ldots$.

2.2.4 Device Design Considering a Strong Field Transport

When we want to design high-speed devices, the main goal is to find the best way of making the electron go as fast as possible along the active region of submicron devices. Generally for a given active region of width d, the problem is to find the optimum electric field configuration and the doping concentration that should characterize the component.

The simplest way to solve this problem is to study the motion of an ensemble of electrons in a uniform bulk semiconductor to a time configuration of the electric field. From the variation of the average drift velocity $v(t)$, the average distance d traveled by the carrier over a time $t$and, consequently, the average velocity over a distance d can

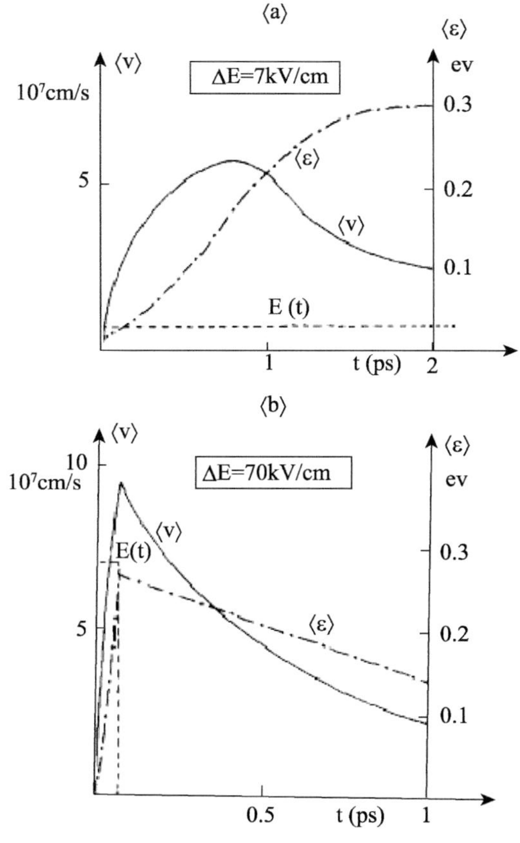

Figure 2.13 Average velocity and average energy of electron in GaAs. The impurity concentration $N_D = 0$ and $T = 77$ K subjected to a time-electric-field step as functions of time. (a) At $t = 0$ an electric-field step $\Delta E = 7$ kV/cm is applied; (b) an electric field pulse $\Delta E = 70$ kV/cm with the duration $\Delta t = 0.06$ ps is applied.

be obtained from the following equations:

$$d = \int_0^T v(t)\,dt, \quad v(d) = \frac{d}{T}. \tag{2.23}$$

Figure 2.13 shows the average velocity and average energy of electrons in GaAs. The impurity concentration $N_D = 0$ and $T = 77$ K are subjected to a time-electric-field step as functions of time [8]. The results are obtained by the Monte Carlo simulation. (a) At $t = 0$

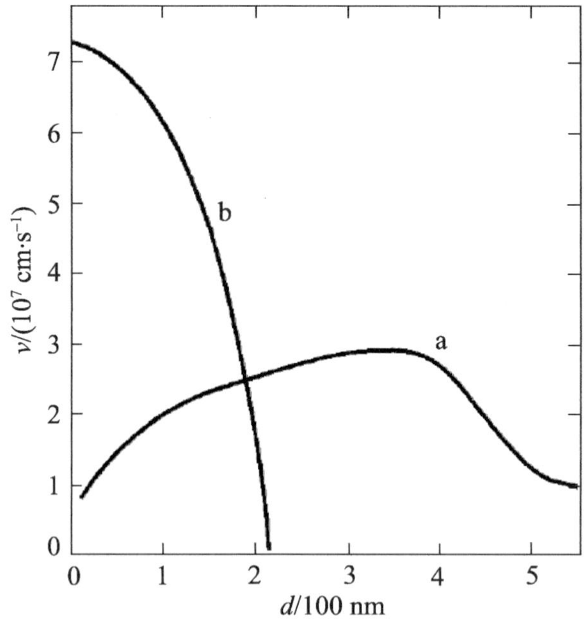

Figure 2.14 Electron average velocities v as functions of traveling distance d: (a) overshoot case and (b) ballistic case.

an electric-field step $\Delta E = 7$ kV/cm is applied; (b) an electric field pulse $\Delta E = 70$ kV/cm with the duration $\Delta t = 0.06$ ps is applied, shown by dashed curves in figures. The former case is called an overshoot, and the latter case is called ballistic by the authors.

When the electron energy is smaller than the polar optical phonon energy (case 1) or smaller than the intervalley energy of Γ and L valleys $\Delta\varepsilon_{\Gamma_L}$ (case 2), the scattering probability is small. Thus, it is possible to obtain very high instantaneous drift velocities 3.5×10^7 and 10^8 cm/s, respectively, in the two cases.

The distance and average velocity of electrons in the time period T can be calculated by Eq. 2.23. For the two cases the calculated $v(d)$ curves are shown in Fig. 2.14. From the figure we see that in the case of an overshoot the maximum average velocity is only half of that in the ballistic case, which occurs at $d = 375$ nm. At this distance the average energy reaches the polar optical phonon energy. In the ballistic case, the average velocity decreases monotonously with the distance, and at $d = 0$ it has the maximum value 7.3×10^7 cm/s. This is because after the electric field pulse the electron does not have energy supplement and the farthest distance is about 200 nm.

2.3 Transport Related to Space

Because the scale of the device becomes smaller and smaller, the role of the velocity overshoot in FET operation is still not entirely clear. The reason is multifold. For one, the overshoot, which may be important in FETs, arises not from the time dependence directly but from the fact that electrons travel across the device in small amounts of time. Transport at each point may, therefore, be time independent. It is the rapid variation (the gradients) of the electric field that results in a velocity overshoot and therefore in increased velocities in FETs even if the transistor is continuously on. Since the spatial gradients of the field are so important, the devices must be very short to exhibit such high gradients.

A high-electron-mobility transistor (HEMT) is shown in Fig. 2.15 [9]. The wider-gap AlGaAs is grown on top of GaAs and is highly doped. The electrons, however, leave their donors and transfer to the lower-gap GaAs, leaving the AlGaAs virtually depleted and creating a high-mobility channel in the GaAs, which is almost free of impurities (modulation doping effect). Electrons move in the high-mobility

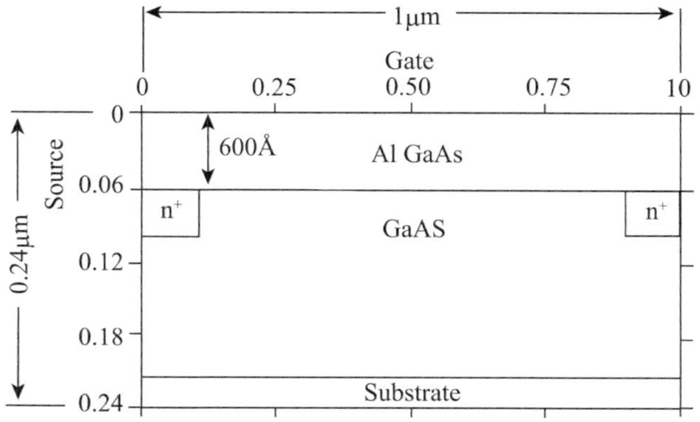

Figure 2.15 Schematic representation of the AlGaAs/GaAs HEMT.

channel, whose length is smaller than 1 μm; hence the electron transport is related to the space.

Here the Monte Carlo simulation needs simultaneously to study the electron motion orbits in the real and momentum spaces. And because the electric field at each point of space is not only decided by the applied electric field but also related to the charge density at that point, the complete self-consistent Monte Carlo simulation should include the solutions of Poisson's equation.

The average velocities of electrons in GaAs after injection in the ⟨100⟩ direction at *k* values are shown in Fig. 2.16 [9]. Figure 2.16a is for a given electric field in the base (10 kV/cm), and in Fig. 2.16b the electric field is varied and the injection energy (*k* value) is fixed. The injected electron is accelerated by a constant electric field and then decelerated due to scattering. Curves a–g in Fig. 2.16a are for injection energies below $\Delta E_{\Gamma L}$ and only curve h is for an energy slightly above $\Delta E_{\Gamma L}$. One can see that within the Γ valley average near-ballistic velocities of up to 8×10^7 cm/s are possible over distances close to 1000 Å. But if the electrons transfer to

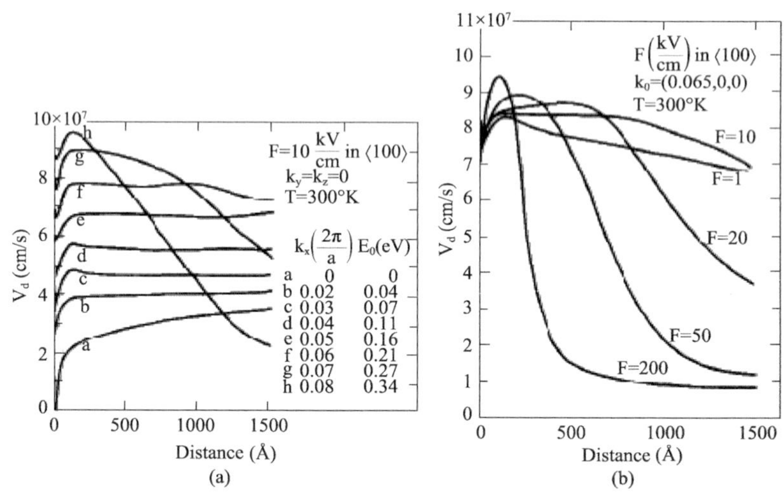

Figure 2.16 Average velocities of electrons in GaAs after injection in the ⟨100⟩ direction at *k* values. (a) Various injection energies for a given electric field in the base (10 kV/cm); (b) the electric field is varied and the injection energy (*k* value) is fixed.

the L valley, the velocity decreases quickly due to the intervalley scattering, as shown by curve h; the ballistic distance decreases to a few hundred angstroms. Figure 2.16b shows the field dependence of the velocity. The higher fields accelerate the electrons above ΔE_{Γ_L} and electron transfer, thus decreasing the velocity. When the electric field is 10 kV/cm, the scattering is compensated by the acceleration of the electric field. The ballistic velocity can be kept to a longer distance. Therefore, the near-ballistic range should extend from a few hundred to thousand angstrom distances depending on the injection energy and the electric field.

Generally, the range in which the electron velocity increases from 0 to the maximum in Fig. 2.16 is called the ballistic range and the range after the maximum to the point of reaching the steady-state velocity is called the overshoot range. In a moderate electric field the steady-state velocity is smaller than 10^7 cm/s. The separation of the ballistic range and the overshoot range is not always as clear as indicated above, and the ballistic range does not always extend to the point of velocity maximum. Especially when the electron injection energy is small (or zero) the velocity maximum is reached only after 4000 Å for an electric field of 10 kV/cm, while the ballistic range is much shorter ($\sim$1000 Å). The velocities as functions of distance for two different injection energies and electric fields are shown in Fig. 2.17 [9].

Figure 2.18 shows the velocity as a function of distance in the channel for a gate voltage of 0.4 eV and various drain voltages of the HEMT device shown in Fig. 2.15 [9], where the heavy horizontal bar marks source, gate, and drain regions. An overshoot is clearly present, particularly at the right side of the gate. There is, however, also a significant portion (close to the source) of the device in which the velocity is below the maximum steady-state value. In this region, the mobility is degraded by hot-electron effects and the field gradients are too small to produce an overshoot. As a consequence, figures of merit such as current switching time, cutoff frequency, and the average draft velocity in the channel, etc., are complicated averages over overshoot regions and regions of mobility degradation and may show no direct trace of overshoot effects.

An alternative method for studying space-dependent phenomena makes use of the balance equation as obtained by the integration

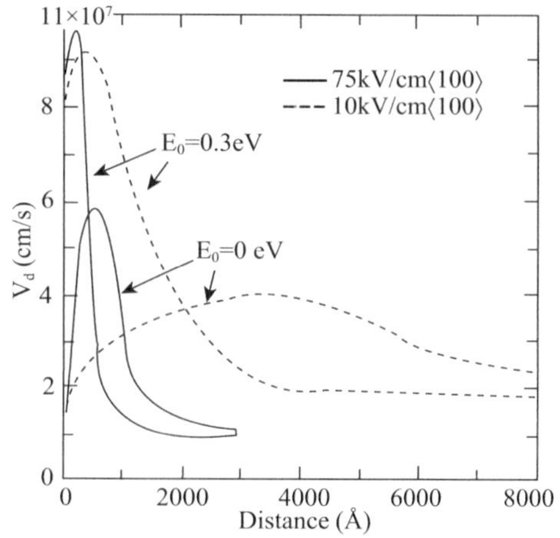

Figure 2.17 Velocities as functions of distance for two different injection energies and electric fields.

of the Boltzmann equation over k space [4]. In the single-valley case, by using a 1D treatment this integration leads to the three fundamental macroscopic equations similar to the balanced equations Eqs. 2.16, which express, respectively, the conversation of the number, momentum, and energy of the particle ensemble,

$$\frac{\partial n}{\partial t}+\frac{\partial\left(n\bar{v}\right)}{\partial x}=0,$$

$$\frac{\partial}{\partial t}\left(nm^{*}\bar{v}\right)=enE-\frac{\partial}{\partial x}\left(nk_{\mathrm{B}}T\right)-\frac{\partial}{\partial x}\left(nm^{*}\bar{v}^{2}\right)-\frac{nm^{*}\bar{v}}{\tau_{\mathrm{p}}\left(\bar{\varepsilon}\right)},$$

and

$$\frac{\partial}{\partial t}\left(n\bar{\varepsilon}\right)=enE\bar{v}-\frac{\partial}{\partial x}\left[n\bar{v}\left(\bar{\varepsilon}+k_{\mathrm{B}}T\right)\right]-n\frac{\bar{\varepsilon}-\varepsilon_{0}}{\tau_{\varepsilon}\left(\bar{\varepsilon}\right)}. \tag{2.24}$$

Solving Eq. 2.24 and Poisson's equation, the spatial dependence of the drift velocity is obtained for a GaAs n^+-n-n^+ structure under static conditions at 77 K and compared with that obtained with the Monte Carlo method. The results are shown in Fig. 2.19 [4]. It is seen that the agreement is good. The non-uniform velocity distribution in space is similar to that in Fig. 2.18.

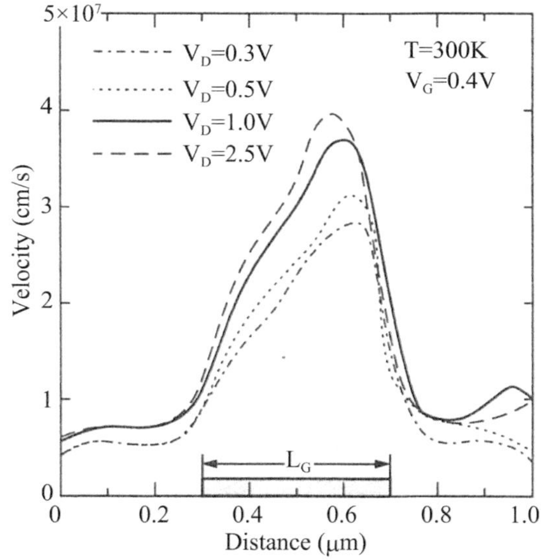

Figure 2.18 Velocity as a function of distance in the channel for a gate voltage of 0.4 eV and various drain voltages of the HEMT device shown in Fig. 2.15.

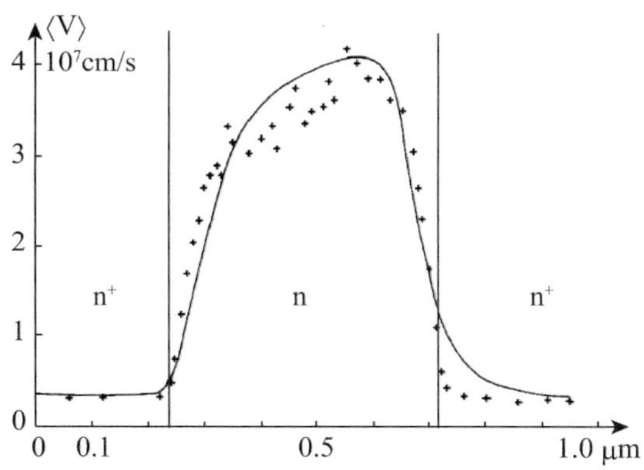

Figure 2.19 Spatial dependence of the drift velocity for a GaAs n^+-n-n^+ structure under static conditions at 77 K obtained from a Monte Carlo simulation (+ + +), and an analytical solution of Eq. 2.24 under static conditions (____).

2.4 Transport in a Si-MOSFET

Silicon metal-oxide semiconductor field-effect transistors (Si-MOSFETs) have been investigated using a large variety of physical models, from full-band Monte Carlo simulations to parabolic-band approximations coupled to Monte Carlo or simplified solutions to the transport equation, but relatively consistent results cannot be obtained. The reason is that the energy band structure of Si is more complicated. It is an indirect bandgap semiconductor, the conduction band bottom is not at Γ point but at X point in the BZ. Thus the various scattering and impact ionization mechanisms are highly anisotropic. Besides, the basic knowledge of transport of the high-energy (>1 eV) electrons in Si is still missed.

For example, due to the anisotropic band structure of Si and the anisotropic intrinsic in ionization processes the spread in the ionization rate is extended over about 1 order of magnitude. Taking into account the realistic joint density of states in Si, and the anisotropic ionization processes, Sano et al. [10] calculated the ionization rate at every point in the first BZ, as shown in Fig. 2.20 by points, with the electron energy measured from the conduction band edge. From the figure we see a large spread in the ionization

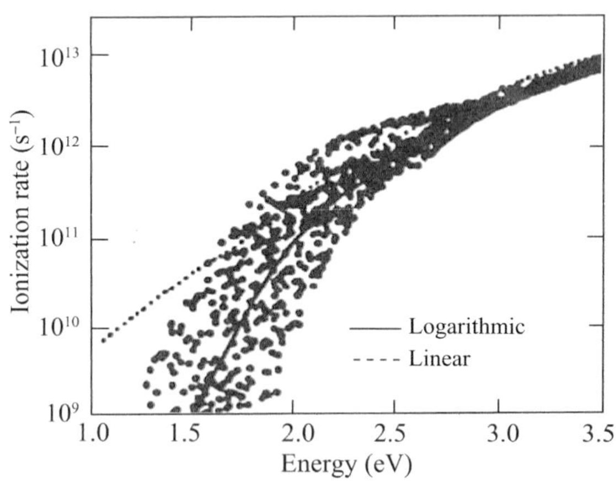

Figure 2.20 Calculated ionization rate as a function of electron energy marked by points, each point for a **k** point in the first Brillouin zone.

rates at low electron energies ($\leq$3 eV), which extends over 2 orders of magnitude. The anisotropy is greatly diminished above 3 eV, where the direct transition across the gap becomes possible. Therefore, a realistic model of the ionization process has to reflect this complexity at low electron energies associated with the realistic joint density of states.

The present anisotropic ionization rate $w_{ii}(\mathbf{k})$ can be transformed into the isotropic energy-dependent ionization rate $w_{ii}(\varepsilon)$ by integrating it over all directions,

$$w_{ii}(\varepsilon) = \frac{\int d\mathbf{k}\delta(\varepsilon - \varepsilon_{\mathbf{k}})\, w_{ii}(\mathbf{k})}{\int d\mathbf{k}\delta(\varepsilon - \varepsilon_{\mathbf{k}})}. \tag{2.25}$$

Here the wave-vector-dependent ionization rate $w_{ii}(\mathbf{k})$ is logarithmically interpolated with the eight nearest corners $(k^{(n)})$ $(n = 1, \ldots, 8)$ of each cubic mesh in the BZ,

$$\ln[w_{ii}(k)] = \sum_{n=1}^{8} \left[\prod_{m=x,y,z} \left(1 - \frac{\left| k_{\mathrm{m}} - k_{\mathrm{m}}^{(n)} \right|}{\Delta k_{\mathrm{m}}} \right) \right] \times \ln\left[w_{ii}\left(k^{(n)}\right) \right]. \tag{2.26}$$

The ionization rates interpolated logarithmically and linearly are shown in Fig. 2. 20 [10] by solid and dotted lines, respectively. From the figure we see that the linear interpolation yields a value larger than all values for the different **k**s, with a given energy near the threshold. This is because when the interpolation is done with a large mesh spacing, the energies at the eight corners in a cubic mesh could vary greatly so that some rates in a cubic mesh could be quite large. As a result, the ionization rate linearly interpolated from those values could be larger than the values for the different *k*s with a given energy.

The present energy-dependent ionization rate averaged over all directions (solid line) is compared in Fig. 2.21 [10] with Kane's ab initio theoretical result (dotted line), the rates extracted from recent experiments of the IBM group (dashed line), and the ab initio calculation result of the Osaka group (black points). The present results closely agree with those of both the IBM group and the Osaka group.

On the basis of the relative consistence of the ionization rates and electron-phonon scattering rates Sano et al. applied the full-band Monte Carlo method to simulate a typical MOSFET structure

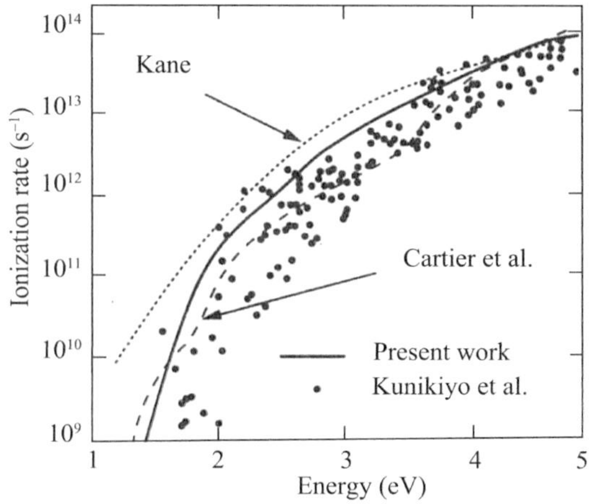

Figure 2.21 Present energy-dependent ionization rate averaged over all directions (solid line) is compared with Kane's ab initio theoretical result (dotted line), the rates extracted from recent experiments of the IBM group (dashed line), and the ab initio calculation result of the Osaka group (black points).

with two different gate lengths $L_g = 250$ nm and 40 nm to present, respectively, submicron and sub-0.1-micron devices. Figure 2.22 shows the electric field profiles as a function of distance along the channel under various applied source-drain voltages V_d [11]. From

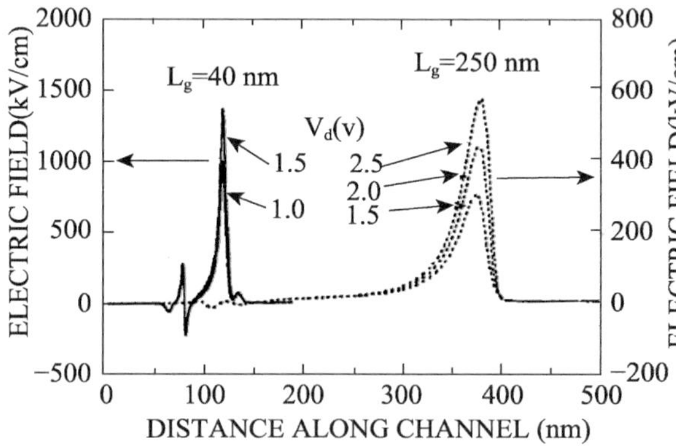

Figure 2.22 Electric field profiles as a function of distance along the channel under various applied source-drain voltages V_d. Notice that the scales of the electric fields of the two MOSFETs are different.

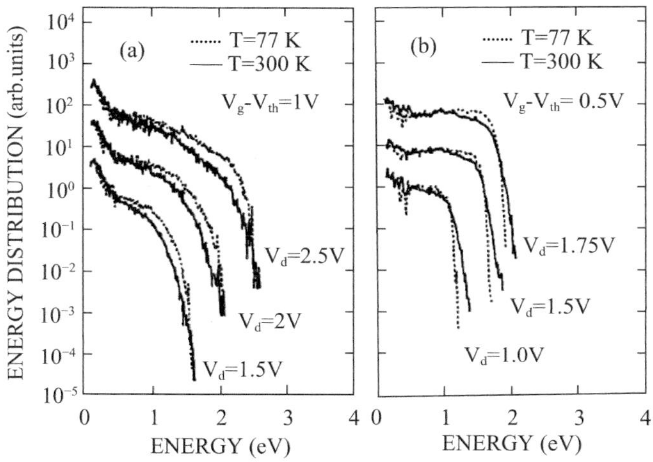

Figure 2.23 Electron energy distributions integrated over the drain regions at 300 K (solid lines) and 77 K (dotted lines) for (a) $L_g = 250$ nm and (b) 40 nm.

the figure we see that there is a very large electric field in the channel of the gate region. Especially in the small-scale device ($L_g = 40$ nm), the electric field can reach 1000 kV/cm.

Figure 2.23 shows the electron energy distributions integrated over the drain regions at 300 K and 77 K [11]. The shapes of the energy distribution are very different for the two devices: in the 250-nm-gate MOSFET, electrons suffer several energy-dissipating scatterings before they reach the drain and, thus, the energy distribution in the regions between several hundred millielectron volts and qV_d smoothly decreases. The number of electrons at 77 K in these energy regions is much larger because of the reduced phonon scattering. This fact demonstrates the diffusive transport property. On the other hand, in a 40-nm-gate MOSFET, the energy distribution does not show significant energy dissipation and the contribution of ballistic electrons is much more pronounced. The electron energy distribution shows a clear transition from a diffusion-energy-dissipating transport regime in a 250-nm-gate MOSFET to a quasi-ballistic transport regime in a 40-nm-gate MOSFET.

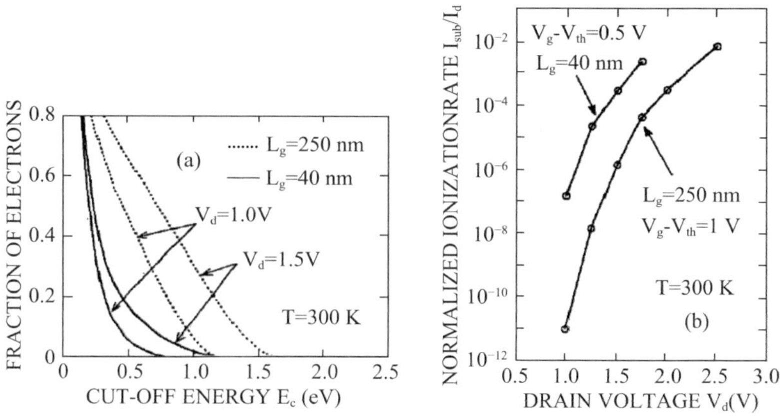

Figure 2.24 (a) Fractions of electrons with energy above the cutoff energy E_c (normalized by the total number of electrons) as a function of $(E - E_c)$ for $L_g = 250$ nm MOSFET (solid lines) and 40 nm MOSFET (dotted lines). (b) Normalized ionization rate for two MOSFETs at 300 K as a function of drain voltage.

Such transition of transport characteristics implies that the fraction of overshoot electrons with respect to the whole electrons in the drain becomes larger as the device size decreases. Figure 2.24a shows the number of electrons with energy above the cutoff energy E_c (normalized by the total number of electrons) as a function of $(E - E_c)$ [11]. The number of electrons with energy above E_c decreases rapidly in the 250-nm-gate MOSFET, whereas it occupies quite a large portion in the 40-nm-gate MOSFET. Therefore, as the device size shrinks, impact ionization tends to be more significant, as shown in Fig. 2.24b. In other words, even if the supply drain voltage is reduced, hot carrier effects could still exist in sub-0.1-micron devices and even more violent than in submicron devices.

2.5 Quantum Simulation Method: Quantum Moment Equations

The classical semiconductor transport theory is based on the Boltzmann transport equation (BTE). The Monte Carlo method provides the most accurate and detailed solution but is limited in practical engineering applications for its computational expenses. One improvement uses a set of quantum moment equations developed from a Wigner function prototype, which preserves explicit quantum corrections as well as the classical hydrodynamic model features [12]. Explicit quantum corrections are built into these equations by using the quantum-mechanical expression of the second moments of the Wigner function, which results in an average electron energy that consists of drift kinetic, thermal, and quantum-potential terms. These equations are then applied to the numerical simulation of ultra-submicron GaAs metal-semiconductor field-effect transistors (MESFETs), which demonstrates expected quantum effects in the devices.

To get explicit quantum corrections into the moment equations, several methods have been proposed. On the basis of the quantum representation of the second moments $\langle p^2 \rangle$, the quantum corrections through the energy representation are explicitly incorporated,

$$w = \frac{1}{2} m^* v^2 + \frac{3}{2} k_{\mathrm{B}} T + U_{\mathrm{q}}, \tag{2.27}$$

where T is the average electron effective temperature; U_{q} is the quantum correction,

$$U_{\mathrm{q}} = -\frac{\hbar^2}{8m^*} \nabla^2 \ln(n); \tag{2.28}$$

and n is the electron density. The quantum correction involves the second-order space derivative of the log density. Thus the correction tends to smooth the electron distribution, especially where the electron density has a sharp change, for example, when approaching a large potential barrier.

A set of quantum moment equations with temperature representation is then written as

$$\frac{\partial n}{\partial t} + \nabla \cdot (n\mathbf{v}) = 0,$$

$$\frac{\partial \mathbf{v}}{\partial t} + \mathbf{v} \cdot \nabla \mathbf{v} = -\frac{q\mathbf{E}}{m^*} - \frac{1}{nm^*} \nabla \left(n k_{\mathbf{B}} T_{\mathrm{q}} \right) - \frac{\mathbf{v}}{\tau_{\mathrm{m}}},$$

and

$$\frac{\partial T}{\partial t} + \frac{1}{3} \mathbf{v} \cdot \nabla \left(T_{\mathrm{q}} \right) = -\frac{2}{3} \nabla \cdot \left(\mathbf{v} T_{\mathrm{q}} \right) + \frac{m^* v^2}{3k_{\mathbf{B}}} \left(\frac{2}{\tau_{\mathrm{m}}} - \frac{1}{\tau_{\mathrm{w}}} \right) - \frac{T - T_0}{\tau_{\mathrm{w}}}, \tag{2.29}$$

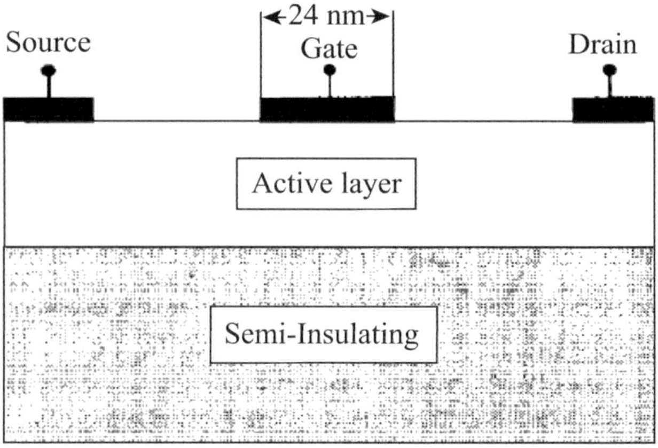

Figure 2.25 MESFET structure for simulation.

where **E** is the electric field, τ_m is the momentum relaxation time, τ_w is the energy relaxation time, T_0 is the lattice temperature, and

$$T_q = T + \frac{2}{3k_B} U_q. \tag{2.30}$$

This set of equations preserves all classical features and gives explicit quantum corrections. As $\hbar$ goes to zero, the equations return to full classical hydrodynamics equations. From Eq. 2.30 one observes that if the thermal energy of the electron is large, the quantum correction has less effect. But as the temperature is lowered, the quantum correction will become dominant. Allying with the Poison equation, Eq. 2.29 forms the basic equations for numerical simulation of ultra-small devices.

The MESFET for simulation is shown in Fig. 2.25 [12], which consists of semi-insulating substrate; an active layer; and three electrodes, source, gate, and drain. The source and drain electrodes are ohmic contact, and the gate electrode is a Schottky contact. The gate voltage V_g produces a depletion region in the channel, whose thickness depends on V_g, then decides the conduction channel between the source and drain.

The parameters of the simulated MESFET device are gate length 24 nm, doping in the channel 1.5×10^{18} cm^{-3}, and lattice temperature 300 K. Assume that the electron property is

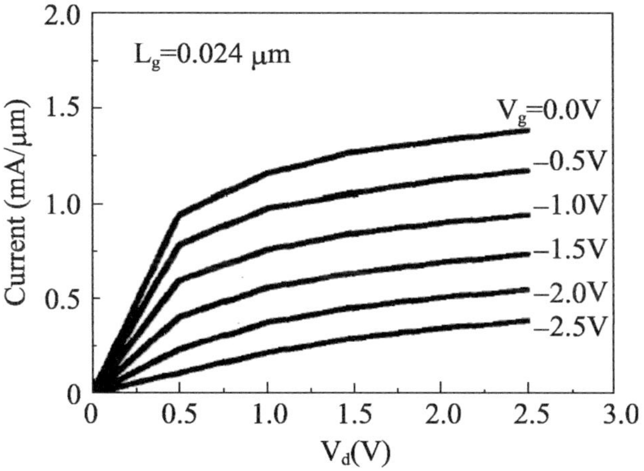

Figure 2.26 I–V characteristics of the 24-nm-gate device.

independent along the transverse coordinate z, the current is along the x direction, and the perpendicular direction is y direction. The finite-difference method is used to discretize Eq. 2.29 and the Poison equation in the $x - y$ two-dimensional (2D) MESFET structure (Fig. 2.25).

The I–V characteristics of the device are shown in Fig. 2.26 [12]. The gate voltage runs from 0 to -2.5 V. The characteristics suggest that the device has a normal working performance. Saturation of the current is observed. When $V_g = -2.5$ V, the electrons are completely depleted in the channel. The remaining current is due to the substrate current as the electrons are pushed into the substrate. Within the saturation region, the maximum transconductance occurs near $V_g = -0.5$ V.

The effect of the gate length on the transconductance is investigated at a drain voltage of 2.0 V, for a fixed active layer depth and doping concentration. Figure 2.27 shows the transconductance characteristics in the range of gate lengths from 24 to 96 nm, for two active layer depths ($a = 30$ nm and 39 nm) [12]. The transconductance for $a = 30$ nm has a maximum value of about 800 mS/mm at a gate length of 60 nm. The transconductance for $a = 39$ nm decreases monotonically for the entire range as gate length

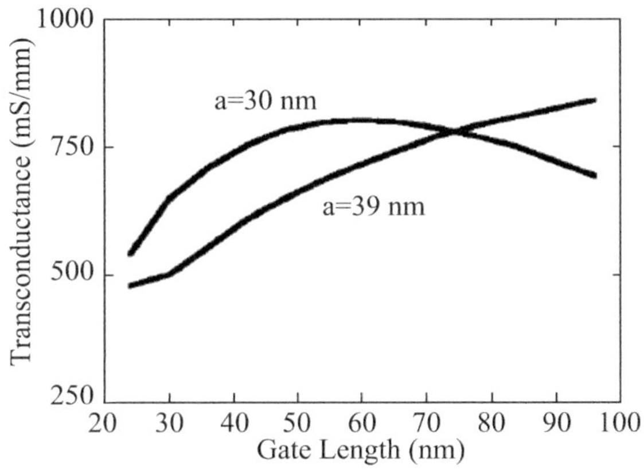

Figure 2.27 Transconductance versus gate length for two active layer depths ($a = 30$ nm and 39 nm).

decreases. The small aspect ratio effect begins at a larger gate length for thicker active devices.

2.6 Simulation of Ultra-Small HEMT Devices

The schematic structure of a HEMT is shown in Fig. 2.15. In a GaAs MESFET the impurity doping in the channel is high and ionized impurity scattering decreases the electron mobility at room temperature from the best value of $\sim$9000 cm^2/(V·s) to 2000 $\approx$ 3000 cm^2/(V·s). In the 1970s the modulation doping heterojunction was proposed. In a wide-gap semiconductor, for example AlGaAs, there is a δ doping layer. At room temperature the impurities are ionized and the electrons from the ionized impurities move into the neighboring GaAs layer. Due to the space charge effect of the ionized impurities a potential well of triangular shape is formed at the interface of AlGaAs and GaAs. The electrons gather in this quantum well and form a two-dimensional electron gas (2DEG) layer. Because the doping layer leaves from the interface a distance, the ionized impurities provide electrons and at the same time do not scatter the electrons in the 2DEG, thus greatly increasing electron mobility, which may even exceed the best value in bulk GaAs.

Zhou et al. [13] used the quantum moment equations (Eq. 2.29) to simulate the HEMT devices, including the carrier degeneracy effects. As the electron density increases to a high concentration level, the Pauli principle ensures that not all electrons can return to the lowest energy level, even at equilibrium. The effect leads to a higher value of the total average energy of the electrons. A direct effect of the degeneracy is on the electron mobility, which is determined by the scattering processes at a given energy level. The momentum and energy relaxation times are computed from Monte Carlo results. If one assumes that the increase of energy due to degeneracy is essentially the thermal energy $3/2\ \gamma k_{\mathrm{B}} T$, then the use of scattering rates from the Monte Carlo results without degeneracy calculated at T instead of γT remains good approximation.

Modulation of the momentum equations (Eq. 2.29) can be made by assuming a drift Fermi–Dirac distribution function, which results in a degeneracy correction factor to the effective electron temperature. The total electron energy can be written as

$$w = \frac{1}{2} m^* v^2 + \frac{3}{2} \gamma k_{\mathrm{B}} T + U_{\mathrm{q}}, \tag{2.31}$$

where γ is the degeneracy factor,

$$\gamma = \gamma\left(\mu / k_{\mathrm{B}} T\right) = \frac{F_{3/2}\left(\mu / k_{\mathrm{B}} T\right)}{F_{1/2}\left(\mu / k_{\mathrm{B}} T\right)}, \tag{2.32}$$

F_{j} is the Fermi–Dirac integral, and μ is the Fermi energy measured from the conduction band bottom. After modification of the equations, the T_{q} in Eq. 2.29 is changed to

$$T_{\mathrm{q}} = \gamma T + \frac{2}{3k_{\mathrm{B}}} U_{\mathrm{q}}. \tag{2.33}$$

The degeneracy factor γ as a function of the factor $n/(k_{\mathrm{B}} T)^{3/2}$ is shown in Fig. 2.28 for GaAs and AlGaAs [13]. The degeneracy factor γ rises significantly only after the ratio of density to thermal energy exceeds 10^{20} $\mathrm{cm^{-3} \cdot eV^{-3/2}}$, that is, the degeneracy effect occurs at a high density and low temperature.

The electron density profile of a 24-nm-gate-length device at the center of the gate as a function of the distance from the surface (see Fig. 2.25) into the bulk at a bias condition of $V_{\mathrm{g}} = -1.5$ V and $V_{\mathrm{d}} = 1.0$ V is shown in Fig. 2.29 [13], where the interface between GaAs and AlGaAs is at 40 nm. The solid line is the result including the

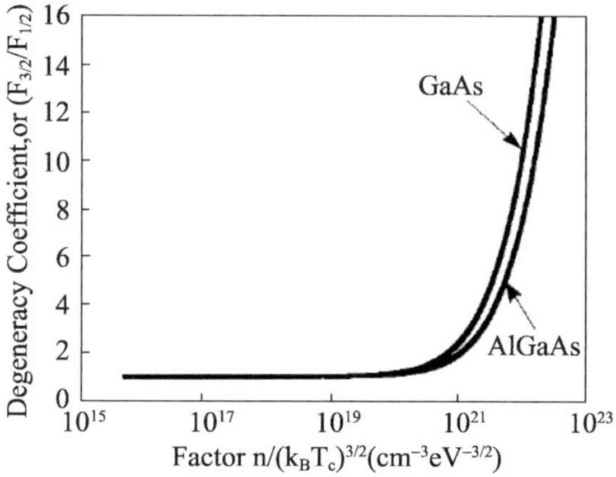

Figure 2.28 Degeneracy factor γ as a function of the factor $n/(k_B T)^{3/2}$ for GaAs and AlGaAs.

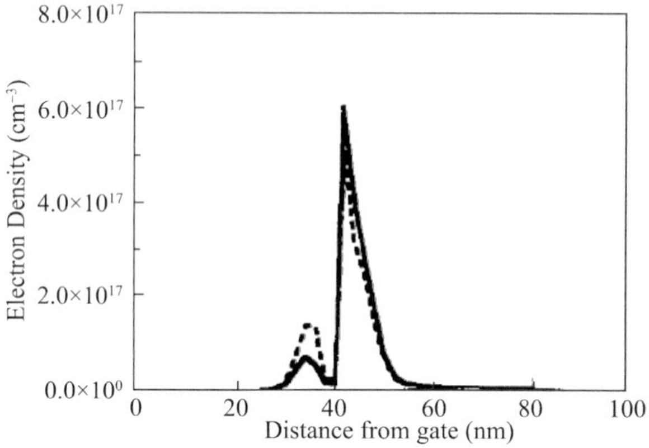

Figure 2.29 Electron density profile of a 24-nm-gate-length device at the center of the gate as a function of the distance from the surface (see Fig. 2.25) into the bulk.

quantum pressure effect, and the dashed line is that neglecting the quantum pressure effect. From the figure one sees that a parallel conduction channel in the AlGaAs is present but is reduced by the quantum effect. The conduction electrons are mostly confined within 10 nm of the interface in the GaAs. The tail of the density distribution in the GaAs (toward the bulk) shows evidence that more

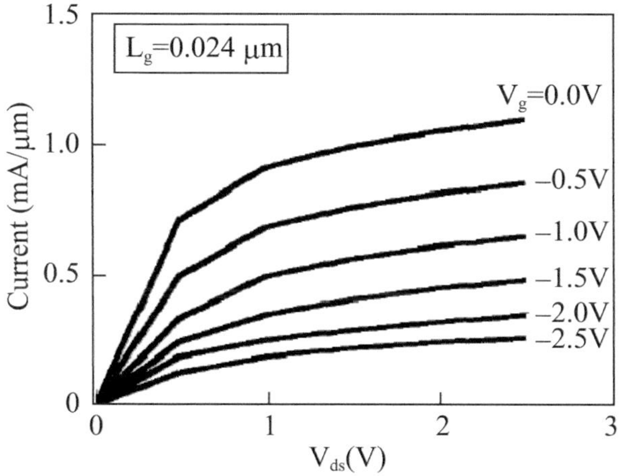

Figure 2.30 *I–V* characteristics for a 24-nm-gate-length device.

electrons are pushed into the substrate when the quantum pressure is included.

The computed *I–V* characteristics for a 24-nm-gate-length device are shown in Fig. 2.30 [13], where the gate bias runs from 0 to −2.5 V, in steps of −0.5 V. The transconductance decreases linearly from about 500 mS/mm to some 250 mS/mm as the gate bias becomes more negative.

References

1. Ye, L. X. (1969). Monte-Carlo Simulation of Small Scale Semiconductor Devices (Academic Press, Beijing) (in Chinese).
2. Rees, H. D. (1969). Calculation of distribution functions by exploiting the stability of the steady state. *J. Phys. Chem. Solids*, **30**, 643–655.
3. Nougier, J. P., Vaissiere, J. C., and Gasquet, D. (1981). Determination of transient regime of hot carriers in semiconductors, using the relaxation time approximations. *J. Appl. Phys.*, **52**, 825–832.
4. Constant, E. (1985). *Hot-Electron Transport in Semiconductors*, eds. Reggiani, L., Chapter 8 "Non-steady-state transport in semiconductors in perspective with submicrometer devices," (Springer-Verlag, Berlin), pp. 227–261.

5. Tomizawa, K., Awano, Y., and Hashizume, N. (1984). Monte Carlo simulation of AlGaAs/GaAs heterojunction bipolar transistors. *IEEE Electron Device Lett.*, **5**, 362–364.
6. Unterrainer, K., Keay, B. J., Wanke, M. C., et al. (1996). *Hot Carriers in Semiconductors*, eds. Hess, K., Leburton, J. P., Rivaioli, U., "Strong terahertz-photocurrent resonances in miniband superlattices at the Bloch frequency" (Plenum, New York), pp. 135–138.
7. Xia, J. B., (1998). Theory of terahertz-photocurrent resonances in miniband superlattices. *Phys. Rev. B*, **58**, 3565–3567.
8. Ghis, A., Constant, E., and Boittiaux, B. (1983). Ballistic and overshoot electron transport in bulk semiconductors and in submicronic devices. *J. Appl. Phys.*, **54**, 214–221.
9. Hess, K., and Iafrate, G. J. (1988). Theory and applications of near ballistic transport in semiconductors. *Proc. IEEE*, **76**, 519–532.
10. Sano, N., and Yoshii, A. (1994). Impact ionization rate near thresholds in Si. *J. Appl. Phys.*, **75**, 5102–5105.
11. Sano, N., Tomizawa, M., and Yoshii, A. (1996). *Hot Carriers in Semiconductors*, eds. Hess, K., Leburton, J. P., Rivaioli, U., "Impact ionization in submicro and sub-0.1 micro Si-MOSFETs" (Plenum, New York), pp. 337–342.
12. Zhou, J. R., and Ferry, D. K. (1992). Simulation of ultra-small GaAs MESFET using quantum moment equations. *IEEE Trans. Electron Devices*, **39**, 473–478.
13. Zhou, J. R., and Ferry, D. K. (1992). Modeling of quantum effects in ultrasmall HEMT devives. *IEEE Trans. Electron Devices*, **40**, 421–427.

Chapter 3

Resonant Tunneling

In 1970 Tsu and Esaki [1] proposed a one-dimensional superlattice. At the same time they proposed the concept of resonant tunneling. They indicated that in such a structure the period is smaller than the electron mean free path and intriguing transport properties such as negative differential conductivity and Bloch oscillation can be realized. Soon afterward Chang et al. [2] observed experimentally the resonant tunneling of electrons in double-barrier structures having a thin GaAs potential well sandwiched between two GaAlAs barriers. The current and conductance characteristics of this structure show the resonant singularities corresponding to the resonant states in the GaAs quantum well. The singularities actually exhibit a decrease in current, giving rise to a negative differential resistance (NDR). Due to the development of the molecular beam epitaxy technology, and the improving of the quality of the sample, the negative resistance can be observed not only at a low temperature but also at room temperature clearly [3–5]. The peak-to-valley ratio (PVR) representing the level of the resonant tunneling structure of 30 and 63 at 300 K and 77 K, respectively, has been reached for an AlAs/InAs/AlAs double potential barrier structure [5].

As for application, the advantage of the resonant tunneling diode (RTD) is that the basic time related to the tunneling process is very

Quantum Waveguide in Microcircuits (Second Edition)
Jian-Bai Xia, Duan-Yang Liu, and Wei-Dong Sheng

ISBN 978-981-4968-30-0 (Hardcover), 978-1-003-65146-8 (eBook)
www.jennystanford.com

short, generally taken as the reciprocal of the quasi-confined state in the quantum well (i.e., the reciprocal of the resonant width). For the high-frequency response of a RTD, it is decided by two time constants: one is the tunneling time through the barrier and the other is the RC time constant associated with the structure capacitance. Optimizing the device structure, these time constants can be minimized. The frequency of the mixed frequency device using the NDR effect has reached 1.8 THz (1.8×10^{12} Hz). When the NDR effect is used as an oscillator, the frequency reached is 0.42 THz [6]. When the NDR effect is used for a high-speed switch, the raising time is shorter than 2 ps [7].

3.1 Single-Barrier Structure

To explain some basic concepts of resonant tunneling (RT), we consider a single potential barrier as shown in Fig. 3.1. Assume that both sides are GaAs material and the middle potential barrier is $Al_xGa_{1-x}As$ material, whose width is $w = 2a$, and the barrier height is V_0 depending on x. The Schrödinger equation of electron motion is

$$\left(-\frac{\hbar^2}{2}\frac{\partial}{\partial z}\frac{1}{m^*(z)}\frac{\partial}{\partial z} + V_{\text{eff}}(z)\right)\varphi(z) = E\varphi(z), \tag{3.1}$$

where $m^*(z)$ is the effective mass, which is generally different for GaAs and $Al_xGa_{1-x}As$, and thus is a function of z.

An electron with energy E enters into the structure from the left 1 region and leaves from the right 3 region. The electron wave

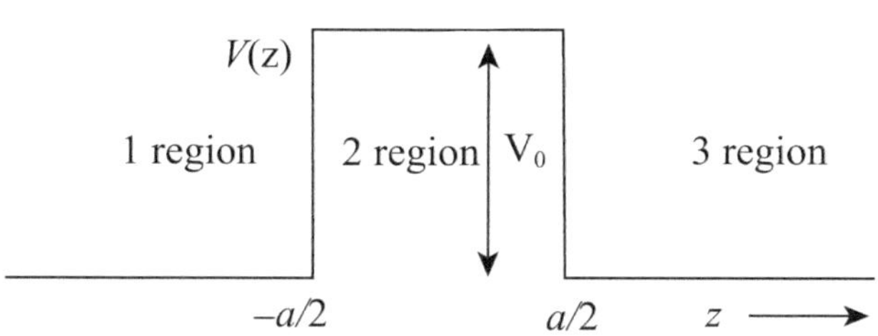

Figure 3.1 Energy band diagram of the single potential barrier.

function in the three regions in Fig. 3.1 are, respectively,

$$\varphi_1(z) = a_1 e^{ikz} + b_1 e^{-ikz}, \qquad z < -a;$$
$$\varphi_2(z) = a_2 e^{Kz} + b_2 e^{-Kz}, \qquad -a < z < a;$$

and

$$\varphi_3(z) = a_3 e^{ikz} + b_3 e^{-ikz}, \qquad z > a, \tag{3.2}$$

where

$$k = \sqrt{\frac{2m_1{}^* E}{\hbar^2}}, \qquad K = \sqrt{\frac{2m_2{}^*(V_0 - E)}{\hbar^2}}. \tag{3.3}$$

where $m_1{}^*$ and $m_2{}^*$ are the electron effective masses in the 1,3 and 2 regions, respectively. In the 1 and 3 regions, the electron energy $E > 0$. Hence the electron waves are plane waves. In the 2 region, $E < V_0$, and the electron wave is an evanescent wave.

By use of the boundary conditions at the interface of the 1 and 2 regions ($z = -a$),

$$\varphi_1(-a) = \varphi_2(-a),$$

and

$$\left.\frac{1}{m_1{}^*}\frac{\partial \varphi_1}{\partial z}\right|_{-a_-} = \left.\frac{1}{m_2{}^*}\frac{\partial \varphi_2}{\partial z}\right|_{-a_+}. \tag{3.4}$$

We obtain

$$a_1 + b_1 = a_2 + b_2,$$

and

$$\frac{ik}{m_1{}^*}(a_1 - b_1) = \frac{K}{m_2{}^*}(a_2 - b_2). \tag{3.5}$$

In calculating Eq. 3.5 we assumed that the coordinate origin is taken at $z = -a$. The choice of the coordinate origin can be as one pleases; the results are only different from a phase factor.

Introduce the concept of the transfer matrix, which connects the coefficients of the wave function at point A to those at point B. For example, the coefficients of the wave function at the 1 region $z = -a_-$ are a_1 and b_1 and those at the 2 region at $z = -a_+$ are a_2 and b_2. They are connected by the transfer matrix M_1,

$$\begin{pmatrix} a_1 \\ b_1 \end{pmatrix} = M_1 \begin{pmatrix} a_2 \\ b_2 \end{pmatrix}. \tag{3.6}$$

The M_1 can be calculated directly from Eq. 3.5,

$$M_1 = \frac{1}{2}\begin{vmatrix} (1-ir) & (1+ir) \\ (1+ir) & (1-ir) \end{vmatrix}, \tag{3.7}$$

where $r = m_1{}^*K/m_2{}^*k$.

In the 2 region, define the transfer matrix,

$$\begin{pmatrix} a_2 \\ b_2 \end{pmatrix} = M_2 \begin{pmatrix} a_2' \\ b_2' \end{pmatrix}, \tag{3.8}$$

where $a_2{}'$ and $b_2{}'$ are the coefficients of $\varphi_2(z)$ with the coordinate origin at $z = a$. Thus

$$M_2 = \begin{vmatrix} e^{-2Ka} & 0 \\ 0 & e^{2Ka} \end{vmatrix}. \tag{3.9}$$

Using the boundary conditions at the interface of the 2 and 3 regions ($z = a$), we obtain

$$a_2' + b_2' = a_3 + b_3,$$

and

$$\frac{K}{m_2{}^*}(a_2' - b_2') = \frac{ik}{m_1{}^*}(a_3 - b_3). \tag{3.10}$$

Define the transfer matrix,

$$\begin{pmatrix} a_2' \\ b_2' \end{pmatrix} = M_3 \begin{pmatrix} a_3 \\ b_3 \end{pmatrix}. \tag{3.11}$$

The transfer matrix M_3 can be calculated from Eq. 3.10,

$$M_3 = \frac{1}{2}\begin{vmatrix} \left(1+\frac{i}{r}\right) & \left(1-\frac{i}{r}\right) \\ \left(1-\frac{i}{r}\right) & \left(1+\frac{i}{r}\right) \end{vmatrix}. \tag{3.12}$$

The total transfer matrix M connects the wave function at the input end and that at the output end. Hence

$$\begin{pmatrix} a_1 \\ b_1 \end{pmatrix} = M \begin{pmatrix} a_3 \\ b_3 \end{pmatrix}, \tag{3.13}$$

where

$$M = M_1 M_2 M_3.$$

Suppose the coefficient of the injected wave function e^{ikz} at the input end $a_1 = 1$; then b_1 represents the amplitude of the reflective wave e^{-ikz}. At the output end the coefficient of the transmitted wave e^{ikz} is a_3 and the coefficient of the returning wave e^{-ikz} $b_3 = 0$. Thus from Eq. 3.13,

$$\begin{pmatrix} 1 \\ b_1 \end{pmatrix} = \begin{pmatrix} M_{11} & M_{12} \\ M_{21} & M_{22} \end{pmatrix} \begin{pmatrix} a_3 \\ 0 \end{pmatrix}. \tag{3.14}$$

We obtain

$$a_3 = 1/M_{11} \quad \text{and} \quad b_1 = M_{21}a_3 = M_{21}/M_{11}. \tag{3.15}$$

The transmission probability and the reflection probability are, respectively,

$$T = |a_3|^2 \quad \text{and} \quad R = |b_1|^2. \tag{3.16}$$

The transfer matrix elements M_{11} and M_{21} in Eq. 3.15 can be calculated from Eq. 3.13 and Eqs. 3.7, 3.9, and 3.12,

$$\begin{aligned} M_{11} &= \frac{1}{4}\left[\left(2 - ir + \frac{i}{r}\right)e^{-2Ka} + \left(2 + ir - \frac{i}{r}\right)e^{2Ka}\right] \\ &= \cosh(2Ka) - \frac{i}{2}\left(r - \frac{1}{r}\right)\sinh(2Ka), \end{aligned}$$

and

$$\begin{aligned} M_{21} &= \frac{1}{4}\left[\left(ir + \frac{i}{r}\right)e^{-2Ka} + \left(-ir - \frac{i}{r}\right)e^{2Ka}\right] \\ &= -\frac{i}{2}\left(r + \frac{1}{r}\right)\sinh(2Ka). \end{aligned} \tag{3.17}$$

Then from Eqs. 3.15 and 3.16 we obtain

$$T = \frac{1}{1 + \left(r + \frac{1}{r}\right)^2 \sinh^2(2Ka)},$$

and

$$R = \frac{\left(r + \frac{1}{r}\right)^2 \sinh^2(2Ka)}{1 + \left(r + \frac{1}{r}\right)^2 \sinh^2(2Ka)}. \tag{3.18}$$

It is deduced from Eq. 3.18 that $T + R = 1$, which means the conservation of particle number.

The problem of the barrier tunneling is a nonstatic problem. We only know the amplitude of the injected wave and need to calculate simultaneously the amplitudes of the reflective wave and transmitted wave. The transfer matrix method is suitable for the two-terminal device with a complicated structure in the middle. For the multiple terminal devices, the scattering matrix method is useful.

Suppose there is an M terminal device, there are injected wave and transmitted wave at each terminal, and their coefficients are b_i and a_i $(i = 1, 2, \ldots, M)$, respectively. The scattering matrix S connects the two sets of coefficients,

$$\begin{pmatrix} a_1 \\ a_2 \\ \vdots \\ a_M \end{pmatrix} = (S(E)) \begin{pmatrix} b_1 \\ b_2 \\ \vdots \\ b_M \end{pmatrix}. \tag{3.19}$$

If the coefficient of the injected wave at the j-th terminal $b_j = 1$, those at other terminals all equal zero. Then the coefficients of the transmitted waves at all other terminals and that of the reflective wave at the j-th terminal can be calculated from Eq. 3.19,

$$a_i = S_{ij} \quad (i = 1, \cdots M, \neq j)$$

and

$$a_j = S_{jj}. \tag{3.20}$$

We take the single-barrier tunneling as an example to calculate the scattering matrix. Define $b_2 e^{-Kz}$ as the wave that travels toward right and $a_2 e^{Kz}$ as the wave that travels toward left. Then the scattering matrix at the boundary $z = -a$ is written as

$$\begin{pmatrix} b_1 \\ b_2 \end{pmatrix} = (S_1) \begin{pmatrix} a_1 \\ a_2 \end{pmatrix}. \tag{3.21}$$

The scattering matrix S_1 can be calculated from Eq. 3.5,

$$S_1 = \frac{1}{1 + ir} \begin{pmatrix} 1 - ir & 2ir \\ 2 & -1 + ir \end{pmatrix}. \tag{3.22}$$

In the barrier 2 region the scattering matrix is

$$\begin{pmatrix} a_2 \\ b_2' \end{pmatrix} = (S_2) \begin{pmatrix} b_2 \\ a_2' \end{pmatrix} \tag{3.23}$$

and

$$S_2 = \begin{pmatrix} 0 & e^{-2Ka} \\ e^{-2Ka} & 0 \end{pmatrix}. \tag{3.24}$$

The scattering matrix at the boundary $z = a$ is

$$\begin{pmatrix} a_2' \\ a_3 \end{pmatrix} = (S_3) \begin{pmatrix} b_2' \\ b_3 \end{pmatrix}. \quad (3.25)$$

S_3 can be calculated from Eq. 3.10,

$$S_3 = \frac{1}{1+ir} \begin{pmatrix} -1+ir & 2 \\ 2ir & 1-ir \end{pmatrix}. \quad (3.26)$$

The total scattering matrix in the single-barrier tunneling case is

$$\begin{pmatrix} b_1 \\ a_3 \end{pmatrix} = (S) \begin{pmatrix} a_1 \\ b_3 \end{pmatrix}. \quad (3.27)$$

But the total scattering matrix is not the multiplication of three matrices S_1, S_2, and S_3 but equals a special multiplication of three matrices introduced by Ando et al. [7],

$$S = S_1 \otimes S_2 \otimes S_3. \quad (3.28)$$

Suppose that there are two scattering matrices,

$$S_1 = \begin{pmatrix} r_1 & t_1' \\ t_1 & r_1' \end{pmatrix} \text{ and } S_2 = \begin{pmatrix} r_2 & t_2' \\ t_2 & r_2' \end{pmatrix}. \quad (3.29)$$

The matrix elements of the multiplication of S_1 and $S_2 S_{12} = S_1 \otimes S_2$ are

$$t_{12} = t_2(1 - r_1' r_2)^{-1} t_1,$$
$$t_{12}' = t_1'(1 - r_2 r_1')^{-1} t_2',$$
$$r_{12} = r_1 + t_1' r_2 (1 - r_1' r_2)^{-1} t_1,$$

and

$$r_{12}' = r_2' + t_2 r_1' (1 - r_2 r_1')^{-1} t_2'. \quad (3.30)$$

The results obtained from the transfer matrix method and the scattering matrix methods are the same. The transfer matrix method is convenient for the two-terminal devices, but for the multi-terminal devices we can only use the scattering matrix method. Besides, if the length of the barrier region a is too large the transfer matrix element (Eq. 3.9) e^{2Ka} from the evanescent wave will approach infinite, resulting in calculation divergence difficulty. While in the scattering matrix (Eq. 3.24) where the two matrix elements are e^{-2Ka}, the divergence difficulty does not appear.

In Eq. 3.16 the transmission probability T and reflection probability R equal the square of the coefficients of the transmitted wave a_3 and the reflective wave b_1, respectively. In the general cases, they should equal

$$T = |a_3|^2 v_2 \quad \text{and} \quad R = |b_1|^2 v_1 . \tag{3.31}$$

The injected electron current,

$$I = |a_1|^2 v_1, \tag{3.32}$$

where v_1 and v_2 are the velocities of the electron waves at the input end and output end, respectively. For the conduction band electron the velocity is proportional to the wave vector, $v = \hbar k/m^*$. In the case of a single barrier (Fig. 3.1) $k_1 = k_2$, then Eq. 3.16 is valid, but is a special case.

To illustrate the meaning of Eq. 3.31 we consider a non-symmetric single potential barrier as shown in Fig. 3.2, where the energy bottom of the output region is lower than those of the input and barrier regions $-V_1$. In this case, the transfer matrices M_1 and M_2 are the same as Eqs. 3.7 and 3.9, while

$$M_3 = \frac{1}{2}\begin{vmatrix} \left(1+\dfrac{i}{r'}\right) & \left(1-\dfrac{i}{r'}\right) \\ \left(1-\dfrac{i}{r'}\right) & \left(1+\dfrac{i}{r'}\right) \end{vmatrix}, \tag{3.33}$$

where

$$r' = \frac{m_1{}^* K}{m_2{}^* k'} \quad \text{and} \quad k' = \sqrt{\frac{2m_1{}^*(E+V_1)}{\hbar^2}}. \tag{3.34}$$

Suppose the injected electron current is 1. From Eqs. 3.31 and 3.32 we define

$$a_1 \sim \frac{1}{\sqrt{k}}, \quad a_3 \sim \frac{t}{\sqrt{k'}}, \quad \text{and} \quad b_1 \sim \frac{r}{\sqrt{k}}. \tag{3.35}$$

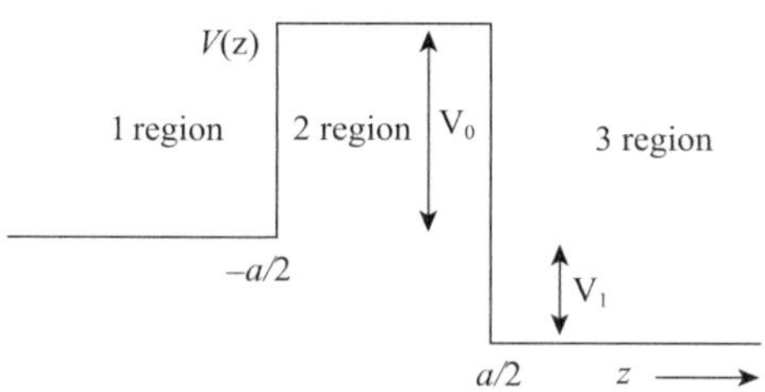

Figure 3.2 Non-symmetric single potential barrier schematic diagram.

The transmission probability and reflection probability are, respectively,

$$T = |t|^2 \quad \text{and} \quad R = |r|^2. \tag{3.36}$$

Equation 3.14 is changed to

$$\begin{pmatrix} 1/\sqrt{k} \\ r/\sqrt{k} \end{pmatrix} = \begin{pmatrix} M_{11} & M_{12} \\ M_{21} & M_{22} \end{pmatrix} \begin{pmatrix} t/\sqrt{k'} \\ 0 \end{pmatrix}, \tag{3.37}$$

$$\begin{aligned} \frac{1}{\sqrt{k}} &= \frac{M_{11}t}{\sqrt{k'}}, \quad t = \sqrt{\frac{k'}{k}}\frac{1}{M_{11}}, \\ \frac{r}{\sqrt{k}} &= \frac{M_{21}t}{\sqrt{k'}}, \quad r = \frac{M_{21}}{M_{11}}. \end{aligned} \tag{3.38}$$

Meanwhile, Eqs. 3.17 and 3.18 are changed as

$$M_{11} = \frac{1}{2}\left(1 + \frac{r}{r'}\right)\cosh 2Ka - \frac{i}{2}\left(r - \frac{1}{r'}\right)\sinh 2Ka,$$

and

$$M_{21} = \frac{1}{2}\left(1 - \frac{r}{r'}\right)\cosh 2Ka - \frac{i}{2}\left(r + \frac{1}{r'}\right)\sinh 2Ka. \tag{3.39}$$

Similarly, the scattering matrix should be correspondingly changed [8]. The wave function in the l-th channel is written as

$$\varphi_l(z) = \frac{1}{\sqrt{k_l}}\left(a_l e^{ik_l z} + b_l e^{-ik_l z}\right). \tag{3.40}$$

The scattering matrix is defined as

$$\begin{pmatrix} a_1 \\ a_2 \\ \vdots \\ a_M \end{pmatrix} = (S(E)) \begin{pmatrix} b_1 \\ b_2 \\ \vdots \\ b_M \end{pmatrix}, \tag{3.41}$$

where $a_l/\sqrt{k_l}$ and $b_l/\sqrt{k_l}$ are the coefficients of the transmitted and injected waves at the l-th channel, respectively.

For the non-symmetric single barrier (Fig. 3.2), the coefficient of the injected wave is $b_1/\sqrt{k_1} = 1/\sqrt{k_1}$ and the amplitudes of the reflective and transmitted waves are, respectively,

$$r = \sqrt{k_1} \cdot \left(\frac{a_1}{\sqrt{k_1}}\right) = a_1 = S_{11}$$

and

$$t = \sqrt{k_2} \cdot \left(\frac{a_2}{\sqrt{k_2}}\right) = a_2 = S_{21}. \tag{3.42}$$

Therefore, the scattering matrix defined in Eq. 3.41 satisfies the unitarity, that is,

$$S^+ S = 1. \tag{3.43}$$

In the two-terminal case, we have

$$|S_{11}|^2 + |S_{21}|^2 = |t|^2 + |r|^2 = 1. \tag{3.44}$$

In the n-terminal case,

$$\sum_l S_{nl} S_{ml}^* = \delta_{nm}. \tag{3.45}$$

The transmission and reflection probabilities are, respectively,

$$T_{nm} = |S_{nm}|^2, \quad m \neq n,$$

and

$$R_n = |S_{nn}|^2. \tag{3.46}$$

If there are no applied magnetic field and spin-orbital coupling, for the real energy E the wave functions ψ and ψ^* are all the solutions of Schrödinger equation and hence the S matrix is symmetric, that is,

$$S_{mn} = S_{nm}. \tag{3.47}$$

Equation 3.47 represents an important result: detailed balance principle $T_{mn} = T_{nm}$, that is, for any structure the transmission probability of the positive injection path equals that of the anti-injection path. It is also indicated from Eq. 3.47 that not only the amplitudes but also the phases are all equal.

In the actual case the chances of using the transfer matrix or scattering matrix to solve analytically the transport problem are less. Even in the single-barrier problem as shown in Fig. 3.1, the tunneling current is measured in the electric voltage bias V, that is, the barrier is under an applied electric field F. The electron motion equation includes a potential energy term,

$$V(z) = V_0(z) - eFz, \tag{3.48}$$

where $V_0(z)$ is the potential without the electric field, as shown in Fig. 3.1, the second term is the additional electric field potential term, which makes the potential energy inclining downward. This time the electron wave function is not a plane wave but a special function $A_i(z)$ and it is nearly impossible to use this function to calculate the transfer or scattering matrix.

Generally one uses the numerical method to solve the electron wave propagation problem in a structure with an arbitrary form of potential [9]. The electron motion Schrödinger equation (in atomic units, length = Born radius a_B and energy = Rydberg R) is written as

$$-\frac{1}{2m^*}\frac{\partial^2}{\partial z^2}\varphi(z) + V(z)\varphi(z) = E\varphi(z) \tag{3.49}$$

or

$$\frac{\partial^2\varphi(z)}{\partial z^2} = -2m^*[E - V(z)]\varphi(z). \tag{3.50}$$

Suppose that the wave function at the output end is $e^{ik'z}$. Then the boundary conditions at the interface between the barrier (2 region) and the output end (3 region) are

$$\varphi(z = a) = 1$$

and

$$\frac{d\varphi}{dz}(z = a) = ik'. \tag{3.51}$$

Starting from the initial condition (Eq. 3.51), integrating Eq. 3.50 from right to left, we obtain the values of the wave function and its differential U and U' at $z = -a$. The wave function at the input end is

$$\varphi(z) = a_1 e^{ikz} + b_1 e^{-ikz}. \tag{3.52}$$

The values of the wave function and its differential at $z = -a$ are

$$\varphi(z = -a) = a_1 + b_1 = U_1$$

and

$$\frac{d\varphi}{dz}(z = -a) = ik(a_1 - b_1) = U_1'. \tag{3.53}$$

The coefficients a_1 and b_1 can be calculated from Eq. 3.53. According to the definition of the transfer matrix (Eq. 3.13),

$$\begin{pmatrix} a_1 \\ b_1 \end{pmatrix} = (M) \begin{pmatrix} 1 \\ 0 \end{pmatrix}. \tag{3.54}$$

Hence,

$$a_1 = M_{11} \quad \text{and} \quad b_1 = M_{21}. \tag{3.55}$$

This numerical method is suitable for arbitrary potential barrier form or hole tunneling (heavy hole and light hole, and there is coupling between them) case [9].

3.2 Resonant Tunneling of Double Potential Barriers

Similar to the symmetry of the scattering matrix S (Eq. 3.47), we give the symmetry of the transfer matrix. In the case of no magnetic field and spin-orbital coupling the wave functions $\varphi(z)$ and $\varphi^*(z)$ are all the solutions of the Schrödinger equation. Thus

$$\varphi_1{}^*(z) = a_1{}^* e^{-ikz} + b_1{}^* e^{ikz}$$

and

$$\varphi_3{}^*(z) = a_3{}^* e^{-ikz} + b_3{}^* e^{ikz}. \tag{3.56}$$

From the definition of the transfer matrix Eq. 3.13,

$$\begin{pmatrix} b_1{}^* \\ a_1{}^* \end{pmatrix} = \begin{pmatrix} M_{11} & M_{12} \\ M_{21} & M_{22} \end{pmatrix} \begin{pmatrix} b_3{}^* \\ a_3{}^* \end{pmatrix}. \tag{3.57}$$

Take the complex number of Eq. 3.57

$$\begin{pmatrix} b_1 \\ a_1 \end{pmatrix} = \begin{pmatrix} M_{11}{}^* & M_{12}{}^* \\ M_{21}{}^* & M_{22}{}^* \end{pmatrix} = \begin{pmatrix} b_3 \\ a_3 \end{pmatrix} \tag{3.58}$$

and rearrange the order,

$$\begin{pmatrix} a_1 \\ b_1 \end{pmatrix} = \begin{pmatrix} M_{22}{}^* & M_{21}{}^* \\ M_{12}{}^* & M_{11}{}^* \end{pmatrix} = \begin{pmatrix} a_3 \\ b_3 \end{pmatrix}. \tag{3.59}$$

Then we obtain the symmetry relation,

$$M_{11}{}^* = M_{22} \quad \text{and} \quad M_{12}{}^* = M_{21}. \tag{3.60}$$

For single potential barrier structure we obtained the transfer matrix from Eq. 3.15,

$$M = \begin{vmatrix} 1/t_1 & r_1^*/t_1^* \\ r_1/t_1 & 1/t_1^* \end{vmatrix}, \tag{3.61}$$

where r_1 and t_1 are amplitudes of the reflective and transmitted waves, respectively. For the double-barrier structure, suppose the width of the middle potential well is w. Then the total transfer matrix is

$$M = \begin{vmatrix} 1/t_1 & r_1^*/t_1^* \\ r_1/t_1 & 1/t_1^* \end{vmatrix} \cdot \begin{vmatrix} e^{-ikw} & 0 \\ 0 & e^{ikw} \end{vmatrix} \cdot \begin{vmatrix} 1/t_1 & r_1^*/t_1^* \\ r_1/t_1 & 1/t_1^* \end{vmatrix}. \tag{3.62}$$

The matrix element

$$M_{11} = e^{-ikw}/t_1^2 + e^{ikw} R_1/T_1, \tag{3.63}$$

where $R_1 = |r_1|^2$ and $T_1 = |t_1|^2$ are the reflection and transmission probabilities, respectively, for the single-barrier structure. The total transmission probability can be calculated from Eq. 3.63,

$$T = 1/|M_{11}|^2 = \left[1 + \frac{4R_1}{T_1^2}\cos^2(kw + \theta)\right]^{-1}, \tag{3.64}$$

where θ is the phase angle of t_1 and $t_1 = |t_1|e^{i\theta}$.

From Eq. 3.64 we see that though the transmission probability T_1 of the single barrier is small, when $kw + \theta = (n + 1/2)\pi$ the transmission probability T of the double barrier equals 1. Physically, because the electron energy equals the energy of the confined state in the quantum well there occurs RT and the transmission probability can reach 1. If the electron energy deviates the resonant energy, the transmission probability descends rapidly, even several orders of magnitude.

Figure 3.3 shows the calculated transmission probabilities of an AlAs/GaAs/AlAs double potential barrier structure as functions of electron energy [10]. The barrier height $V_0 = 1.355$ eV, barrier width $L_B = 2.3$ nm, and the three curves correspond to GaAs potential well widths $L_W = 5$ nm, 7 nm, and 9 nm, respectively. When the electron energy equals the energy of a quasi-confined state there is a RT peak and the transmission probability reach 1. For a narrow quantum well $L_W = 5$ nm the confined states are less and their energies are

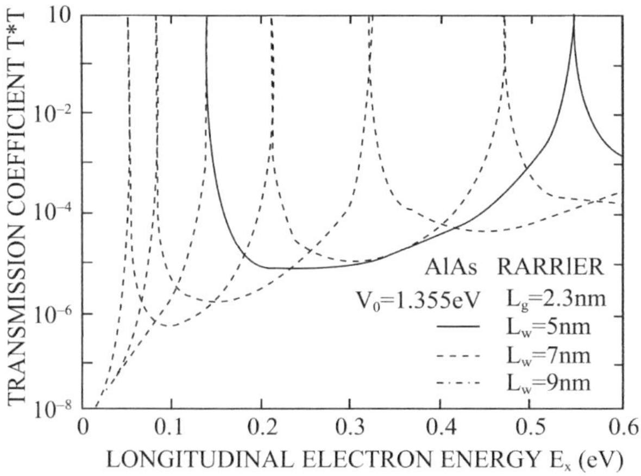

Figure 3.3 Calculated transmission probabilities of an AlAs/GaAs/AlAs double potential barrier structure as functions of electron energy.

higher. There are only two resonant peaks in the energy range of $E = 0 \approx 0.6$ eV. For a wide quantum well $L_W = 9$ nm there are three resonant peaks at lower energies.

Expanding the formula of the RT probability Eq. 3.64 near the resonant energy E_n, we obtain the Lorentz line-shape,

$$T(E) = \frac{\Gamma_n^2/4}{\Gamma_n^2/4 + (E - E_n)^2}, \tag{3.65}$$

where Γ_n is the half-height width of the resonant peak,

$$\Gamma_n = \left(\frac{\hbar^2 E_n T_1^2}{2m^* w^2 R_1}\right)^{1/2}. \tag{3.66}$$

The confined state in the quantum well of a double-barrier structure is a quasi-confined state. If an electron stays at the state, it will escape from the well and thus the energy of the quasi-confined state is a complex number,

$$E = E_0 - i\Gamma_n/2. \tag{3.67}$$

The wave function of electron state,

$$\begin{aligned}\varphi(t) = e^{-i\frac{E}{\hbar}t} &= e^{-i\frac{t}{\hbar}(E_0 - i\Gamma_n/2)} \\ &= e^{-i\frac{E_0}{\hbar}t - \frac{\Gamma_n}{2\hbar}t}.\end{aligned} \tag{3.68}$$

From Eq. 3.68 the electron density obtained is

$$|\varphi(t)|^2 = e^{-\frac{\Gamma_n}{\hbar}t}. \tag{3.69}$$

It means that in the well the electron density decays with time. Then we define the electron lifetime

$$\tau = \frac{\hbar}{\Gamma_n}. \tag{3.70}$$

The above calculation results are obtained by assuming that the single electron has only momentum k in the z direction. Actually the electrons at the input end have momentum $\mathbf{k}$ in the three-dimensional (3D) space and obey the Fermi distribution,

$$f(\mathbf{k}) = \frac{1}{1 + e^{(E - E_F)/kT}}. \tag{3.71}$$

The tunneling current equals the integration of all electron states,

$$J = +e\frac{2}{(2\pi)^3}\int T(E_z)f(\mathbf{k})v_z(\mathbf{k})d\mathbf{k}, \tag{3.72}$$

where $v_z(\mathbf{k})$ is the velocity in the z direction,

$$v_z = \frac{1}{\hbar}\frac{\partial E(\mathbf{k})}{\partial k_z} = \frac{\hbar k_z}{m^*}. \tag{3.73}$$

The total electron energy,

$$E = E_z + E_t = \frac{\hbar^2 k_z^2}{2m^*} + \frac{\hbar^2 k_t^2}{2m^*}, \tag{3.74}$$

where

$$k_t^2 = k_x^2 + k_y^2.$$

The integration of Eq. 3.72 can be divided into integration over k_z and k_t,

$$J = \frac{em^*}{2\pi^2\hbar^2}\int_0^\infty dE_z T(E_z)\int_0^\infty dE_t\left[f_l(E_z, E_t) - f_r(E_z, E_t)\right], \tag{3.75}$$

where f_l and f_r are the Fermi distribution function (Eq. 3.71) of electrons at the left side and the right side of the barrier, respectively. If there is no bias voltage on the structure, $f_l = f_r$ and $J = 0$, there is no tunneling current. If a bias voltage is applied to the structure, the potential energy curve inclines toward right (reverse voltage), so $f_l > f_r$ and tunneling current is produced.

Integrating Eq. 3.75 for E_t,

$$\int_0^\infty dE_t\,[f_l(E_z, E_t) - f_r(E_z, E_t)]$$

$$= \int_0^\infty dE_t \left[\frac{1}{1+e^{(E_z+E_t-E_F)/kT}} - \frac{1}{1+e^{(E_z+E_t-E_F+eV)/kT}}\right]$$

$$= kT \ln\left(\frac{1+e^{(E_F-E_z)/kT}}{1+e^{(E_F-eV-E_z)/kT}}\right), \tag{3.76}$$

where E_F is the Fermi energy of the left side and that of the right side is $E_F - eV$ due to the bias voltage. Inserting Eq. 3.76 into Eq. 3.75 we obtain

$$J = \frac{em^*kT}{2\pi^2\hbar^2}\int_0^\infty dE_z T(E_z) \ln\left(\frac{1+e^{(E_F-E_z)/kT}}{1+e^{(E_F-eV-E_z)/kT}}\right). \tag{3.77}$$

Equation 3.77 was derived first by Tsu et al. [1].

When $T = 0$, the electron energy distribution is step-like,

$$\ln\left[1+e^{(E_F-E_z)/kT}\right] = \begin{cases} \dfrac{E_F - E_z}{kT}, & E_z < E_F \\ 0. & E_z > E_F \end{cases} \tag{3.78}$$

Thus Eq. 3.77 is written as

$$J = \frac{em^*}{2\pi^2\hbar^2}\left[\int_0^{E_F} dE_z T(E_z)(E_F - E_z) - \int_0^{E_F-eV} dE_z T(E_z)(E_F - eV - E_z)\right]. \tag{3.79}$$

If $E_F - eV < 0$, the Fermi energy level descends to the bottom of the energy band at the left side ($E = 0$). The right side is completely empty for the tunneling electron from the left side, so the second term in Eq. 3.79 equals zero.

At the neighbor of the resonant energy E_n the tunneling probability can be expanded as in Eq. 3.65. In the limit that Γ_n approaches zero, it changes to be a δ function,

$$\delta(E_z - E_n) = \frac{1}{\pi}\lim_{\Gamma_n\to 0}\frac{\Gamma_n/2}{(\Gamma_n^2/4) + (E_z - E_n)^2}. \tag{3.80}$$

Thus

$$T(E_z) \approx \pi \cdot \frac{\Gamma_n}{2}\delta(E_z - E_n). \tag{3.81}$$

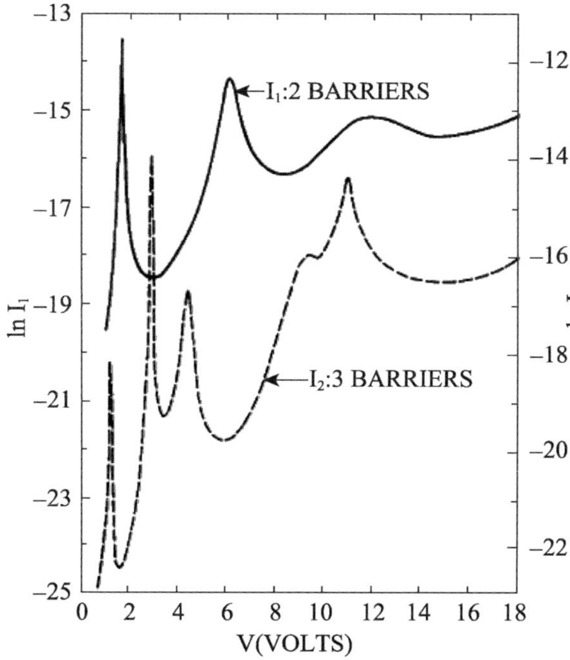

Figure 3.4 Calculated current density at 0 K for the double- and triple-barrier cases as a function of applied voltage.

Inserting Eq. 3.81 into Eq. 3.79, and assuming that $E_F - eV < 0$, we obtain

$$J = \frac{em^* T_{res} \Gamma_n}{4\pi \hbar^2}(E_F - E_n), \tag{3.82}$$

where T_{res} is a constant. From Eq. 3.82 we see that the tunneling current is proportional to $E_F - E_n$. When the E_n is far away from E_F, the tunneling current is larger, and when E_n approaches E_F, the tunneling current approaches zero.

Figure 3.4 shows the calculated current density at 0 K for the double- and triple-barrier cases, without the constant factor $em^*/2\pi^2\hbar^2$ as a function of applied voltage [1]. Note that the I–V characteristics indicate a fine structure having differential negative conductivities. The first peak for the double-barrier case is located approximately at 0.16 eV, which is about twice the transmission peak of 0.082 eV. This is because the bottom of the well shifts

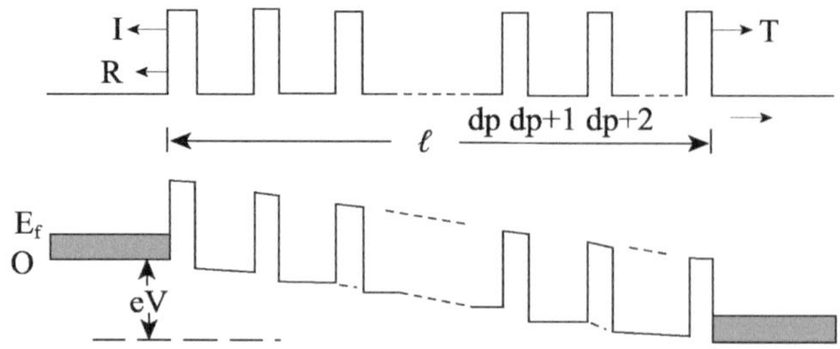

Figure 3.5 Schematic energy band of the finite superlattice: the upper part is without bias and the lower part is with bias.

only half the amount of the applied voltage for this case. When the voltage increases continuously, E_F approaches the energy of the second confined state of the well and there appears the second resonant current peak. For the triple-barrier case, the situation is more complicated, E_F can be resonant with the energy of the confined state in the first well or in the second well, and thus multiple resonant peaks appear.

Tsu et al. [1] calculated the tunneling probabilities of the superlattices composed of multi-barriers with the transfer matrix method. The schematic energy band of the finite superlattice is shown in Fig. 3.5: the upper part is without bias, and the lower part is with bias. Figure 3.6 shows in the case of no bias the transmission probabilities as functions of electron energy for the double, triple, and quintuple barriers. The barrier and well widths are 2 nm and 5 nm, respectively. The barrier height is 0.5 eV. Note that the resonant energies for the triple barrier case is split into doublets and those for the quintuple barrier case are split into quadruplets. The line widths are roughly determined by the tunneling probability of the barrier width. Generally, for n barriers, there will be an $(n - 1)$-fold splitting. This splitting is caused by the coupling of the energy levels of the neighboring wells.

Using the negative resistance property produced by the RT effect one can design and fabricate the two-terminal RT oscillator. Figure 3.7 shows the simplest equivalent circuit of a two-terminal RT oscillator [11], where $-R_D$ and C_D represent the resistance and capacitance of the device, respectively. R_S and L_S are parasitic resistance and inductance of the lead, respectively. This equivalent

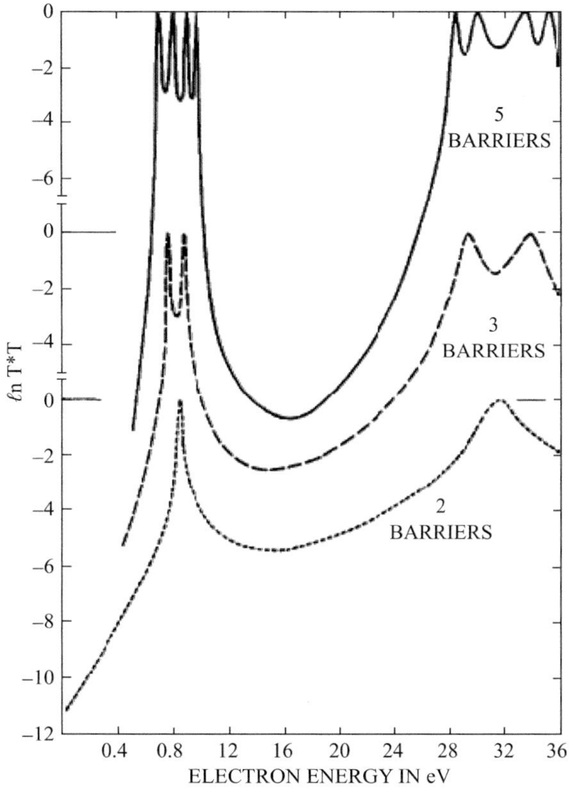

Figure 3.6 Transmission probabilities as functions of electron energy for the double, triple, and quintuple barriers.

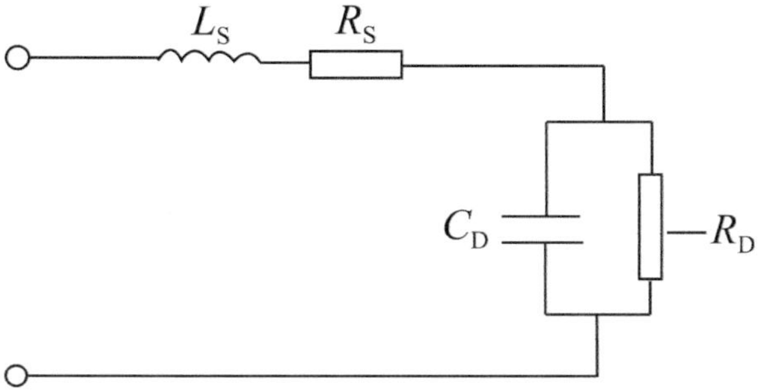

Figure 3.7 Equivalent circuit of a two-terminal RT oscillator.

circuit has been successfully applied to analyze the tunneling diodes, and the calculated I–V characteristics are in agreement with experimental results. The real part of the impedance in the equivalent circuit R_{eg} is

$$R_{eg} = R_s + \frac{-R_D}{1 + (\omega R_D C_D)^2}. \qquad (3.83)$$

For steady oscillation R_{eg} should be negative; then the cutoff frequency can be calculated from Eq. 3.83,

$$f_{max} = \frac{1}{2\pi R_D C_D}\sqrt{\frac{R_D}{R_s} - 1}. \qquad (3.84)$$

For raising the f_{max} the parasitic resistance R_s and the diode capacitance C_D should be minimum. Differentiating Eq. 3.84 to R_D, we obtain the condition that f_{max} is maximum and

$$R_D = 2R_s. \qquad (3.85)$$

That means that the absolute value of the negative resistance R_D is also very small. Therefore, a steep drop of the current and high peak-valley ratio in the I–V characteristic curve will be helpful to raise the f_{max}. Another demand for the RT oscillator is large power output, generally,

$$P_{max} \approx (V_p - V_v)(I_p - I_v), \qquad (3.86)$$

where V_p, I_p, V_v, and I_v are the voltage and current of the current peak and valley, respectively. Thus large peak-valley ratio and high current density are necessary for a high performance oscillator.

Brown et al. obtained oscillations at frequency 712 GHz in InAs/AlSb double-barrier resonant tunneling diodes (RTDs) at room temperature [12]. The InAs/AlSb materials system has several advantages over GaAs/AlAs for making high-speed RTDs. First, the InAs/AlSb band offset (staggered type II at the Γ pint) allows an electron to tunnel more easily through the AlSb barrier than the AlAs barrier (type I band offset). This leads to a larger difference of the peak and valley currents $\Delta I = I_p - I_v$ and a smaller R_D. A second advantage is that the electron effective mass in InAs (0.023 m_0) is smaller than that in GaAs (0.067 m_0), so the electrons will drift across a given depletion layer much more rapidly in InAs than

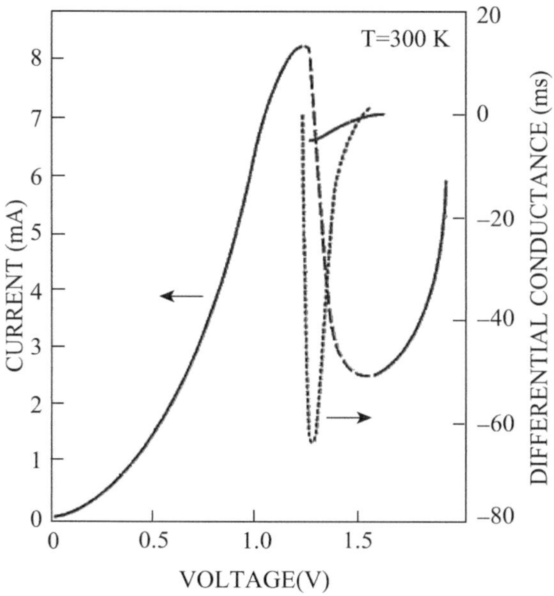

Figure 3.8 DC $I-V$ characteristic (solid line) and differential conductance (dotted line) of a 1.8 μm diameter diode.

in GaAs. A third advantage is that InAs RTDs have a lower total series resistance R_S.

Figure 3.8 shows the direct current (DC) I–V characteristic and differential conductance of a 1.8 μm diameter diode with a negative voltage applied to the top contact [12]. The I–V curve shows a peak-to-valley current ratio of about 3.4 at room temperature and a peak current density of 2.8×10^5 A cm^{-2}, corresponding to $\Delta I \cong 2.0 \times 10^5$ A cm^{-2}. The discontinuous nature of the experimental I–V curve in the NDC region is caused by self-rectification of the oscillations.

Figure 3.9 shows the highest power density as a function of the frequency obtained in the InAs/AlSb RTDs (dotted line) and GaAs/AlAs RTDs (solid line) [12]. At 360 GHz an absolute power of 3 μW was obtained, corresponding to a power density of 90 Wcm^{-2}. This is 50 times the power density obtained previously from GaAs/AlAs RTDs at 370 GHz. At 712 GHz a power of 0.3 μW was

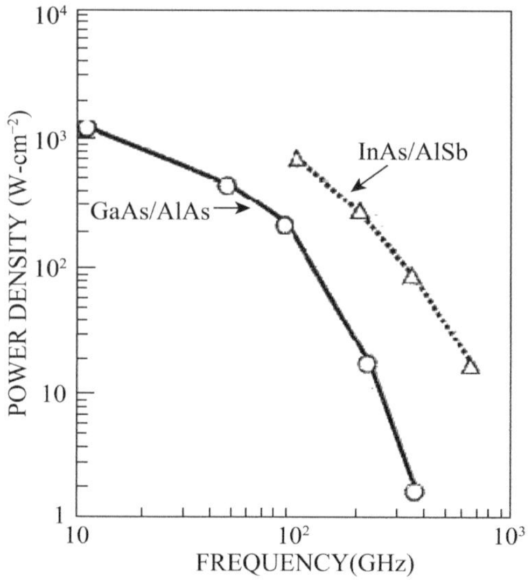

Figure 3.9 Highest power density as a function of the frequency obtained in the InAs/AlSb RTDs (dotted line) and GaAs/AlAs RTDs (solid line).

obtained, corresponding to a power density of 15 Wcm^2. The power density decreases with the oscillation frequency increasing.

The RTD oscillators have not applied on a large scale; the difficulties are operation at room temperature and large-scale integration.

3.3 Hole Resonant Tunneling

Mendez et al. first observed the hole RT in the GaAs/AlAs double-barrier structure [13]. The conductance (dI/dV) versus voltage bias of an undoped AlAs/GaAs/AlAs double-barrier structure for four temperatures is shown in Fig. 3.10 [13]. There are many conductance minima representing the positions of resonant peaks. On the basis of the temperature dependence behavior and the relative strength of the resonances, the authors assigned the high T set of structures to RT of light holes and the low T set to heavy hole tunneling. To compare with experiments they assumed that

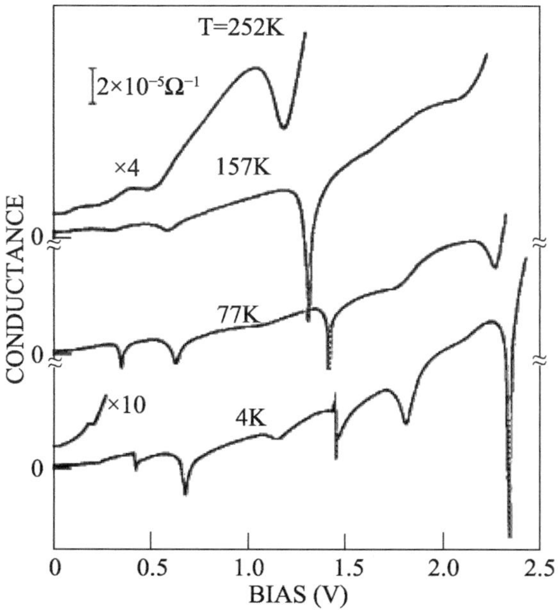

Figure 3.10 Conductance (dI/dV) versus voltage bias of an undoped AlAs/GaAs/AlAs double-barrier structure for four temperatures.

the effective masses of the light and heavy holes in GaAs are 0.1 m_0 and 0.6 m_0, respectively. The width of the GaAs and AlAs layers are 5 nm and the barrier height is 550 meV. Assuming that there is no coupling between the light hole and heavy hole, they calculated the tunneling probabilities as functions of energy separately for two kinds of holes; the results are shown in Fig. 3.11 [13]. It is considered approximately that the bias voltages to which the tunneling peaks correspond multiplied by the electron charge e are twice the energy to which the tunneling probability peaks correspond. From Fig. 3.11 the light hole resonant peaks should be at 0.168 and 0.634 V, which contrasts with the experimental values of 0.43, 0.67, and 1.45 V. The heavy hole resonant peaks should be at 0.038, 0.154, 0.344, 0.604, and 0.915 V, while experimentally only four peaks, at 0.20, 1.15, 1.81, and 2.33 V, were observed. The authors believed that non-parabolicity effects and band mixing (coupling between light hole and heavy hole) are probably the main reasons for the discrepancies found.

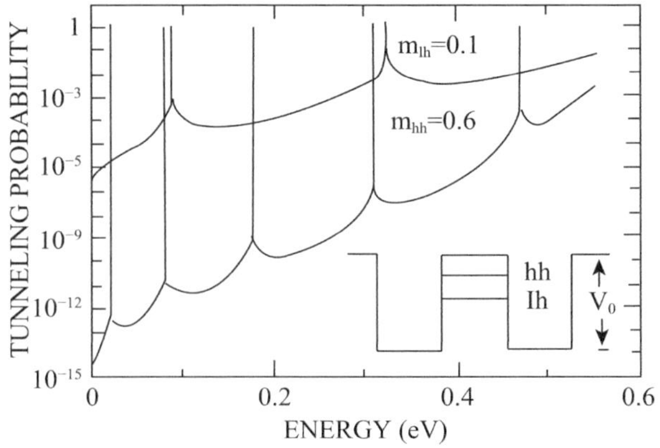

Figure 3.11 Calculated tunneling probabilities as functions of energy separately for two kinds of holes.

The authors also studied the effect of a strong magnetic field on the positions of resonant current peaks. The magnetic field is applied along the current direction. If the hole energy band is a simple parabolic band, the problem is simpler. The hole energy band in the plane perpendicular to the magnetic field direction is split into a series of Landau energy levels. The basis state energy equals $\hbar\omega_c/2 \approx B$. In the strong magnetic field the RT energy will increase linearly with the magnetic field. The surprising results are that with the magnetic field increasing some peak energy increases, but some peak energy decreases, which proves the complexity of the valence bands. We should consider the coupling between light and heavy holes and the non-parabolicity.

To solve the above problem, Xia proposed the hole RT theory [9]. In the double-barrier structure, the effective mass Hamiltonian is written as

$$H = H_{\mathrm{L}} + V(z), \tag{3.87}$$

where $V(z)$ is the effective potential of the structure,

$$H_{\mathrm{L}} = \frac{1}{2}\begin{vmatrix} P_1 & R & Q & 0 \\ R^* & P_2 & 0 & -Q \\ Q^* & 0 & P_2 & R \\ 0 & -Q^* & R^* & P_1 \end{vmatrix}, \tag{3.88}$$

$$
\begin{aligned}
P_1 &= (\gamma_1 + \gamma_2)P_{\parallel}^2 + (\gamma_1 - 2\gamma_2)P_z^2, \\
P_2 &= (\gamma_1 - \gamma_2)P_{\parallel}^2 + (\gamma_1 + 2\gamma_2)P_z^2, \\
Q &= -i2\sqrt{3}\gamma_3 P_z(P_x - iP_y), \\
R &= \sqrt{3}\left[\gamma_2(P_x^2 - P_y^2) - i2\gamma_3 P_x P_y\right],
\end{aligned}
\tag{3.89}
$$

where $P_\perp$ and P_z are the momentum operators perpendicular and parallel to the z direction, respectively. γ_1, γ_2, and γ_3 are the Luttinger parameters.

For the superlattice or double-barrier structures only P_z is the operator; P_x and P_y can be replaced by their eigen-values k_x and k_y. To simplify the calculation of the transfer matrix, we transform the Hamiltonian (3.88) by a unitary transform proposed by Briodo et al. [14],

$$
U = \begin{vmatrix}
\frac{1}{\sqrt{2}}e^{-i\phi} & 0 & 0 & -\frac{1}{\sqrt{2}}e^{i\phi} \\
0 & \frac{1}{\sqrt{2}}e^{-i\eta} & -\frac{1}{\sqrt{2}}e^{i\eta} & 0 \\
0 & \frac{1}{\sqrt{2}}e^{-i\eta} & \frac{1}{\sqrt{2}}e^{i\eta} & 0 \\
\frac{1}{\sqrt{2}}e^{-i\phi} & 0 & 0 & \frac{1}{\sqrt{2}}e^{i\phi}
\end{vmatrix},
\tag{3.90}
$$

where ϕ and η are determined by k_x and k_y. The transformed Hamiltonian becomes two independent 2×2 matrices,

$$
UH_{\mathrm{L}}U^+ = \frac{1}{2}\begin{vmatrix}
P_1 & i|Q|P_2 + |R| & 0 & 0 \\
-i|Q|P_2 + |R| & P_2 & 0 & 0 \\
0 & 0 & P_2 & -i|Q|P_2 + |R| \\
0 & 0 & i|Q|P_2 + |R| & P_2
\end{vmatrix},
\tag{3.91}
$$

where

$$
|R| = \sqrt{3}\left[\gamma_2^2(k_x^2 - k_y^2) + 4\gamma_3^2 k_x^2 k_y^2\right]^{1/2}
$$

and

$$
|Q| = 2\sqrt{3}\left[k_x^2 + k_y^2\right]^{1/2}. \tag{3.92}
$$

Even with the 2×2 matrix the tunneling problem still cannot be solved analytically. We will use the numerical integration method [9]. We first consider the upper-left 2×2 matrix of Eq. 3.91; the result of the lower right 2×2 matrix is the same.

$$
H_1 = \frac{1}{2}\begin{vmatrix}
P_1 & i\,|Q|\,p_z - |R| \\
-i\,|Q|\,p_z - |R| & P_2
\end{vmatrix}. \tag{3.93}
$$

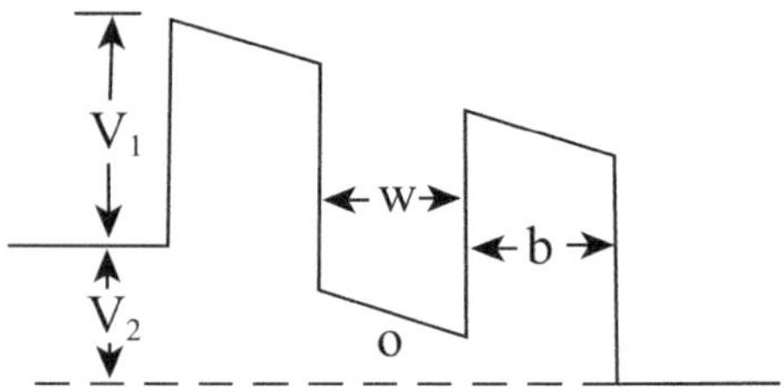

Figure 3.12 Potential barrier region for hole tunneling.

The potential barrier region is illustrated in Fig. 3.12. On the left side of the potential barrier region $V(z)$ is assumed to be zero. The hole wave function is of the form

$$\psi = \begin{pmatrix} a \\ b \end{pmatrix} e^{i\mathbf{k}_\parallel \cdot \mathbf{r}_\parallel + ik_z z}. \tag{3.94}$$

As the holes enter into the potential barrier region, where $V(z)$ is not a constant (see Fig. 3.12), the hole wave function becomes

$$\psi = \begin{pmatrix} U_1(z) \\ U_2(z) \end{pmatrix} e^{i\mathbf{k}_\parallel \cdot \mathbf{r}}, \tag{3.95}$$

where $\mathbf{k}_\parallel$ is still a good quantum number.

From the effective mass equation with the Hamiltonian (3.93) we obtain the differential equations satisfied by U_1 and U_2,

$$U_1'' = \frac{1}{\gamma_1 - 2\gamma_2}\left\{(\gamma_1 + \gamma_2)k_\parallel^2 U_1 - |R|U_2 + |Q|U_2' - 2[E - V(z)]U_1\right\},$$

and

$$U_2'' = \frac{1}{\gamma_1 + 2\gamma_2}\left\{(\gamma_1 - \gamma_2)k_\parallel^2 U_2 - |R|U_1 - |Q|U_1' - 2[E - V(z)]U_2\right\}. \tag{3.96}$$

From Eq. 3.96 we see that there is coupling between the heavy and light holes, so we cannot deal with the heavy hole and light hole independently.

First we calculate the hole wave function on the left side and right side of the barrier region. Because $V(z) = V_0$ is a constant on the left and right sides, inserting the hole wave function (Eq. 3.94) into the effective mass equation with the Hamiltonian (Eq. 3.93) we obtain the eigen-energy

$$E - V_0 = \frac{1}{2}\gamma_1 k^2 \pm \left[r_2^2 k^4 + 3(\gamma_3^2 - \gamma_2^2)(k_x^2 k_y^2 + k_y^2 k_z^2 + k_z^2 k_x^2)\right]^{1/2}, \tag{3.97}$$

where the sign $\pm$ corresponds to the light and heavy holes, respectively. For given k_x and k_y, the k_z can be calculated from Eq. 3.97. The coefficients a and b of the hole wave function (3.94) can be equally calculated.

When the energy and the parallel wave vectors k_x and k_y are given, there are generally four independent hole states on the left-hand side of the potential barrier region: the heavy hole states ψ_{h,k_h} and $\psi_{h,-k_h}$ with perpendicular wave vectors k_h and $-k_h$ and the light hole states ψ_{l,k_l} and $\psi_{l,-k_l}$ with perpendiculars k_l and $-k_l$. The k_h and k_l are calculated from Eq. 3.97 for given E, V_0, k_x, and k_y. At the left side and right side, the wave functions are, respectively,

$$\psi_l = \alpha\psi_{h,k_h} + \beta\psi_{h,-k_h} + \gamma\psi_{l,k_l} + \delta\psi_{l,-k_l}$$

and

$$\psi_r = \alpha'\psi_{h,k'_h} + \beta'\psi_{h,-k'_h} + \gamma'\psi_{l,k'_l} + \delta'\psi_{l,-k'_l}. \tag{3.98}$$

The coefficients $(\alpha, \beta, \gamma, \delta)$ and $(\alpha', \beta', \gamma', \delta')$ are connected by a transfer matrix,

$$\begin{pmatrix} \alpha \\ \beta \\ \gamma \\ \delta \end{pmatrix} = (M) \begin{pmatrix} \alpha' \\ \beta' \\ \gamma' \\ \delta' \end{pmatrix}. \tag{3.99}$$

Suppose at the right side a heavy hole state goes out; then the boundary conditions at the right barrier of the barrier U_1, U'_1 and U_2, U'_2 can be determined. Numerically integrating Eq. 3.96 from right to left (e.g., by the Adams method), we obtain the values of wave functions and their derivatives U_1, U'_1 and U_2, U'_2 at the left boundary of the barrier region.

The hole wave functions are given by Eqs. 3.98 and 3.94; thus their coefficients α, β, γ, and δ are determined by the following set of equations:

$$a_{h,k_h}\alpha + a_{h,-k_h}\beta + a_{l,k_l}\gamma + a_{l,-k_l}\delta = U_1,$$
$$b_{h,k_h}\alpha + b_{h,-k_h}\beta + b_{l,k_l}\gamma + b_{l,-k_l}\delta = U_2,$$
$$ik_h a_{h,k_h}\alpha - ik_h a_{h,-k_h}\beta + ik_l a_{l,k_l}\gamma - ik_l a_{l,k_l}\delta = U'_1,$$

and

$$ik_h b_{h,k_h}\alpha - ik_h b_{h,-k_h}\beta + ik_l b_{l,k_l}\gamma - ik_l b_{l,k_l}\delta = U'_2, \tag{3.100}$$

where (a_{hk_h}, b_{hk_h}) and (a_{lk_l}, b_{lk_l}) are the coefficients of the wave function (Eq. 3.94) for the heavy hole and light hole, respectively.

If there is a heavy hole at the right side ($k_z' > 0$), the transfer matrix equation (Eq. 3.99) will be

$$\begin{pmatrix} \alpha \\ \beta \\ \gamma \\ \delta \end{pmatrix} = (M) \begin{pmatrix} 1 \\ 0 \\ 0 \\ 0 \end{pmatrix}. \tag{3.101}$$

From Eq. 3.101 we obtain

$$M_{11} = \alpha, \quad M_{21} = \beta, \quad M_{31} = \gamma, \quad \text{and} \quad M_{41} = \delta\,. \tag{3.102}$$

Similarly if there is a light hole at the right side ($k_z' > 0$), we obtain

$$M_{13} = \alpha, \quad M_{23} = \beta, \quad M_{33} = \gamma, \quad \text{and} \quad M_{43} = \delta. \tag{3.103}$$

Knowing the transfer matrix (doesn't need all matrix elements), the transmission amplitudes T and the reflection amplitudes R can be calculated from Eq. 3.99,

$$\begin{aligned}
T_{\text{hh}} &= \frac{M_{33}}{D}, \quad T_{\text{hl}} = \frac{-M_{31}}{D}, \\
R_{\text{hh}} &= \frac{M_{21}M_{33} - M_{23}M_{31}}{D}, \quad R_{\text{hl}} = \frac{M_{41}M_{33} - M_{43}M_{31}}{D}, \\
T_{\text{lh}} &= -\frac{M_{13}}{D}, \quad T_{\text{ll}} = \frac{M_{11}}{D}, \\
R_{\text{lh}} &= \frac{-M_{21}M_{13} + M_{23}M_{11}}{D}, \quad R_{\text{ll}} = \frac{-M_{41}M_{13} + M_{43}M_{11}}{D},
\end{aligned} \tag{3.104}$$

and

$$D = M_{11}M_{33} - M_{13}M_{31}, \tag{3.105}$$

where T_{hh} represents the transmission amplitude from heavy hole state to heavy hole state, T_{hl} that from heavy hole state to light hole state, etc.

Figure 3.13 shows the transmission probabilities of the heavy and light hole $\log_{10}(T^*T)$ as functions of energy E at $F = 0$ and $k_\parallel = 0$ [9]. Since for $k_\parallel = 0$ there is no coupling between heavy and light holes and the resonant peaks correspond to the independent heavy and light hole states in the quantum well, similar to Fig. 3.11.

Figures 3.14 and 3.15 show the transmission probabilities $(T^*T)_{\text{hh}}$ and $(T^*T)_{\text{hl}}$ as well as $(T^*T)_{\text{ll}}$ and $(T^*T)_{\text{lh}}$ as functions of

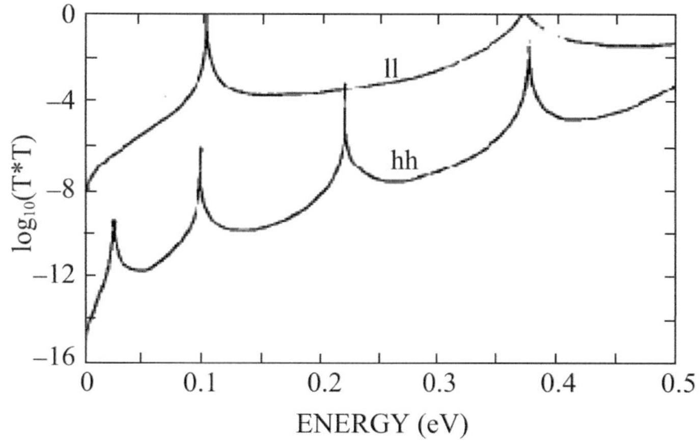

Figure 3.13 Transmission probabilities of the heavy and light hole $\log_{10}(T^*T)$ as functions of energy E at $F = 0$ and $k_{\|} = 0$.

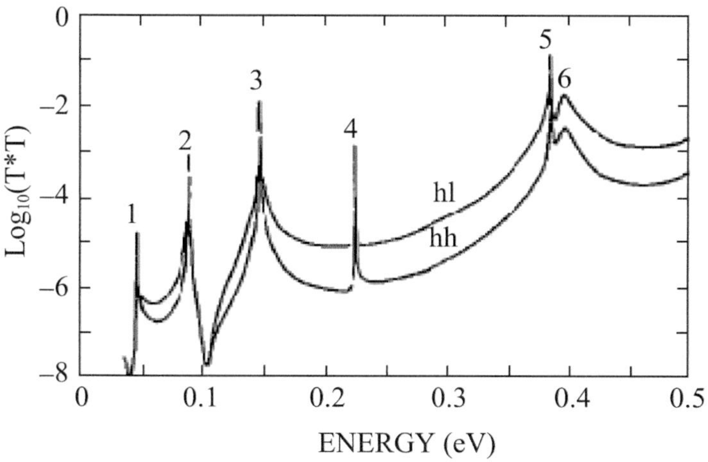

Figure 3.14 Transmission probabilities $(T^*T)_{\text{hh}}$ and $(T^*T)_{\text{hl}}$ as functions of energy E at $F = 0$ and $k_{\|} = 0.3$ $(2\pi/7$ nm).

energy E at $F = 0$ and $k_{\|} = 0.3$ $(2\pi/7$ nm), respectively [9]. From Fig. 3.14 we see that $(T^*T)_{\text{hh}}$ and $(T^*T)_{\text{hl}}$ contain almost all the resonant peaks of the heavy and light holes, which means that there is strong mixing between them. One peak is an exception: the fourth

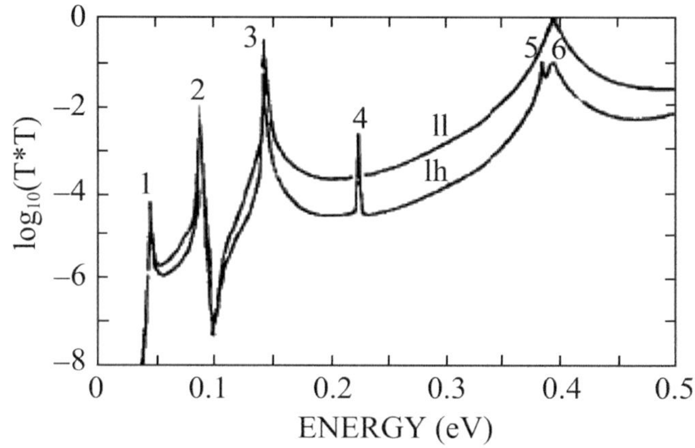

Figure 3.15 Transmission probabilities $(T^*T)_{\text{ll}}$ and $(T^*T)_{\text{lh}}$ as functions of energy E at $F = 0$ and $k_{\parallel} = 0.3\ (2\pi/7\ \text{nm})$.

peak appears in the hh curve but not in the hl curve. Similarly, in Fig. 3.15 the fourth peak appears only in the lh curve. It indicates that this peak is derived from a heavy hole resonance.

3.4 Resonant Tunneling in Dilute Magnetic Semiconductors

When magnetic ions, for example, Mn^{2+}, are doped in the semiconductor, then in the magnetic field the electronic energy levels with different spin orientations will split and the splitting energy is large, which is called giant Zeeman splitting energy. These kinds of semiconductors are called dilute magnetic semiconductors (DMSs). The giant Zeeman splitting energy

$$\Delta E = N_0 \alpha x_{\text{eff}} \langle S_z \rangle \sigma_z, \tag{3.106}$$

where $N_0\alpha$ is the s-d exchange interaction energy. For $Zn_{1-x}Mn_xSe$ DMS, $N_0\alpha = 0.27$ eV. x_{eff} is the effective concentration of magnetic ions, $\langle S_z \rangle$ is the thermal average magnetic moment of magnetic ions:

$$\langle S_z \rangle = S B_J \left(\frac{S g u_{\text{B}} B}{k_{\text{B}}(T + T_0)} \right), \tag{3.107}$$

where B_J is the Brillouin function and J is the angular momentum quantum number of magnetic ion basis state. For Mn^{2+} ion, $S = J = 5/2$. σ_z is the spin of conduction electrons. For holes in valence band, Eqs. 3.106 and 3.107 are still valid, except the parameters are changed to those of the valence band.

If the magnetic ion concentration x is several percents, in the moderate magnetic field the giant Zeeman splitting energy can reach several even several tens of millielectron volt orders of magnitudes. This energy level splitting will have different effects on the electrons with positive and negative spin orientations. Chang et al. studied the RT of electrons through a DMS potential barrier structure [15].

When a small bias is applied across the structure, a non-equilibrium electron population will be generated. The current density $J = \Sigma_\sigma J^\sigma$ can be calculated,

$$J^\sigma = \frac{e^2 V}{2\pi^2 l_{\mathrm{B}}^2} \sum_n \left(\frac{1}{2\pi}\right) \int_0^{k_{\mathrm{n}}^F} dk_z \left(-\frac{\partial f_0}{\partial E}\right) T^\sigma v_z^\sigma, \tag{3.108}$$

where $T^\sigma(n, E_{\mathrm{F}})$ is the transmission probability at the Fermi energy for different spin orientations, $v_z^\sigma = \hbar k_z^\sigma/m$ is the group velocity, $(1/2)\pi l_{\mathrm{B}}^2$ is the density of state of each Landau level, $l_{\mathrm{B}} = \sqrt{\hbar/eB}$ is the magnetic length, and f_0 is the equilibrium distribution function. We use the approximation $-\partial f_0/\partial E \approx \delta(E - E_{\mathrm{F}})$ when $k_{\mathrm{B}} T \ll E_{\mathrm{F}}$. The low-temperature conductance is given by

$$\sigma^\sigma = \sigma_0 \sum_n T^\sigma(k_{nF}^\sigma), \tag{3.109}$$

where

$$k_{nE_{\mathrm{F}}}^\sigma = \sqrt{\frac{2m}{\hbar^2}\left[E_{\mathrm{F}} - \left(E_{\mathrm{n}} \pm \frac{\Delta}{2}\right)\right]}, \tag{3.110}$$

$\sigma_0 = e^2/2\pi l_{\mathrm{B}}^2 h$, $\Delta = J_{s-d}\langle S_z\rangle$ is the giant Zeeman splitting, and $E_{\mathrm{n}} = (n + 1/2)\hbar\omega_c$ is the energy of the Landau level. The total conductivity is the sum of the conductivity of each Landau level at the Fermi energy and for each spin state.

The degree of spin polarization (SP) of the current density is defined by

$$P = \frac{J^\downarrow - J^\uparrow}{J^\downarrow + J^\uparrow}, \tag{3.111}$$

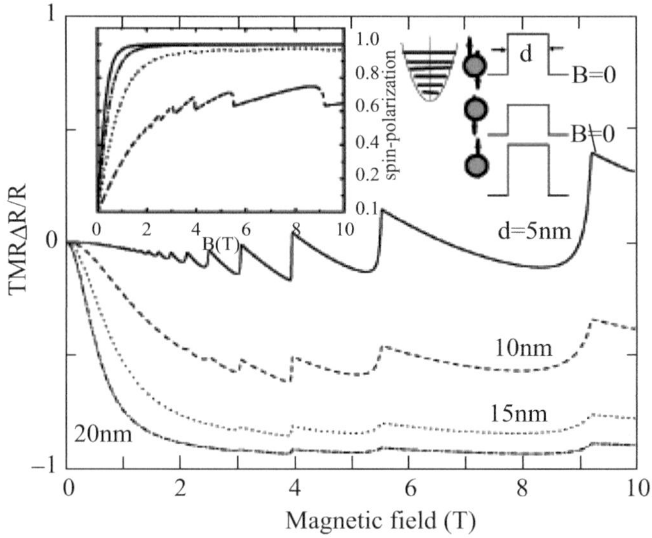

Figure 3.16 Electron tunneling through a single $ZnSe/Zn_{1-x}Mn_{0.07}Se$ ($x = 0.07$) barrier, TMR $\Delta R/R$, and spin polarization as functions of magnetic field for different DMS barrier thicknesses.

where $J^{\uparrow}$ ($J^{\downarrow}$) is the spin-up (spin-down) current density of the spin-polarized current. The tunneling magnetoresistance (TMR) as a result of tunneling through the NMR/DMS structures is defined by

$$\frac{\Delta R}{R} = \frac{R(B) - R(0)}{R(0)} = \frac{R(B)}{R(0)} - 1 = \frac{\sigma(0)}{\sigma(B)} - 1. \tag{3.112}$$

Figure 3.16 shows the TMR $\Delta R/R$ as a function of the magnetic field for different single DMS barrier thicknesses [15]. The inset gives the SP versus the magnetic field. The potential profiles for $B = 0$ and $B \neq 0$ are also plotted. The solid, dashed, dotted, and dash-dotted curves correspond to different thicknesses of the DMS layer: 5, 10, 15, 20 nm, respectively. The other parameters for $ZnSe/Zn_{1-x}Mn_xSe$ are $m^* = 0.16\ m_0$, $T_0 = 1.4$ K, $N_0\alpha = 0.17$ eV, and potential barrier height $V_b = 10$ meV ($x = 0.07$). From Fig. 3.16 we see that the TMR decreases and oscillates with an increasing magnetic field. The oscillations of the TMR and the SP are weakened by increasing the DMS barrier thickness. These oscillations are mainly attributed to the oscillation of the spin-down conductivity component $\sigma^{\downarrow}/\sigma_0$,

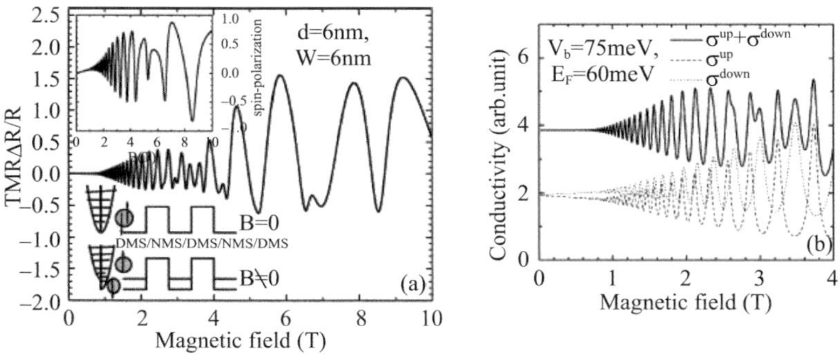

Figure 3.17 (a) Tunneling magnetoresistance (TMR) $\Delta R/R$ of a NMS double-barrier structure with DMS contacts as a function of magnetic field. The inset shows the spin polarization as a function of magnetic field. (b) Total conductivity (thick curve), spin-up (dotted), and spin-down (dashed) components as functions of magnetic field.

which is enhanced by an increasing magnetic field and weakened by the thickness of the DMS layer. The $\sigma^{\downarrow}/\sigma_0$ is larger than that of spin-up component $\sigma^{\uparrow}/\sigma_0$, since the barrier height seen by the spin-up electron is higher than that seen by the spin-down electron due to the magnetic field–induced s–p exchange interaction. Therefore the SP increases and saturates with an increasing magnetic field and barrier thickness.

Figure 3.17 shows the TMR $\Delta R/R$ of a non-magnetic semiconductor (NMS) double-barrier structure with a DMS well and contacts as a function of magnetic field [15]. The inset in Fig. 3.17a shows the SP as a function of magnetic field. The parameters are $E_F = 60$ meV and $V_b = 75$ meV for $Cd_{1-y}Mg_yTe/Cd_{1-x}Mn_xTe$ ($x = 0.04$, $y = 0.08$). From Fig. 3.17a we observe a peculiar beat pattern in the TMR and SP curves, which is directly related to the superposition of the s–p spin-split transmission channels. When the magnetic field increases, the landau levels are swept across the Fermi surface one by one, resulting in oscillations in the magnetoresistance, similar to those of Shubnikov de Haas (SdH) oscillations. And the s-d exchange interaction leads to a giant Zeeman splitting. The beating is a result of the fact that the total current is composed of spin-up and spin-down components. The interplay between the spin-up and spin-down channels results in the beat pattern in

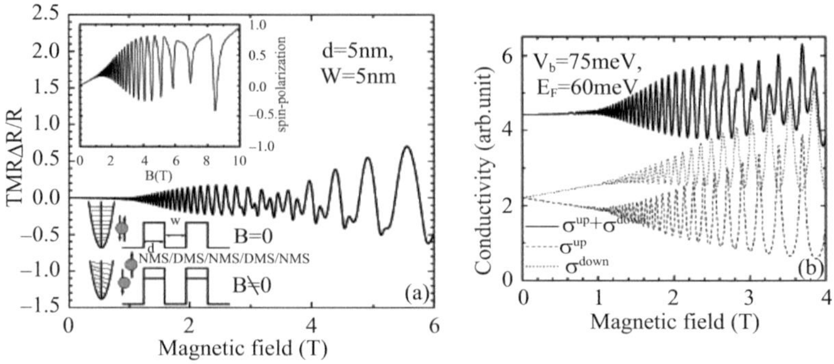

Figure 3.18 (a) Tunneling magnetoresistance (TMR) $\Delta R/R$ of a DMS double-barrier structure with NMS contacts as a function of magnetic field. The inset shows the spin polarization as a function of magnetic field. (b) Total conductivity (thick curve), spin-up (dotted), and spin-down (dashed) components as functions of magnetic field.

the magnetoresistance. This is clearly demonstrated in Fig. 3.17b, which shows the total conductivity (thick curve), spin-up (dotted), and spin-down (dashed) components. The phase difference of both currents results in the beat pattern of the total TMR and SP.

Figure 3.18a shows the TMR $\Delta R/R$ of a DMS double-barrier structure with a NMS well and contacts as a function of magnetic field [15]. The inset shows the SP as a function of magnetic field. The parameters are $E_F = 60$ meV and $V_b = 75$ meV for ZnSe/$Zn_{1-x}Mn_xSe$ ($x = 0.2$). Similar oscillating and beat behaviors as in Fig. 3.17a are found. Figure 3.18b shows the total conductivity and two components as functions of magnetic field. The left-down inset is the schematic diagram of the potential energy; in the magnetic field the potential heights for the spin-up and spin-down electrons are different, resulting in different tunneling probabilities $T^{\uparrow\downarrow}$. From Fig. 3.18a we see that the beat of the magnetoresistance is obvious. The SP oscillates with the magnetic field and is obviously larger than 1, that is, the spin-down current is dominant.

Grube et al. fabricated a BeTe/$Zn_{1-x}Mn_xSe$/BeTe double-barrier RTD and observed a SP of up to 80% in the tunneling current with a semimagnetic layer of only 3.5 nm thickness [16]. Figure 3.19 shows the schematic diagram of the potential energy of this device; the middle potential well is DMS $Zn_{1-x}Mn_xSe$, and the magnetic field

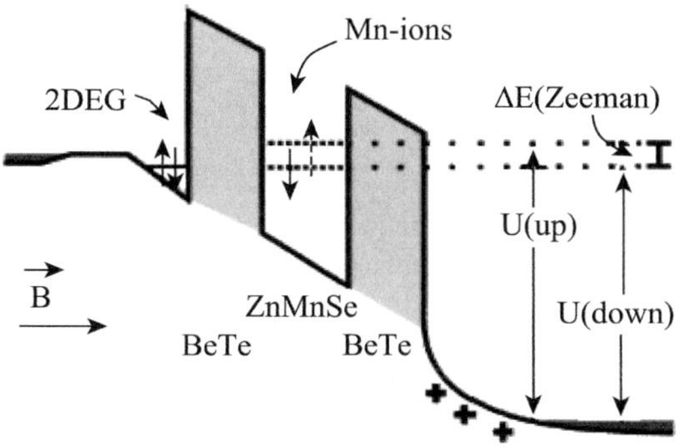

Figure 3.19 Schematic diagram of the potential energy of BeTe/$Zn_{1-x}Mn_xSe$/BeTe double-barrier resonant tunneling diode (RTD).

is applied in the z direction (perpendicular to the junction). Due to the giant Zeeman splitting effect the energy levels of the spin-up and spin-down electrons split. In the magnetic field $B = 2.5$ T the splitting energy is about 20 meV. When the bias is applied to the diode, if the Fermi energy level of the electron in the injecting region and the spin-down energy level in the well region coincide, there will be a tunneling current with spin-down polarization. At the same time the spin-up electron cannot tunnel through the device. Therefore in the collecting region we obtain electrons with only one spin orientation; the SP degree approaches 1 or -1.

To measure the SP degree a III-V $Al_{0.07}Ga_{0.93}As$/GaAs LED is grown directly on the bottom of the RTD. The tunneling current from the RTD is directly injected into the LED, and the spin-polarized electrons combine with holes in the quantum well. Measuring the circular polarization degree of the luminescent light, one can determine the SP degree according to the selected rule of the electron-hole transition [17].

The SP degree of injected current can be derived from the circular polarization degree of the LED luminescence [17],

$$P_{\text{opt}} = \frac{(3n^{\uparrow} + n^{\downarrow}) - (3n^{\downarrow} + n^{\uparrow})}{(3n^{\uparrow} + n^{\downarrow}) + (3n^{\downarrow} + n^{\uparrow})} = \frac{2n^{\uparrow} - 2n^{\downarrow}}{4n^{\uparrow} + 4n^{\downarrow}} = \frac{1}{2}P_{\text{J}}, \qquad (3.113)$$

where $n^{\uparrow}$ and $n^{\downarrow}$ are the occupied number of spin-up and spin-down electrons, respectively. P_{J} is the SP degree of the current; thus the SP

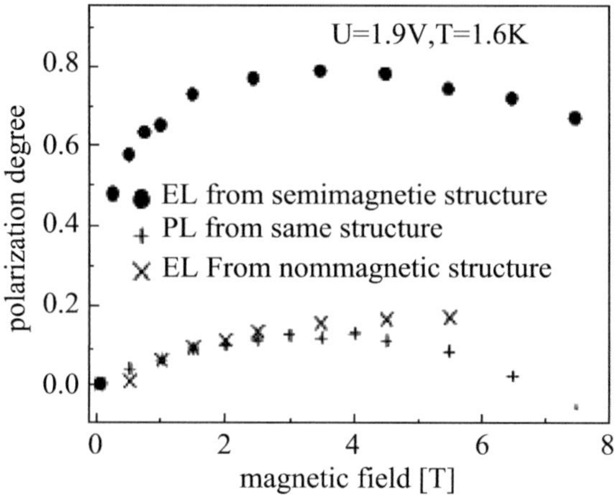

Figure 3.20 Magnetic dependence of the spin polarization degree (•) at a constant voltage of 1.9 V at the beginning of the first resonance in the AlGaAs/GaAs LED. For comparison the results for the nonmagnetic RTD are also shown (+, ×).

degree of the current is double of the circular polarization degree of luminescence.

Figure 3.20 shows the magnetic dependence of the SP degree (•) at a constant voltage of 1.9 V at the beginning of the first resonance in the AlGaAs/GaAs LED [16]. For comparison the results for the nonmagnetic RTD are also shown (+, ×). The results prove that the magnetic RT structure is the origin of the spin injection. From Fig. 3.20 we see that the SP degree increases with the magnetic field until it saturates at 80% for fields above 2–3 T, coinciding very well with the saturation of the giant Zeeman splitting in DMSs. Test measurements with optical excitation of nonpolarized carriers in the same structure with nonmagnetic BeTe-ZnSe tunneling structures reveal only a small polarization degree below 15%.

Figure 3.21 shows the voltage dependence of the SP degree and I–V curve at a constant magnetic field of 2.5 T around the first resonance of the 3.5 nm/3.0 nm BeTe/$Zn_{0.94}Mn_{0.06}Se$ RTD [16]. From Fig. 3.21 we see a clear reduction of the polarization degree over the resonance, which can be explained by the offset of the resonant spin-down level in the well (see Fig. 3.19). Although a

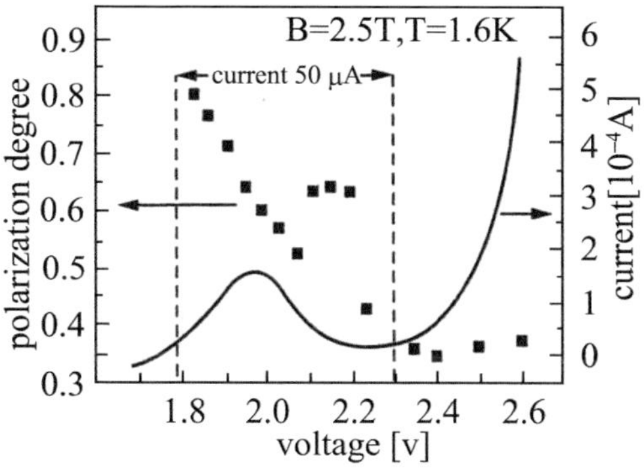

Figure 3.21 Voltage dependence of the spin polarization degree and I–V curve at a constant magnetic field of 2.5 T around the first resonance of the 3.5 nm/3.0 nm BeTe/$Zn_{0.94}Mn_{0.06}Se$ RTD.

reversal of the spin orientation has not been observed, the SP at a constant drive current of 50 μA can be from 80% down to 38% by merely changing the external voltage from 1.8 to 2.3 V. The reason that the spin-up polarization has not been observed is because the transit time is well above 10^{-7} s in the present RTD, whereas the spin-relaxation times in diluted magnetic semiconductors are in the range of 10^{-12} s. Thus, electrons that enter the ZnMnSe quantum well in resonance with the higher energy spin-split subband level will very likely relax into the lower energy spin-split level before they are injected into the III-V material.

Xia et al. studied theoretically the spin-polarized current through a RTD made of a semimagnetic semiconductor [18]. The structure is shown in Fig. 3.19; the parameters are the same as in Ref. [16]: $B = 2.5$ T, $T = 1.6$ K, $T_0 = 1.4$ K, and concentration of Mn^{2+} ions $x = 0.06$. Because the temperature is very low, we take $T = 0$ K in the calculation. Thus the tunneling current is given by Eq. 3.79. Because the RTD is in the magnetic field along the z direction (current direction), one should use the density of states of Landau energy levels instead of the two-dimensional density of states in Eq. 3.79,

$$J = \frac{e}{2\pi^2\hbar}\left(\frac{eB}{\pi\hbar c}\right)\int_0^{E_\mathrm{F}} dE_z N(E_\mathrm{F} - E_z)T(E_z), \tag{3.114}$$

where $eB/(\pi\hbar c)$ is the density of states of each Landau energy level,

$$N(E_F - E_z) = \text{int}\left(\frac{E_F - E_z}{\hbar\omega_c} - \frac{1}{2}\right). \tag{3.115}$$

The transmission probability $T(E_z)$ is calculated by the numerical method.

The parameters of the RTD are well width 3.5 nm, barrier width 3 nm, and potential barrier height 1890 meV. The electron effective masses are assumed to be equal for the well and barrier, $m^* = 0.16$ m_0. For the BeTe/ZnTe quantum well there are three confined states in the well, whose energies are 132, 521, and 1135 meV, respectively. The electron density n in the injection region is 10^{19} cm^{-3}. Using the relation $E_F = \hbar^2(3\pi^2 n)^{2/3}/(2m^*)$, we obtain the Fermi energy $E_F = 105.8$ meV. In the calculation we take $E_F = 100$ meV.

Figure 3.22 shows the spin-polarized tunneling current (down ∇, up Δ), total current ($\Box$), and SP degree ($\bullet$) as functions of bias voltage at $B = 2.5$ T when the Fermi level crosses over the first quasi-confined state [18]. From Fig. 3.22 we see that when the electric field is such that the Fermi level is near the first confined state there is only one kind of spin orientation, that is, spin-down, and the polarization degree reaches 1. As the electric field (voltage) increases, the current of the other kind of spin orientation, that is, spin-up, increases and the polarization degree decreases gradually. These results are basically consistent with

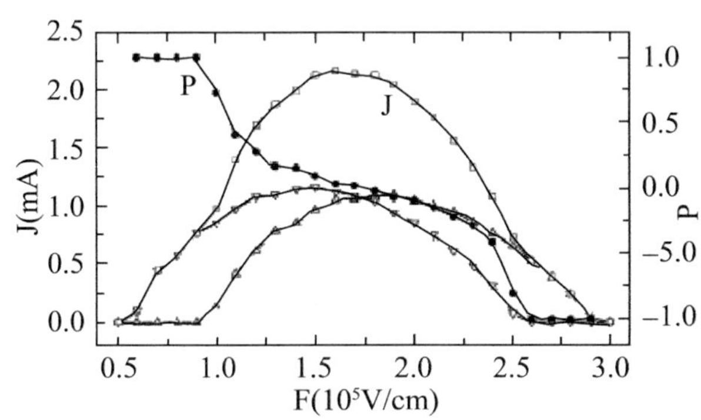

Figure 3.22 Spin-polarized tunneling current (down ∇, up Δ), total current ($\Box$), and spin polarization degree ($\bullet$) as functions of bias voltage at $B = 2.5$ T when the Fermi level crosses over the first quasi-confined state.

the experimental results [16]. When the electric field increases continuously the opposite process occurs. The current of the spin-up becomes larger than that of the spin-down, and consequently, the polarization degree becomes negative. But this behavior has not been observed experimentally; the reason has been indicated above.

References

1. Tsu, R., and Esaki, L. (1973). Tunneling in a finite superlattice. *Appl. Phys. Lett.*, **22**, 562–564.
2. Chang, L. L., Esaki, L., and Tsu, R. (1969). Resonant tunneling in semiconductor double barriers. *Appl. Phys. Lett.*, **24**, 593–595.
3. Morkoc, H., Chen, J., and Reddy, U. K. (1986). Observation of a negative differential resistance due to tunneling through a single barrier into a quantum well. *Appl. Phys. Lett.*, **49**, 70–72.
4. Goodhue, W. D., Sollner, T. C. L. G., Le, H. Q., Brown, E. R., and Vojak, B. A. (1986). Large room-temperature effects from resonant tunneling through AlAs barriers. *Appl. Phys. Lett.*, **49**, 1086–1088.
5. Broekaert, T. P. E., Lee, W., and Fonstad, C. G. (1988). Pseudomorphic $In_{0.53}Ga_{0.47}As$/AlAs/InAs resonant tunneling diodes with peak-to-valley current ratios of 30 at room temperature. *Appl. Phys. Lett.*, **53**, 1545–1547.
6. Brown, E. R., Sollner, T. C. L. G., Parker, C. D., Goodhue, W. D., and Chen, C. L. (1989). Oscillations up to 420 GHz in GaAs/AlAs resonant tunneling diodes. *Appl. Phys. Lett.*, **55**, 1777–1779.
7. Tamura, H., and Ando, T. (1991). Conductance fluctuations in quantum wires. *Phys. Rev. B*, **44**, 1792–1800.
8. Price, P. J. (1993). Transmission and reflection peaks in multichannel ballistic transport. *Phys. Rev. B*, **48**, 17301–17307.
9. Xia, J. B. (1988). Theory of hole resonant tunneling in quantum well structures. *Phys. Rev. B*, **38**, 8365–8370.
10. Tsuchiya, M., and Sakaki, H. (1986). Dependence of resonant tunneling current on well widths in AlAs/GaAs/AlAs double barrier diode structures. *Appl. Phys. Lett.*, **49**, 88–90.
11. Luryi, S., and Zaslavsky, A. (1998). *Modern Semiconductor Device Physics*, ed. Sze, S. M., Chapter 5 "Quantum-effect and hot-electron devices" (John Wiley & Sons, Inc.).

12. Brown, E. R., Söderström, J. R., Parker, C. D., et al. (1991). Oscillations up to 712 GHz in InAs/AlSb resonant-tunneling diodes. *Appl. Phys. Lett.*, **58**, 2291–2293.
13. Mendez, E. E., Wang, W. I., Ricco, B., and Esaki, L. (2008). Resonant tunneling of holes in AlAs-GaAs-AlAs heterostructures. *Appl. Phys. Lett.*, **47**, 415–417.
14. Broido, D. A., and Sham, L. J. (1985). Effective masses of holes at GaAs-AlGaAs heterojunctions. *Phys. Rev. B*, **31**, 888–892.
15. Chang, K., Xia, J. B., and Peeters, F. M. (2002). Oscillating magnetoresistance in dilute magnetic semiconductor barrier structures. *Phys. Rev. B*, **65**, 115209 (1–5).
16. Gruber, Th., Keim, M., Fiederling, R., et al. (2001). Electron spin manipulation using semimagnetic resonant tunneling diodes. *Appl. Phys. Lett.*, **78**, 1101–1103.
17. Fiederling, R., Keim, M., Reuscher, G., et al. (1999). Injection and detection of a spin-polarized current in a light-emitting diode. *Nature*, **402**, 787–790.
18. Xia, J. B., Hai, G. Q., and Wang, J. N. (2003). Spin-polarized current produced by a double barrier resonant tunneling diode. *Solid State Commun.*, 489–492.

Chapter 4

Longitudinal Transport of Superlattices

Longitudinal transport in a superlattice is transport along the growth direction of the superlattice; the resonant tunneling discussed in the preceding chapter is one of the forms of longitudinal transports. The path of the longitudinal transport is very short, only several or several tenths of nanometers. The effect of scattering on electron motion is not obvious. Electron motion is mainly controlled by the quantum effect. At the same time, electron motion velocity is generally very higher than the thermal balance velocity of carriers; therefore these kinds of devices are called quantum effect devices or hot-electron devices [1]. The advantage of such a device is high velocity, and it can be used to fabricate high-frequency devices. But due to the delay of the circuit, for example, the RC time delay, the extremely high velocity in the active region cannot be represented in the whole device performance. Now people use the strong or modulated nonlinear current-voltage characteristics or special electrode symmetry to realize a more complex circuit performance and to replace the circuit unit constructed by a large number of transistors or passive circuits; for example, they use a few tunneling devices to realize a multiple value memory and logic circuit.

The largest obstacles of the practical application of quantum effect devices and hot-electron devices are the difficulties of room

Quantum Waveguide in Microcircuits (Second Edition)
Jian-Bai Xia, Duan-Yang Liu, and Wei-Dong Sheng

ISBN 978-981-4968-30-0 (Hardcover), 978-1-003-65146-8 (eBook)
www.jennystanford.com

temperature operation and large-scale integration, resulting in that they are now only used in some special cases.

4.1 Miniband Transport of a Superlattice

When the potential barrier width of a superlattice becomes gradually smaller, the confined states of neighboring wells will interact and the energy of a single confined state becomes a broad band, called a miniband. These neighboring confined states in quantum wells are similar to the atoms of a one-dimensional (1D) atomic chain. Assume that the energy of each atomic state is E_0, the interaction energy of neighboring atoms is V_0, and the distance between neighboring atoms is d; then the tight binding wave function of the atomic chain is

$$\psi_k = \frac{1}{\sqrt{N}} \sum_i e^{ikR_i} \phi_i(r), \tag{4.1}$$

where k is the wave vector, R_i is the coordinate of the i-th atom, $\phi_i(r)$ is the wave function of the i-th atom, and N is the total number of atoms. The energy of the atomic chain,

$$\begin{aligned} E = \psi^* H \psi &= \frac{1}{N} \sum_{ij} e^{-ikR_j} \phi_j^*(r) H e^{ikR_i} \phi_i(r) \\ &= E_0 + V_0 e^{ikd} + V_0 e^{-ikd} \\ &= E_0 + 2V_0 \cos kd. \end{aligned} \tag{4.2}$$

Equation 4.2 can be also written as

$$E_z(k) = E_0 - \frac{\Delta}{2} \cos kd, \tag{4.3}$$

where $\Delta = 4|V_0|$ is the miniband width, which is proportional to the interaction energy V_0 of confined states in neighboring wells, that is, proportional inversely to the width of the barrier. Figure 4.1 shows Δ as functions of the barrier width for a fixed well width $w = 3.5$ nm in a GaAs/AlAs superlattice.

The miniband given by Eq. 4.3 is different from the ordinary energy band of the semiconductor in the following points: (i) The semiconductor energy band is three-dimensional isotropic, while the superlattice miniband is highly anisotropic; in the x and y

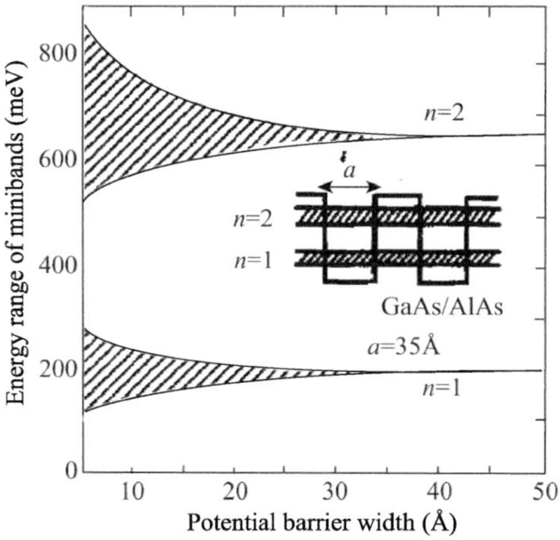

Figure 4.1 Δ as functions of the barrier width for a fixed well width $w = 3.5$ nm in a GaAs/AlAs superlattice.

directions it is same as the bulk energy band, but in the z direction it is a band with a small width Δ (generally several tenths of millielectron volts). (ii) A semiconductor energy band is defined in the Brillouin zone $-\pi/a < k < \pi/a$ for each direction, where a is the lattice constant. The miniband is defined in the Brillouin zone of z direction $-\pi/d < k < \pi/d$, where d is the period of the superlattice in the z direction. $d \gg a$, so the Brillouin zone of the miniband is very smaller than that of the bulk energy band. (iii) In the applied electric field the electrons in the bulk material move near the center of the Brillouin zone and are scattered by impurities or other scattering centers. The derivation of distribution function from the balance state is very small, so the transport can be treated by the perturbation method, for example, solving the Boltzmann equation. In the superlattice due to the small Brillouin zone in the z direction, in the electric field the electron can reach the edge of the Brillouin zone ($k = \pm\,\pi/d$) before it is scattered. At the edge of Brillouin zone the electron effective mass is negative, so negative conductance is possible.

The electron motion equation in a semiconductor is

$$\hbar\frac{d\mathbf{k}}{dt} = e\mathbf{F}, \tag{4.4}$$

where **k** is the wave vector and **F** is the applied electric field. In the electric field along the z direction the electron moves from the center of the Brillouin zone and after time τ reaches the edge of the Brillouin zone π/d. Then

$$\hbar k = \hbar \left(\frac{\pi}{d}\right) = eF\tau. \tag{4.5}$$

If the scattering relaxation time is longer than τ, then the electrons can reach the edge of the Brillouin zone and produce negative conductance. Esaki et al. predicted the Bloch oscillation is probably observed in superlattices on the basis of this idea [2]. Equation 4.5 becomes a necessary condition of occurrence of negative differential conductance (NDC). In the following we accept the relaxation approximation proposed by Huang et al. [3].

From Eq. 4.3 the electron velocity in the z direction

$$v = \frac{1}{\hbar}\frac{\partial E_z(k)}{\partial k} = \frac{\Delta d}{2\hbar}\sin kd. \tag{4.6}$$

From Eq. 4.4 the variation of the electron wave vector with time is

$$k = k_0 + \frac{eF}{\hbar}t, \tag{4.7}$$

where k_0 is the initial wave vector. Suppose the electron scattering time is τ; then the average drift velocity,

$$\begin{aligned} v_{\mathrm{d}}(k_0) &= \frac{1}{\tau}\int_0^\infty v(k_0, t)e^{-t/\tau}\,dt \\ &= \frac{\Delta eFd^2\tau}{2\hbar^2} \cdot \frac{\cos k_0 d + \frac{\hbar}{eFd\tau}\sin k_0 d}{1 + \left(\frac{eFd\tau}{\hbar}\right)^2}. \end{aligned} \tag{4.8}$$

The electron total energy

$$E = E_z(k) + \frac{\hbar^2 k_{||}^2}{2m^*}. \tag{4.9}$$

Assuming that the electrons satisfy the Boltzmann distribution, the electron density

$$\begin{aligned} n &= \frac{C}{4\pi^2}\int_{-\pi/d}^{\pi/d} dk \exp\left(-\frac{E_z(k)}{k_{\mathrm{B}}T}\right)\int_0^\infty dk_{||}^2 \exp\left(-\frac{\hbar^2 k_{||}^2}{2m^* k_{\mathrm{B}}T}\right) \\ &= C \cdot \frac{m^* k_{\mathrm{B}}T}{\pi\hbar^2 d} I_0\left(\frac{\Delta}{2k_{\mathrm{B}}T}\right), \end{aligned} \tag{4.10}$$

where I_0 is the zero-order Bessel function of an imaginary argument.

The calculation of the current density is similar to Eq. 4.10. Calculate the statistical average of the drift velocity $v_{\mathrm{d}}(k)$,

$$\begin{aligned} j &= C\frac{ev_{\mathrm{d}0}}{4\pi^2}\cdot\frac{2m^*k_{\mathrm{B}}T}{\hbar^2}\int_{-\pi/d}^{\pi/d}dk\cos kd\exp\left(-\frac{E_z(k)}{k_{\mathrm{B}}T}\right) \\ &= C\frac{ev_{\mathrm{d}0}m^*k_{\mathrm{B}}T}{\pi\hbar^2 d}I_1\left(\frac{\Delta}{2k_{\mathrm{B}}T}\right) \\ &= nev_{\mathrm{d}0}\frac{I_1(\Delta/2k_{\mathrm{B}}T)}{I_0(\Delta/2k_{\mathrm{B}}T)}, \end{aligned} \tag{4.11}$$

where

$$v_{\mathrm{d}0} = \frac{\Delta eFd^2\tau}{2\hbar^2}\cdot\frac{1}{1+\left(\frac{eFd\tau}{\hbar}\right)^2}. \tag{4.12}$$

In calculating Eq. 4.11 the term sin kd in $v_{\mathrm{d}}(k)$ (Eq. 4.8) is odd symmetric to k. Hence its contribution to the integration for k equals zero.

Therefore the current composes two parts: one is related to electric field strength F, which is included in v_{d0} (Eq. 4.12); the other is related to temperature, which is included in I_1/I_0. When the electric field is very small, $eFd\tau/h \ll 1$, then

$$v_{\mathrm{d}0} = \frac{\Delta eFd^2\tau}{2\hbar^2} = \frac{eF\tau}{m_z{}^*}, \tag{4.13}$$

where $m_z{}^*$ is the effective mass in the z direction,

$$\frac{1}{m_z{}^*} = \frac{1}{\hbar^2}\frac{\partial^2 E_z(k)}{\partial k^2}\bigg|_{k=0} = \frac{1}{\hbar^2}\frac{\Delta d^2}{2}. \tag{4.14}$$

Equation 4.13 is the result of the linear transport theory. If the electric field F and the relaxation time τ are large enough so $eFd\tau/h \approx 1$, then nonlinear conductance will appear and the current will not be proportional to the electric field.

Sibille et al. observed first the nonlinear conductance caused by the miniband transport [4]. They grew some GaAs/AlAs superlattice samples: the width of the GaAs layer is 10~20 monolayers (2.83 Å) and that of the AlAs layer is 3~7 monolayers. The I–V curves of three samples at 300 K are shown in Fig. 4.2 [4], where (12/4) represents that the monolayer number of GaAs and AlAs, 12 and 4, respectively. From Fig. 4.2 we see that when the bias voltage is very

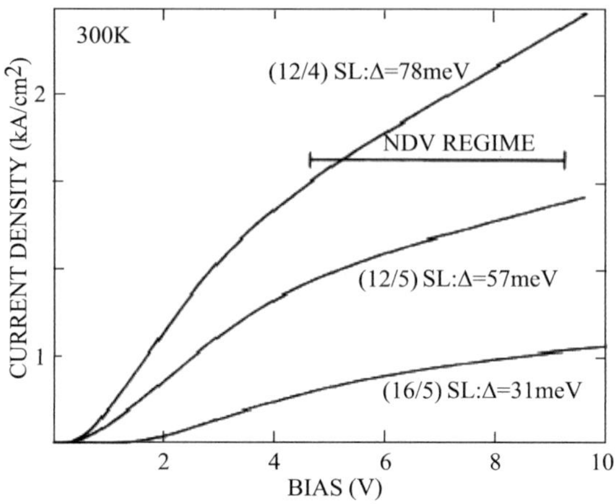

Figure 4.2 I–V curves of three samples at 300 K.

small (3 V) there appears nonlinear conductance. From Eq. 4.12 the drift velocity can be phenomenally written as

$$v_\mathrm{d}(F) = \frac{\mu F}{1 + (F/F_0)^\beta}, \tag{4.15}$$

where β is a parameter. When $\beta = 2$, Eq. 4.15 is just Eq. 4.12. When $\beta > 1$, $v_\mathrm{d}(F)$ reaches maximum if F equals a critical field F_c. When $F > F_\mathrm{c}$, $v_\mathrm{d}(F)$ decreases and exhibits a negative differential velocity (NDV). When $\beta = 1$, $v_\mathrm{d}(F)$ approaches a saturated value; there is no maximum, as shown in Fig. 4.2. When $\beta = 2$, $v_\mathrm{d}(F)$ reaches the maximum $v_\mathrm{p} = \mu F_0/2$ at $F = F_0$.

Sibille et al. measured the I–V curves for another set of samples that have larger miniband widths Δ. The I–V curves for samples with $\Delta = 133$, 81, and 52 meV are shown in Fig. 4.3 [5]. Different from Fig. 4.2 the I–V curves in Fig. 4.3 show obvious NDC behavior.

A shown in Fig. 4.3 the samples are ohmic at a low bias and exhibit NDC beyond a critical voltage. The peak electron velocity v_p can be estimated from the peak current density J_p by applying $J_\mathrm{p} = env_\mathrm{p}$, where $n = 6 \times 10^{16}$ cm^{-3} is the electron density. For the (17/4) sample, $J_\mathrm{p} = 40$ kA/cm^2 and $v_\mathrm{p} = 4 \times 10^6$ cm/s; for the (13/3) sample, $J_\mathrm{p} = 140$ kA/cm^2 and $v_\mathrm{p} = 10^7$ cm/s. Because the electron velocity is very high, these samples can be used to fabricate high-frequency devices. For the highest value, a simple calculation leads

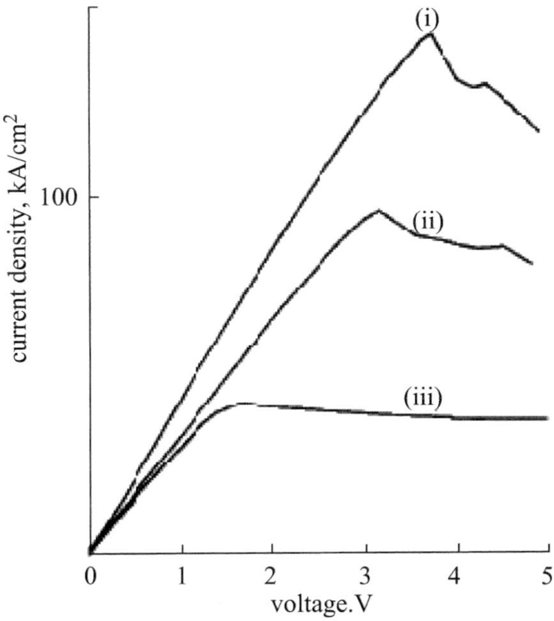

Figure 4.3 I–V curves for samples with $\Delta = 133$ (i), 81 (ii), and 52 (iii) meV.

to an expected maximum fundamental frequency $f_{\max} = v_p/L$ of 200 GHz, where L is the device thickness $L = 0.5$ μm.

Another NDC mechanism was proposed, relying on a transition between Bloch miniband transport and hopping transport between Wannier–Stark quantized levels (WSL). To differentiate between the negative effective mass (NEM) and WSL mechanisms, the edF_c and v_p/d as a function of Δ for a series of samples are shown in Fig. 4.4 [4]. According to the NEM mechanism, Eqs. 4.12 and 4.15, ($\beta = 2$),

$$F_c = \frac{\hbar}{ed\tau}, \quad edF_c = \frac{\hbar}{\tau}, \tag{4.16}$$

and

$$v_p = \frac{\mu F_c}{2} = \frac{\Delta d}{4\hbar}, \quad \frac{v_p}{d} = \frac{\Delta}{4\hbar}. \tag{4.17}$$

On the other hand, the WSL theory predicts $edF_c \approx \Delta$ and a superlinear dependence of v_p/d on Δ. From Fig. 4.4 we see that the experimental values of edF_c rapidly and strongly deviate from the

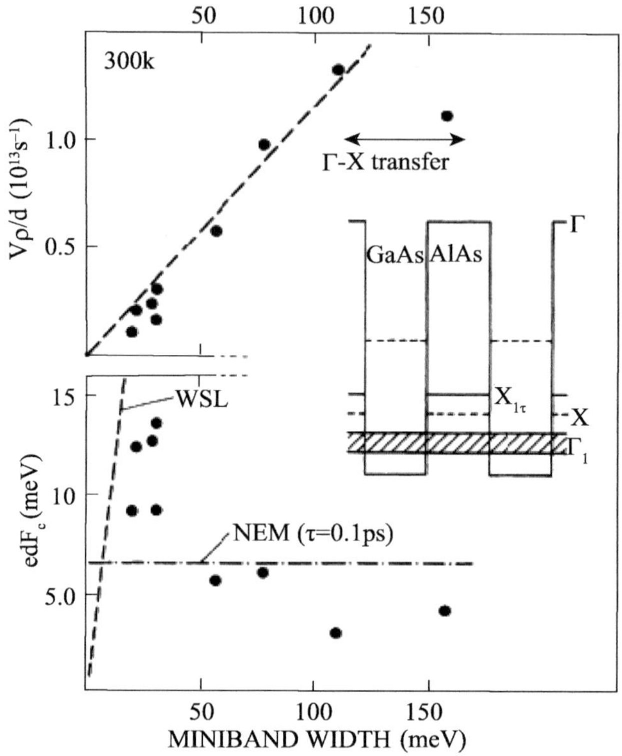

Figure 4.4 edF_c and v_p/d as a function of Δ for a series of samples. Inset: Conduction band diagram of the relevant levels.

line predicted by the WSL mechanism. Furthermore, v_p/d appears to vary linearly with Δ at least to $\Delta = 100$ meV. These two features strongly argue against the WSL mechanism and in favor of the NEM one.

Figure 4.5 shows the frequency dependence of the device conductance for several applied biases [6]. Both curves (two samples) exhibit sharp resonances, resulting in a large negative conductance at peak frequency f_m. The resonance frequencies are 2.7 GHz and 4.9 GHz for the samples of total thickness $L = 1.2$ μm and 0.6 μm, respectively. The f_m indeed scales almost linearly with the inverse of the square loop (SL) thickness, a logical consequence of a transit time effect.

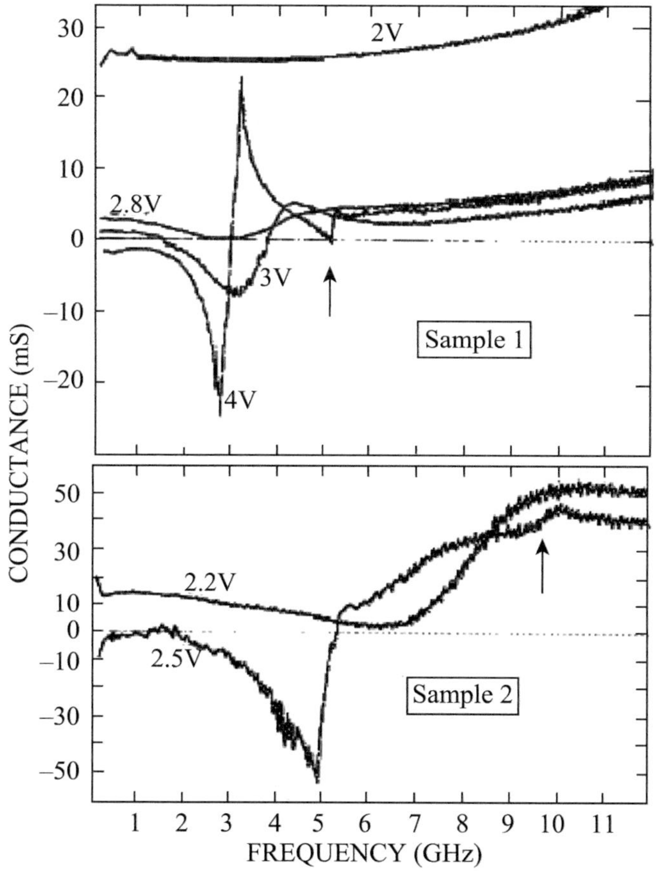

Figure 4.5 Microwave conductance as a function of frequency for several applied biases.

4.2 Bloch Oscillation in Superlattices

When Esaki et al. proposed the miniband transport of a superlattice, they predicted that it is possible to observe the Bloch oscillation in the superlattice [2]. If the collision relaxation time in the superlattice τ satisfies Eq. 3.5, then the Bloch oscillation can occur, whose period equals the time that the electron travels one circle along the Brillouin zone.

$$\hbar\frac{\Delta k}{T} = \hbar\left(\frac{2\pi}{d}\right)\frac{1}{T} = eF$$

$$T = \frac{\hbar}{eF}\left(\frac{2\pi}{d}\right), \quad \omega_{\mathrm{B}} = \frac{2\pi}{T} = \frac{eFd}{\hbar}, \tag{4.18}$$

where ω_{B} is called Bloch oscillation frequency. To observe Bloch oscillation experimentally, people have made many years efforts, until 1996, when the strong terahertz photoelectric resonance at Bloch oscillation frequency was observed [7].

The experimental sample in the superlattice structure consists of 40 periods of 8 nm wide GaAs wells and 2 nm AlGaAs barriers (miniband width $\Delta = 22$ meV). The superlattice is homogeneously Si doped with a concentration of $n = 3 \times 10^{15}$ cm^{-3}. In the University of California, Santa Barbara (UCSB), free electron laser (FEL) were used as a terahertz radiation source, giving pulses of intense terahertz radiation with a pulse of several microseconds. Therefore the electric fields, including direct and alternating, are applied to the superlattice sample longitudinally.

$$F = F_0 + F_1 \cos\omega t. \tag{4.19}$$

Figure 4.6 shows the DC current-voltage curve for increasing FEL intensity. The FEL frequency was fixed to be 0.6 THz (Fig. 4.6a) and 1.5 THz (Fig. 4.6b) [7]. The curves are shifted downward for increasing laser intensity (F_1). From the figures we see that besides the NDC, there are a series of resonance peaks at the Bloch frequency and their harmonics. When the laser intensity is increased further the first peak starts to decrease and a second peak at about twice the voltage of the first peak is observed due to a two-photon resonance. At the highest intensities up to a four-photon resonance was observed. The authors attributed the first additional peak to a resonance of the external laser field where the Bloch frequency $\omega_{\mathrm{B}} = \omega$. The other peaks are higher resonances, where $\omega_{\mathrm{B}} = n\omega$.

To explain this behavior we still use the relaxation time approximation [3] used in the last section to calculate the current density in the presence of an alternating field. Similar to Eqs. 3.6 and 3.7, in the electric field (Eq. 4.19) we have

$$k = k_0 + \frac{eF_0}{\hbar}t + \frac{eF_1}{\hbar\omega}\sin\omega t. \tag{4.20}$$

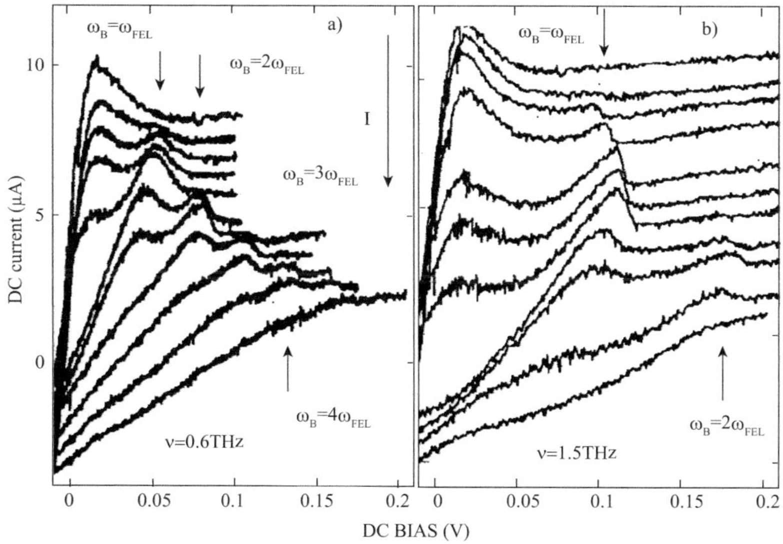

Figure 4.6 DC current-voltage curve for increasing FEL intensity. The FEL frequency was fixed to 0.6 THz (a) and 1.5 THz (b).

$$\begin{aligned} v(t) &= \frac{\Delta d}{2\hbar} \sin kd \\ &= \frac{\Delta d}{2\hbar} \sin\left(k_0 + \frac{eF_0 t}{\hbar} + \frac{eF_1}{\omega} \sin\omega t\right) d \\ &= \frac{v_0}{2i}\left[e^{i(\alpha+\beta\sin\omega t)} - e^{-i(\alpha+\beta\sin\omega t)}\right]. \end{aligned} \tag{4.21}$$

Expanding Eq. 4.21 to t, we obtain

$$\begin{aligned} v(t_0) = v_0 \Bigg\{ &\sin\alpha \left[J_0(\beta) + 2\sum_{k=1}^{\infty} J_{2k}(\beta)\cos 2k\omega t\right] \\ &+ \cos\alpha \left[2\sum_{k=0}^{\infty} J_{2k+1}(\beta)\sin(2k+1)\omega t\right]\Bigg\}. \end{aligned} \tag{4.22}$$

where

$$v_0 = \frac{\Delta d}{2\hbar}, \quad \alpha = \frac{eF_0 d}{\hbar} t + k_0 d, \quad \text{and} \quad \beta = \frac{eF_1 d}{\hbar\omega}. \tag{4.23}$$

When the collision relaxation is taken into account, the electron drift velocity equals

$$v_d(k_0) = \frac{1}{\tau}\int_0^{\infty} v(k_0, t) e^{-t/\tau} dt. \tag{4.24}$$

Inserting Eq. 4.23 into Eq. 4.24, at last we obtain

$$v_{d0} = \frac{\Delta eFd^2\tau}{2\hbar^2}\sum_{n=0}^{\infty} J_n\left(\frac{eF_1 d}{\hbar\omega}\right)Q_n,$$

$$Q_n = \begin{cases} \dfrac{1}{1+\left(\dfrac{eF_0 d\pi}{\hbar}\right)^2}, & n=0 \\ \dfrac{1+\dfrac{n\omega\hbar}{eF_0 d}}{1+\left(\dfrac{eF_0 d}{\hbar}+n\omega\right)^2\tau^2} \pm \dfrac{1-\dfrac{n\omega\hbar}{eF_0 d}}{1+\left(\dfrac{eF_0 d}{\hbar}-n\omega\right)^2\tau^2}. & \begin{array}{l} n=2k \\ n=2k+1 \end{array} \end{cases} \tag{4.25}$$

Phenomenally we can write Eq. 4.25 as follows:

$$v_{d0} = \sum_{n=0}^{\infty} v_{dn}(F_0),$$
$$v_{dn}(F_0) = \frac{\mu_n F_0}{1+\left(\frac{F_0}{F_c} \pm n\omega\tau\right)^2}. \tag{4.26}$$

Equation 4.26 can be used to explain the peak positions in Fig. 4.6. Differentiating $v_{dn}(F_0)$ in Eq. 4.26 to F_0, we obtain

$$F_0 = n\omega\tau F_c = n\omega\tau\cdot\frac{\hbar}{ed\tau}, \quad \text{where} \quad n\omega = \frac{F_0 ed}{\hbar} = \omega_B. \tag{4.27}$$

v_{dn} (F_0) have maximum values. This explains the peaks corresponding to $n\omega_{FEL}$ in the I–V curve of Fig. 4.6, that is, the terahertz laser excites the Bloch oscillation in the superlattice. When the laser intensity F_1 increases, the amplitudes of peaks change according to the expanding coefficients in Eq. 4.25 $J_n(eF_1 d/\hbar\omega)$, which oscillates as F_1 increases.

We calculated the direct current (DC)'s conductance with the above relaxation time approximation theory. In the following we will use the simplified balance equation theory [8] to study the dynamic behavior of the superlattice in the DC and alternating current (AC) electric fields,

$$\frac{dv_d}{dt} = \frac{eF(t)}{m_z{}^*(\varepsilon)} - \frac{v_d}{\tau_v},$$
$$\frac{d\varepsilon}{dt} = eF(t)v_d - \frac{(\varepsilon-\varepsilon_T)}{\tau_\varepsilon}, \tag{4.28}$$

where v_d is the drift velocity; ε is the electron energy; $m_z{}^*(\varepsilon)$ is the electron effective mass,

$$\frac{1}{m_z{}^*(\varepsilon)} = \frac{1}{\hbar^2}\frac{\partial^2 E_z(k)}{\partial k^2} = \frac{\Delta d^2}{2\hbar^2}\cos kd = \frac{1}{m_z{}^*(0)}\left(1 - \frac{2\varepsilon}{\Delta}\right); \quad (4.29)$$

ε_T is the electron average thermal energy,

$$\varepsilon_T = \frac{\Delta}{2}\left[1 - \frac{I_1(\Delta/2kT)}{I_0(\Delta/2kT)}\right]; \quad (4.30)$$

and τ_v and τ_ε are the electron momentum and energy scattering relaxation times, respectively.

To check the simplified balance equation (Eq. 4.28), we calculate the steady solution of Eq. 4.28 in the DC bias voltage. Let $eF(t) = eF$ and $dv_d/dt = d\varepsilon/dt = 0$. We obtain

$$\varepsilon = \frac{C + \varepsilon_T}{1 + 2C/\Delta}, \quad C = \frac{(eF)^2 \tau_v \tau_\varepsilon}{m_z^*(0)}. \quad (4.31)$$

$$v_d = \frac{1}{eF\tau_\varepsilon}[\varepsilon - \varepsilon_T] = \frac{1}{eF\tau_\varepsilon}C \cdot \frac{I_1/I_0}{1 + \frac{2}{\Delta}C} = v_{d0} \cdot \frac{I_1}{I_0}, \quad (4.32)$$

$$v_{d0} = \frac{\Delta eF d^2 \tau_v}{2\hbar^2} \cdot \frac{1}{1 + \left(\frac{eFd\tau}{\hbar}\right)^2}, \quad \tau = \sqrt{\tau_v \tau_\varepsilon}. \quad (4.33)$$

Equation 4.32 is just Eq. 4.11 and Eq. 4.12; only the relaxation time τ is replaced by τ_v and τ_ε.

If the peak value of the drift velocity v_p and the critical electric field E_c are known from the experiment, then the τ_v and τ_ε can be calculated from Eq. 4.32. For example, for the GaAs/AlAs superlattice, $T = 300$ K, $\Delta = 110$ meV, $v_p/d = 1.4 \times 10^{12}\ \mathrm{s}^{-1}$, and $eE_c d = 3.1$ meV and we obtain $\tau_\varepsilon = 2.2 \times 10^{12}\ \mathrm{s}^{-1}$, and $\tau_v = 1.1 \times 10^{13}\ \mathrm{s}^{-1}$ [8].

Introduce the efficiency of oscillator η,

$$\eta(f) = \frac{P_{alt}}{P_{bias}} = \frac{\int v(t) F_1 \cos\omega t dt}{\int v(t) F_0 dt}, \quad (4.34)$$

where P_{bias} and P_{alt} are the powers absorbed from the bias source and the alternating field, respectively. The integration is taken over an oscillation period, and $f = \omega/2\pi$ is the frequency of the alternating field. If $v(t)$ and alternating field $F_1 \cos\omega t$ have the same phases, then $P_{alt} > 0$ ($\eta > 0$) energy of the alternative field

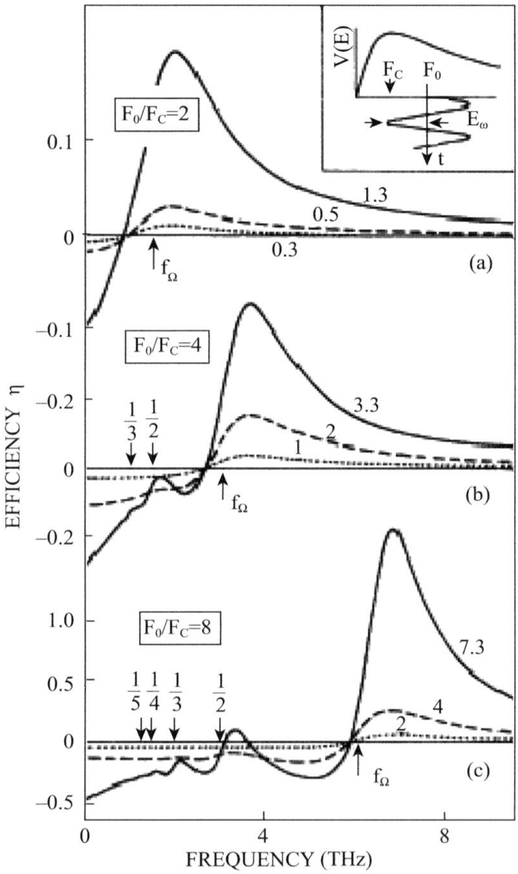

Figure 4.7 Calculated efficiencies as a function of frequency for the mentioned superlattice parameters [8]; (a), (b), and (c) correspond to different DC biases $F_0/F_c = 2$, 4, and 8.

is absorbed by the superlattice. If $v(t)$ and $F_1 \cos\omega t$ have opposite phases $P_{alt} < 0$ ($\eta < 0$) energy is transferred to the alternating field, that is, the alternating field is amplified and, in principle, oscillation of the system is possible.

Figure 4.7 shows the calculated efficiencies as a function of frequency for the mentioned superlattice parameters [8]; (a), (b), and (c) correspond to different DC biases $F_0/F_c = 2$, 4, and 8. The number marked at each curve is the strength of the alternating field F_1/F_c. From the figure we see that the frequency range of amplification ($\eta < 0$) extends from almost DC to about the Bloch frequency $f_\Omega = eF_0d/h$ that lies in the terahertz frequency range. With an increasing bias field the $\eta(f)$ curves show a fine structure near the subharmonics $f_n = f_\Omega/n$, where $n = 2, 3, \ldots$

Experimentally Wankle et al. have observed that when a microwave of 600 GHz irradiates on the GaAs/$Al_{0.3}Ga_{0.7}As$ superlattice, a strong third-order harmonic wave related to the injection power can be excited [9]. The transfer efficiency can reach 0.1%.

4.3 Hopping Conduction between Wannier–Stark States

Wannier [11] predicted early that when a quasi-continuous band is under an applied electric field, it can split into a serious of discrete energy states and the wave function changes from the expanded state to the local state. Later these states are called Wannier–Stark states. In the preceding section we discussed the longitudinal miniband transport in a superlattice. When the applied electric field along the z direction F is large enough, so that $eFd > \Delta$ (miniband width), that is, the potential difference between neighboring wells is larger than Δ, then the miniband doesn't exist, which splits into a series of discrete energy levels. The electron state is localized in a quantum well, and becomes the Wannier–Stark state. In this case the conduction is determined by the hopping conduction of electron states of neighboring quantum wells, as shown in Fig. 4.8 [12]. Figure 4.8 shows the energy bands of neighboring two quantum wells with s distance d in the applied electric field F. The potential difference is eFd. If eFd is such that level 1 coincides with level $2'$, electrons may tunnel resonantly from 1 to $2'$, marked by (a), followed by an inelastic scattering process marked by (b) to level $1'$, in order to repeat the process onto the next cell. Another process is the direct inelastic scattering process from 1 to $1'$, marked by (c). These two processes all need the participation of phonons; thus the hopping conductance is obviously lower than that within one band, resulting in NDC.

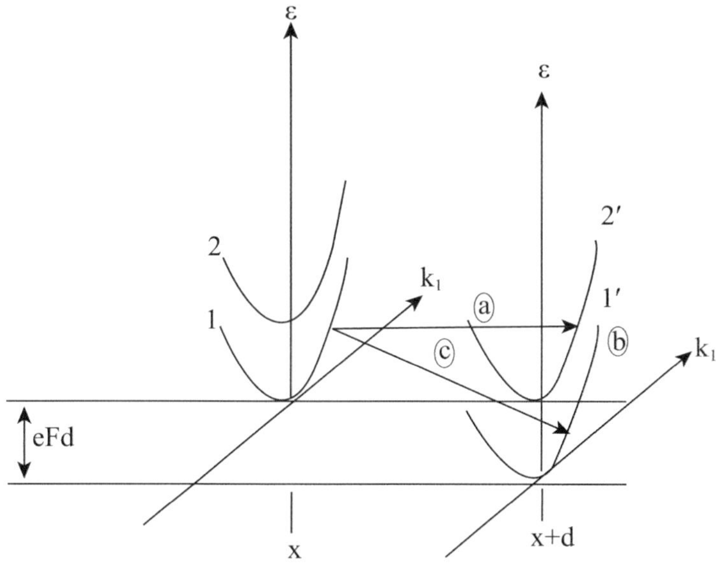

Figure 4.8 Energy bands of two neighboring quantum wells with a distance d in the applied electric field F.

To study the hopping conductance of Wannier–Stark states, we used the one-band model to calculate the Wannier–Stark state [13]. The effective mass Hamiltonian of electron in the electric field along the z direction is

$$H = E(k) + eFz, \tag{4.35}$$

where $E(k)$ is the 1D energy band dispersion relation as shown in Eq. 4.3. We write the Schrödinger equation in the momentum k space,

$$\left[E(k) + eFi\frac{d}{dk}\right]\psi(k) = E\psi(k), \tag{4.36}$$

where the coordinate z becomes an operator in the momentum space:

$$z = i\frac{d}{dk}. \tag{4.37}$$

The solution of Eq. 4.36 is easily obtained,

$$\psi(k) = C\exp\left\{-\frac{i}{eF}\int_0^k [E - E(k')]dk'\right\}, \tag{4.38}$$

where C is a normalization constant. Because $E(k)$ is a periodic function of k, $\psi(k)$ must be a periodic function in momentum space with period $2\pi/d$, which demands that

$$\frac{1}{eF}\int_{-\pi/d}^{\pi/d}[E - E(k)]dk = 2\pi n. \tag{4.39}$$

The eigenenergy is obtained from Eq. 4.39,

$$E = eFnd + \frac{d}{2\pi}\int_{-\pi/d}^{\pi/d} E(k)dk. \tag{4.40}$$

Equation 4.40 is just the Wannier–Stark ladder formula first derived by Wannier [12]. The second term in Eq. 4.40 is the average energy of the energy band.

The energy band $E(k)$ can be expanded by a series of cosine functions,

$$E(k) = \sum_{n=0}^{N} \lambda_n \cos(nkd). \tag{4.41}$$

Then the eigenenergy in Eq. 4.40 is

$$E = eFnd + \lambda_0. \tag{4.42}$$

Equation 4.42 represents the energy of the electron state in each quantum well, called the Wannier–Stark ladder.

The normalization constant $C = \sqrt{d}$ in a wave function (Eq. 4.38) is obtained from the normalization condition,

$$\frac{1}{2\pi}\int_{-\pi/d}^{\pi/d} |\psi(k)|^2 dk = 1. \tag{4.43}$$

With the eigenenergy (Eq. 4.42) we transform the wave function $\psi(k)$ (Eq. 4.38) into the coordinate representation,

$$\begin{aligned}\psi(z) &= \frac{\sqrt{d}}{2\pi}\int dk\varphi_k(z)\exp\left\{-i\frac{1}{eF}\int_0^k [E - E(k')]dk'\right\}\\ &= \frac{\sqrt{d}}{2\pi}\int_{-\pi/d}^{\pi/d} dkU_k(z')\exp\left\{iz'k + \frac{1}{eFd}\sum_{n=1}^{N}\frac{\lambda_n}{n}\sin(ndk)\right\},\end{aligned} \tag{4.44}$$

where $z' = z\text{-}nd$ and $\varphi_k(z)$ is the Bloch function with wave vector k in the single band,

$$\varphi_k(z) = e^{ikz}U_k(z). \tag{4.45}$$

Equation 4.44 gives a general formula of the wave function for any form of band energy $E(k)$. If we take only two terms in the band energy expansion (Eq. 4.41),

$$E(k) = \lambda_0 + \lambda_1 \cos kd, \tag{4.46}$$

and let $z'/d = \nu - n$ (i.e., the coordinate at the center of the well) and let $U_k(z) = U_0$ be a constant, from Eq. 4.44 we obtain

$$\psi(z) = \frac{1}{\sqrt{d}} J_{n-\nu}\left(\frac{\lambda_1}{eFd}\right) U_0. \tag{4.47}$$

Equation 4.47 is just the result obtained by Bastard et al. [14] with the one-band tight-binding model.

At the same time Xia et al. used a finite Kronig–Penny model and series expansion method to calculate the eigenenergies and eigenstates of the Wannier–Stark state. Because the Wannier–Stark state is a local state, we take a large potential well with a finite number of quantum wells and an infinite high barrier at the two edges. We expand the wave function in a series of sine functions,

$$\psi(z) = \sqrt{\frac{2}{L}} \sum_n C_n \sin\left(\frac{n\pi z}{L}\right), \tag{4.48}$$

where L is the width of the large well, the sine function guarantees that the wave function equals zero at the edges, and the origin of the z coordinate is taken at the left edge. Inserting the wave function (Eq. 4.48) into the equation,

$$-\frac{1}{2m^*}\frac{d^2\psi}{dz^2} + [V(z) + eFz]\psi = E\psi, \tag{4.49}$$

where m^* is the effective mass and $V(z)$ is the Kronig–Penny potential (finite superlattice potential), we obtain the secular equation for the expansion coefficients C_n in an expanded form (Eq. 4.48).

In calculation we take the parameters for the GaAs/$Al_{0.35}Ga_{0.65}As$ superlattice: $m^* = 0.085\ m_0$, potential barrier height 0.26 eV, and widths of the well and the barrier 3 nm and 3.5 nm, respectively. Figure 4.9 shows the distribution of the wave functions along the z direction for the electric field 2×10^4 Vcm^{-1} (curve 1) and 5×10^4 Vcm^{-1} (curve 2) [13]. The unit of the abscissa is the superlattice d. From Fig. 4.9 we see that when the electric field is small (curve 1) the

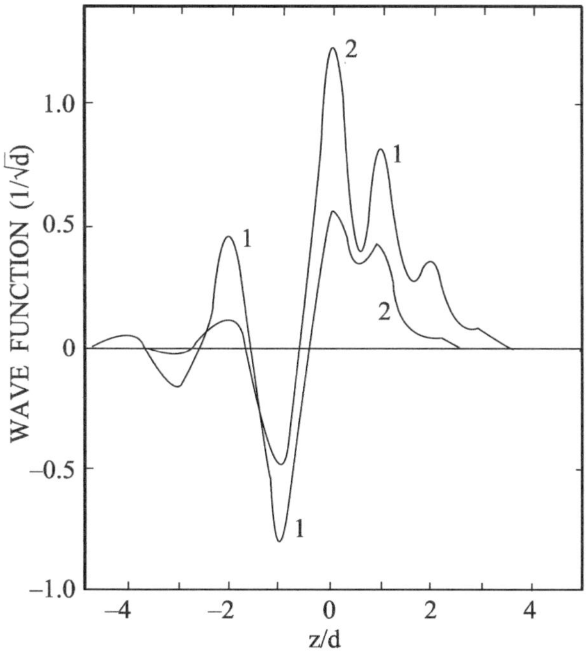

Figure 4.9 Distribution of the wave functions along the z direction for the electric field 2×10^4 Vcm^{-1} (curve 1) and 5×10^4 Vcm^{-1} (curve 2).

wave function extends over 10 periods and when the electric field is large (curve 2) the wave function extends over 7 periods, more local than the former. It is also to be noted that the wave function is not symmetrical with respect to the origin: on the left-hand side (lower potential) the wave function is oscillatory, taking positive and negative values at successive wells, while on the right-hand well (higher potential) the wave function is always positive, with the maxima at the successive wells. It is also found that the wave function values calculated from the numerical method and Eq. 3.47 are in good agreement.

Tsu et al. calculated inelastic scattering probabilities of process (c) in Fig. 4.8 for phonon emission at 300 K as a function of F, at fixed transverse energies $\varepsilon_\perp = \hbar^2 k_\perp^2/2m^* = 0.1$ eV [12]. Three curves in Fig. 4.10 correspond to the total energy of the first well $\varepsilon_1 = 0.01$, 0.02, and 0.1 eV (in the figure $\varepsilon_1 = 0.2$ eV is wrong; it should be

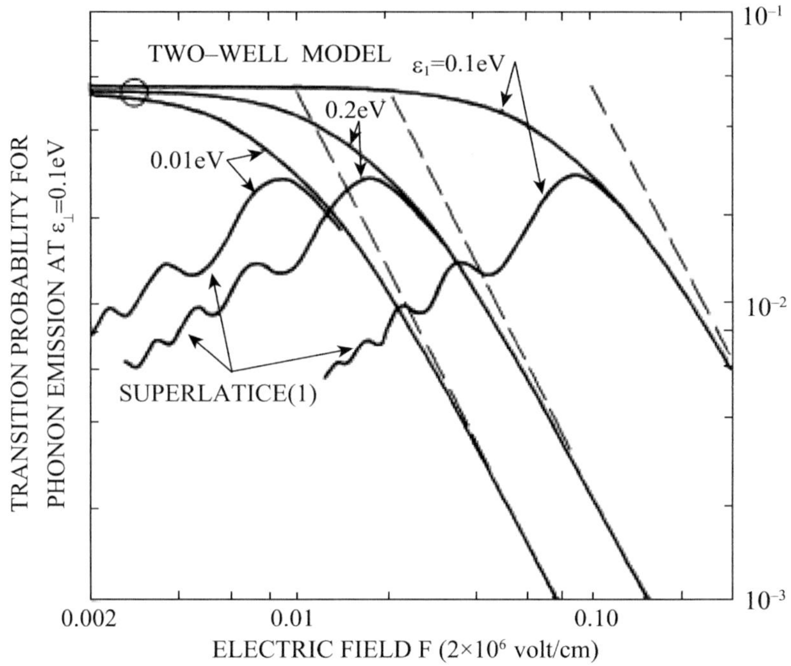

Figure 4.10 Transition probabilities of process (c) in Fig. 4.8 for phonon emission at 300 K as a function of F, at fixed transverse energies $\varepsilon_\perp = \hbar^2 k_\perp^2/2m^* = 0.1$ eV. Three curves correspond to the total energy of the first well $\varepsilon_1 = 0.01$, 0.02, and 0.1 eV.

0.02 eV).

$$\varepsilon_1 = E_0 + \left[V^2 + \alpha^2\right]^{1/2} + \frac{\hbar^2 k_\perp^2}{2m^*}, \tag{4.50}$$

where E_0 is the energy of the confined state in the quantum well, V is the interaction energy of neighboring wells, and 2α is the energy-level difference of the two wells. From Fig. 4.10 we see that the transition probability increases with increasing electric field F, reaches a maximum, and decreases as F increases continuously. The larger is the ε_1, the larger is the critical field F_c, at which the transition probability reaches the maximum. Thus F_c is proportional approximately to Δ, which corresponds to the formation of the Wannier–Stark state, $F_c \approx \Delta/ed$. When the electric field F is small, the transition probability oscillates with F. It is caused by the Bessel function factor in the Wannier–Stark wave function (4.47).

References

1. Luryi, S., and Zaslavsky, A. (1998). *Modern Semiconductor Device Physics*, ed. Sze, S. M., Chapter 5 "Quantum-effect and hot-electron devices" (John Wiley & Sons, Inc.).
2. Esaki, L., and Tsu, R. (1970). Superlattice and negative differential conductivity in semiconductors. *IBM J. Res. Dev.*, **14**, 61–65.
3. Huang, K., and Zhu, B. F. (1992). Temperature dependence of the low-field mobility of miniband conduction in superlattices. *Phys. Rev. B*, **45**, 14404–14406.
4. Sibille, A., Palmier, J. F., Wang, H., and Mollot, F. (1990). Observation of Esaki-Tsu negative differential velocity in GaAs/AlAs superlattices. *Phys. Rev. Lett.*, **64**, 52–55.
5. Hadjazi, M., Sibille, A., and Palmier, J. F. (1991). Negative differential conductance in GaAs/AlAs superlattices. *Electron. Lett.*, **27**, 1101–1103.
6. Sibille, A., Palmier, J. F., Wang, H., Esnault, J. C., and Mollot, F. (1990). dc and microwave negative differential conductance in GaAs/AlAs superlattices. *Appl. Phys. Lett.*, **56**, 256–258.
7. Unterrainer, K., Keay, B. J., Wanke, M. C., et al. (1996). *Hot Carriers in Semiconductors*, eds. Hess, K., Leburtor, J. P., and Rivaioli, U., "Strong terahertz-photocurrent resonances in miniband superlattices at the Bloch frequency" (Plenum, New York), pp. 135–138.
8. Ignatov, A. A., Renk, K. F., and Dodin, E. P. (1993). Esaki-Tsu superlattice oscillator: Josephson-like dynamics of carriers. *Phys. Rev. Lett.*, **70**, 1996–1999.
9. Wanke, M. C., Mankelz, A. G., Unterrainer, K., Allen, S. J., and Bhatt, R. (1996). *Hot Carriers in Semiconductors*, eds. Hess, K., Leburtor, J. P., and Rivaioli, U., "Third harmonic generation in a Gaas/Algaas superlattice in the Bloch oscillator regime" (Plenum, New York), pp. 161–163.
10. Xia, J. B. (1998). Theory of terahertz-photocurrent resonances in miniband superlattices. *Phys. Rev. B*, **58**, 3565–3567.
11. Wannier, G. H. (1962). Dynamics of band electrons in electric and magnetic fields. *Rev. Mod. Phys.*, **35**, 645–655.

12. Tsu, R., and Döhler, G. (1975). Hopping conduction in a "superlattice". *Phys. Rev. B*, **12**, 680–686.
13. Xia, J. B., and Huang, K. (1991). Wannier quantization of a superlattice subband under an electric field. *J. Phys. C*, **3**, 4639–4644.
14. Bleuse, J., Bastard, G., and Voisin, P. (1988). Electric-field induced localization and oscillatory electro-optical properties of semiconductor superlattices. *Phys. Rev. Lett.*, **60**, 220–223.

Chapter 5

Mesoscopic Transport

5.1 Contact Resistance

It is found experimentally that when we measure a sample of high mobility and small size, the current is always finite, though the transport is ballistic. There occurs a problem: where does the resistance come from?

Consider an experimental installation as shown in Fig. 5.1a [1]. A conductor of length L and width W is placed between contacts 1 and 2. According to the classical theory, the conductance of this conductor will equal $G = \sigma W/L$, where σ is the conductivity, independent of the sample dimensions. When the length L decreases, then we would expect the conductance to grow infinitely. Experimentally, however, it is found that the measured conductance approaches a limiting value G_c when the length of the conductor becomes much shorter than the mean free path ($L \ll L_m$).

Theory analysis indicates that the resistance arises from the interface between the conductor and the contact; hence the resistance G_c^{-1} is called contact resistance. This contact resistance is not due to the difference of two materials. The current is carried in the contacts by infinitely transverse modes, but inside the conductor only a few modes. This requires a redistribution of the current

Quantum Waveguide in Microcircuits (Second Edition)
Jian-Bai Xia, Duan-Yang Liu, and Wei-Dong Sheng

ISBN 978-981-4968-30-0 (Hardcover), 978-1-003-65146-8 (eBook)
www.jennystanford.com

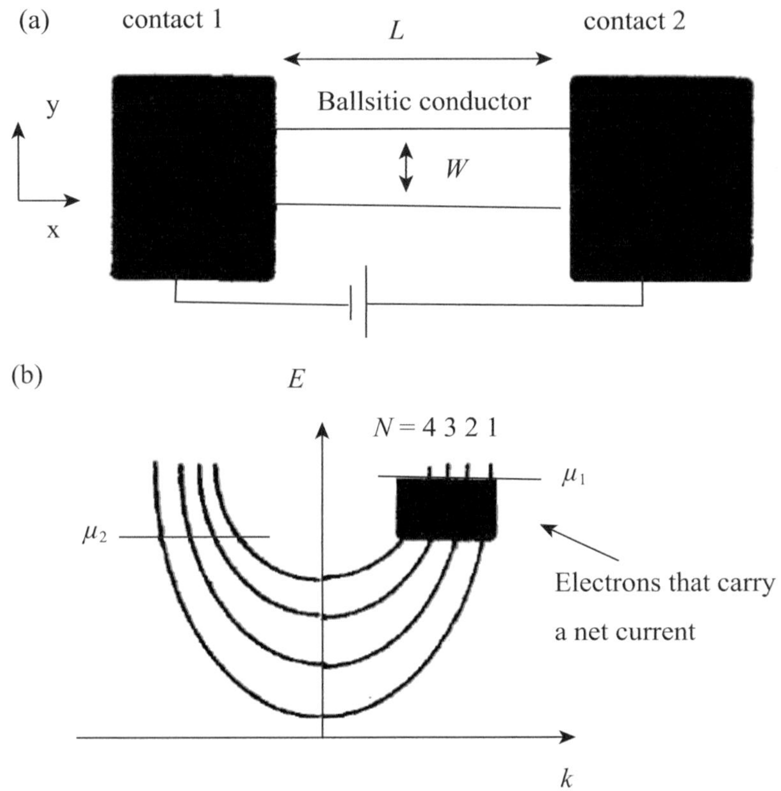

Figure 5.1 (a) A conductor connected with two contacts. (b) Conductor energy band and electron filling state.

among the current-carrying modes at the interface, leading to interface resistance.

To calculate the contact resistance G_c^{-1} we consider a ballistic conductor and calculate the current through it for a given applied bias $\mu_1 - \mu_2$. Numerical calculations indicate that an electron can exit from a narrow conductor into a wide contact with negligible probability of reflection [2]. Therefore, the reflection is negligible only when being transmitted from the narrow conductor to the wide contact. Going the other way, from the contact to the conductor, the reflections can be quite large.

The $+k$ state in the conductor is occupied by electrons coming from the left contact until μ_1, while the $-k$ state is empty because electrons are reflectionless through the right contact. The $-k$ state is occupied only by electrons coming from the right contact until μ_2, and the $+k$ state is empty because electrons are reflectionless through the left contact. The quasi-Fermi energy levels of $+k$ and $-k$ states are μ_1 and μ_2, respectively, when a bias $V = \mu_1 - \mu_2$ is applied (see Fig. 5.1b). At a low temperature the current equals that carried by the electrons in the states between μ_1 and μ_2 (shaded part of Fig. 5.1b).

The states in the narrow conductor belong to different transverse modes or subbands, as shown in Fig. 5.1b. Each mode has a dispersion relation $E(N, k)$ with a cut-off energy

$$\varepsilon_N = E(N, k = 0), \tag{5.1}$$

below which the current cannot propagate. The number of transverse modes at an energy E is obtained by counting the number of modes having cut-off energies smaller than E:

$$M(E) = \sum_N \theta(E - \varepsilon_N). \tag{5.2}$$

Consider a single transverse mode whose $+k$ state is occupied according to some distribution function $f^+(E)$. A uniform electron gas with n electrons per unit length moving with a velocity v carries a current equal to $en\mathbf{v}$. Since the electron density associated with a single k state in a conductor of length L is $(1/L)$ we can write the current I^+ carried by the $+k$ states as

$$I^+ = \frac{e}{L}\sum_k vf^+(E) = \frac{e}{L}\sum_k \frac{1}{\hbar}\frac{\partial E}{\partial k} f^+(E). \tag{5.3}$$

Changing the summation over k to the integration, we obtain

$$I^+ = \frac{e}{L}\cdot\frac{L}{\pi}\int dk \cdot \frac{1}{\hbar}\frac{\partial E}{\partial k} f^+(E) = \frac{2e}{h}\int_{\varepsilon}^{\infty} f^+(E)dE, \tag{5.4}$$

where ε is the cut-off energy of the transverse mode. If there are multi-modes then Eq. 5.4 can be written as

$$I^+ = \frac{2e}{h}\int_{-\infty}^{\infty} f^+(E)M(E)dE, \tag{5.5}$$

where $M(E)$ is given by Eq. 5.2. The general result given by Eq. 5.4 is that the current carried per mode per unit energy by an occupied state equals $2|e|/h$ (which is about 80 nA/meV).

At a low temperature the total current

$$I = I^{+} - I^{-} = \frac{2e}{h}\left[\int_{-\infty}^{\mu_1} f^{+}(E)M(E)dE - \int_{-\infty}^{\mu_2} f^{-}(E)M(E)dE\right]. \tag{5.6}$$

The number of modes M is constant over the energy range $\mu_1 < E < \mu_2$, we can write

$$\begin{aligned} I &= \frac{2e}{h}M\left[\int_{-\infty}^{\mu_1} f^{+}(E)dE - \int_{-\infty}^{\mu_2} f^{+}(E)dE\right] \\ &= \frac{2e^2}{h}M\frac{(\mu_1-\mu_2)}{e}, \end{aligned} \tag{5.7}$$

$$G = \frac{2e^2}{h}M. \tag{5.8}$$

Equation 5.8 is a universal result, independent of the length and width of the conductor, the energy band dispersion, etc. The conductance is quantized, which is integral times of $2e^2/h$. The contact resistance is given by

$$G_c^{-1} = \frac{h}{2e^2 M} = \frac{12.9\ k\Omega}{M}. \tag{5.9}$$

The contact resistance goes down inversely with the number of modes. The contact resistance of a single-mode conductor is ~12.9 kΩ, which is certainly not negligible. Only if the size of the sample reaches macro-size, the M approaches infinite, and the contact resistance can be neglected. Assuming that the width of the sample is W and the electron energy is E, we can estimate the number of modes M,

$$E = \frac{\hbar^2}{2m^*}\left(\frac{n\pi}{W}\right)^2$$

and

$$M = \text{Int}\left[\frac{W}{\pi}\sqrt{\frac{2m^*E}{\hbar^2}}\right] = \text{Int}\left[\frac{W}{\lambda_f/2}\right], \tag{5.10}$$

where λ_f is the Fermi wavelength and $\text{Int}(x)$ represents the integer that is just smaller than x. If the typical electron density $n_s = 3.6 \times 10^{11}\ \text{cm}^{-2}$, then $M = 1$ for $W = 22$ nm and $M = 47$ for $W = 1\ \mu\text{m}$.

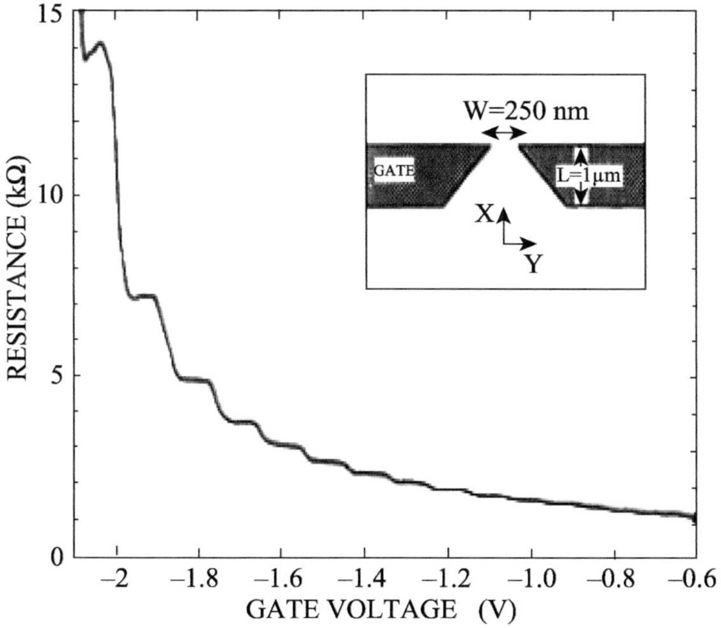

Figure 5.2 Point-contact resistance as a function of gate voltage at 0.6 K. Inset: Point-contact layout.

Wees et al. reported the first measurements of the conductance of single ballistic contacts in a two-dimensional electron gas (2DEG) [3]. A novel quantum effect is found: the conductance is quantized in units of $2e^2/h$, as shown in Eq. 5.8. A high-mobility 2DEG is formed on the interface of GaAs/AlGaAs heterostructure grown by molecular beam epitaxy. The electron density is 3.56×10^{11} cm^{-2}, and the mobility 85,000 cm^2/Vs (at 0.6 K). Using electron beam lithography, a metal gate is made on top of the heterostructure, with an opening 250 nm wide (inset in Fig. 5.2). The point contacts are defined by the application of a negative voltage to the gate. At $V_g = -0.6$ eV the electron gas underneath the gate is depleted, the conduction taking place through the point contact only. At this voltage the point contacts have their maximum width, about equal to the opening between the gates. By a further decrease of the gate voltage, the width of the point contacts can gradually be reduced, until they are fully pinched off at $V_g = -2.2$ eV.

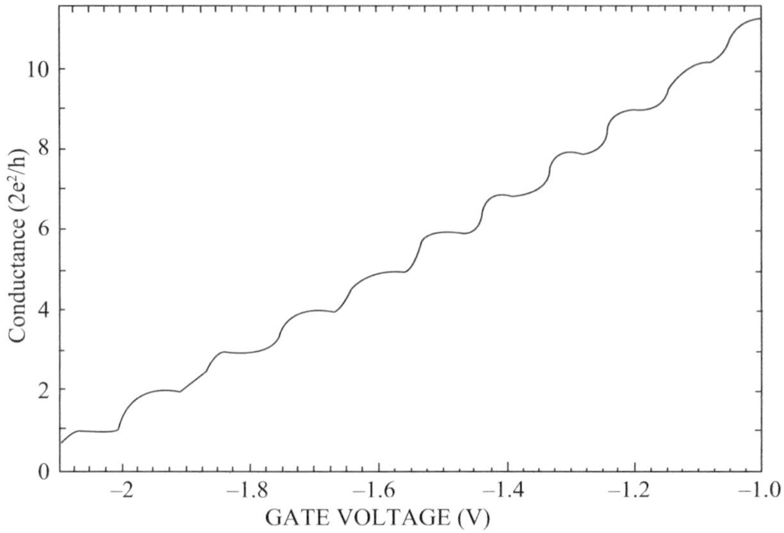

Figure 5.3 Point-contact conductance as a function of gate voltage, obtained from the data of Fig. 5.2 after subtraction of the lead resistance.

Figure 5.2 shows the measured resistance of a point contact as a function of gate voltage [3]. In total, 16 plateaus are observed when the gate voltage is varied from -0.6 to -2.2 V. The measured resistance consists of the resistance of the point contact, and a constant series resistance from the 2DEG leads to the point contact. The conductance calculated from the measured resistance after subtraction of a lead resistance of 400 Ω as a function of gate voltage is shown in Fig. 5.3 [3], which shows clear plateaus at integer multiples of $2e^2/h$, verifying the theoretical result of Eq. 5.8.

Equations 5.8 and 5.9 are general theoretical results. It is still not clear how is contact resistance is formed. Kirczenov calculated the conductance of a short, narrow ballistic channel in a 2DEG [4]. The model is of a heterostructure in the $x - y$ plane with a 2DEG occupying the left (L) and right (R) half-spaces, $x < -d$ and $x > d$, respectively, and a narrow channel (C) of length $2d$ centered on the x axis connecting the 2D regions; see Fig. 5.4 lower-right inset [4]. The shaded part is the infinite high potential barrier region produced by the metallic gate.

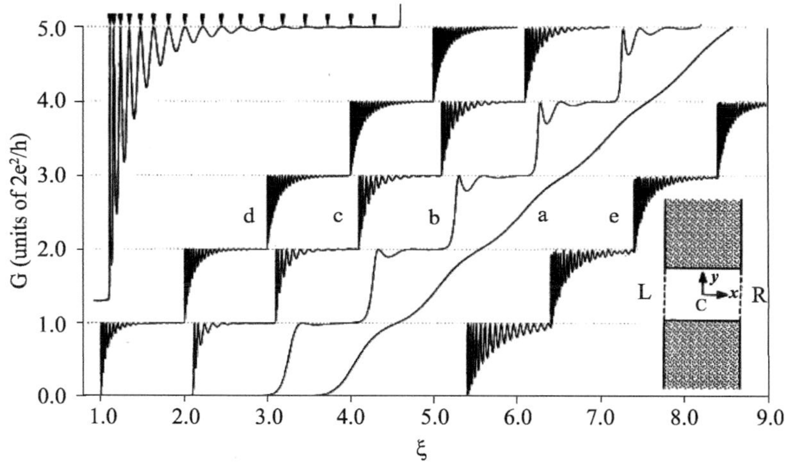

Figure 5.4 Conductance G as a function of the electron Fermi level ξ at $T = 0$ K. $U = 0$, $d = 0$, 1, 5, and 10 for curves a, b, c, and d. Curve e: $U = 2.5$ and $d = 10$.

Suppose that an electron moves from the left region into the channel region. Its energy is ε, wave vector is $\mathbf{k} = (k, K)$, and k and K are the components along the x and y directions, respectively. The wave function in the left region can be written as

$$\psi_k^L(\mathbf{r}) = e^{ikx}\phi_k(y) + \sum_{k'} a_{k'}^L e^{-ik'x}\phi_{k'}(y), \qquad (5.11)$$

where $\phi_k(y) = e^{iky}$ and $k' = (2m^*\varepsilon_k/\hbar^2 - K'^2)^{1/2}$. The second term in Eq. 5.11 represents the reflection wave. The sum is over all transverse momentum K' so that imaginary values of k' (evanescent partial waves) are included. In the channel the wave function is

$$\psi_k^c(\mathbf{r}) = \sum_n \left(a_n^+ e^{iq_n x} + a_n^- e^{-iq_n x}\right)\phi_n(y), \qquad (5.12)$$

where $\phi_n(y)$ is the eigenfunction of the n-th transverse eigenstate of the confining potential $U(y)$, $\phi_n = \sqrt{2/w}\sin(n\pi y/w)$, and $q_n = \left[2m^*\varepsilon_n/\hbar^2 - (n\pi/w)^2\right]^{1/2}$. In the right-hand region the transmitting wave (not including the reflection wave) is

$$\psi_k^R(\mathbf{r}) = \sum_{k'} a_{k'}^R e^{ik'x}\phi_{k'}(y). \qquad (5.13)$$

Using the boundary conditions at $x = -d$ and $x = d$, the continuities of the wave function and its differential, we obtain four sets of equations of the coefficients $a_{k'}^{L}$, a_n^+, a_n^-, and $a_{k'}^{R}$. Eliminating $a_{k'}^{L}$ and $a_{k'}^{R}$ from the sets of equations yields the equations of a_n^+ and a_n^- [4],

$$\sum_n \left[(T_{mn} + q_n\delta_{mn})e^{-iq_nd}a_n^+ + (T_{mn} - q_n\delta_{mn})e^{ig_nd}a_n^-\right] = 2ke^{-ikd}M_{Km}$$

and

$$\sum_n \left[(T_{mn} - q_n\delta_{mn})e^{iq_nd}a_n^+ + (T_{mn} + q_n\delta_{mn})e^{-ig_nd}a_n^-\right] = 0$$

$$(m = 1, 2, \cdots), \tag{5.14}$$

where

$$T_{mn} = \sum_{k'} k' M_{k'm} M_{-k'n}$$

and

$$M_{kn} = \int_{-w/2}^{w/2} \phi_k(y)\phi_n(y)dy. \tag{5.15}$$

In terms of a_n^+ and a_n^-, the electric current carried through the channel by ψ_k is

$$i_x(k) = <\psi_k|j_x|\psi_k> = \frac{i\hbar e}{2m^*}\int_{-w/z}^{w/z}\left(\psi_k^*\frac{\partial\psi_k}{\partial x} - \psi_k\frac{\partial\psi_k^*}{\partial x}\right)dy$$

$$= -\frac{\hbar e}{m^*}\left[\mathrm{Re}\sum_n q_n(|a_n^-|^2 - |a_n^+|^2) + \mathrm{Im}\sum_n q_n(a_n^{+*}a_n^- - a_n^+a_n^{-*})\right]. \tag{5.16}$$

Under the electric potential V, the electron Fermi circle shifts dk along the x direction. The total current J through the channel at $T = 0$ is given by the sum of the contributions of all states ψ_k in the k space, which is the difference of the Fermi circles before and after shift, and in the energy interval eV near the Fermi energy E_F, where V is the potential difference between the two 2DEG regions. The conductance is then

$$G = \left|\frac{J}{V}\right| = -2\int_{-a}^{a}\frac{m^*e}{h^2k}i_x(k)\,dK, \tag{5.17}$$

where $a = \left(2m^*E_F/\hbar^2\right)^{1/2}$ and $k = \left(2m^*E_F/\hbar^2 - K^2\right)^{1/2}$.

Figure 5.4 shows the calculated conductance as a function of electron Fermi energy [4], where the energy unit $\Delta = h^2/8m^*W^2$,

$\hat{E}_F = E_F/\Delta$, $\hat{U} = U_0/\Delta$, U_0 is the potential inside the channel, the aspect ratio $\hat{d} = 2d/W$, and the abscissa $\xi = \left(\hat{E}_F - \hat{U}\right)^{1/2}$. $\hat{U} = 0$ and $\hat{d} = 0, 1, 5$, and 10 for curves a, b, c, and d. Curve e: $\hat{U} = 2.5$ and $\hat{d} = 10$.

The horizontal scale is for curve d; the other curves are offset to the right by multiples of 1.1. Curve a is the limiting case of zero aspect ratio, $\hat{d} \to 0$. For this "ideal" point contact there is no quantization of conductance; a channel of nonzero length is necessary for quantization. In the cases of curves b, c, and d the conductance is close to $G = \nu 2e^2/h$ near the right-hand side of each plateau. But the rise to each plateau is oscillatory, the strength of the oscillations increasing with d and with ν. The $\nu = 1$ step of curve d is shown enlarged in the upper left. The resonance peaks correspond to the condition $n\lambda_f/2 = 2d$, where λ_f is the electron Fermi wavelength. This indicates that the open-ended channel supports longitudinal resonant states, i.e. the quasi-standing wave in the channel and at this time the conductance decreases. The curve e is for $\hat{U} = 2.5$ and $\hat{d} = 10$. The main effect is to increase the amplitude of the oscillations.

5.2 Landauer Formula

The last section discussed the case of an ideal conductor. Now if there is a scatterer in the conductor, as shown in Fig. 5.5, the carriers have a probability T for traveling of the scatterer and a probability R of being reflected. The left reservoir injects carriers into the perfect wire up to a quasi-Fermi energy μ_1, and the right reservoir emits carriers up to a quasi-Fermi energy μ_2 (see Fig. 5.4b). The net current, for two injection directions, is given by the sum of the contributions of all states in the energy range between μ_2 and μ_1,

$$I = ev\left(\frac{\partial n}{\partial E}\right)(\mu_1 - \mu_2), \tag{5.18}$$

where v is the Fermi velocity and $\partial n/\partial E$ is the density of states for two spin directions and for carriers with a positive velocity. In one dimension $\partial n/\partial E = 1/\pi$ and $\partial n/\partial E = 1/\pi\hbar v$. Thus, the total current emitted by the left reservoir due to the difference in the

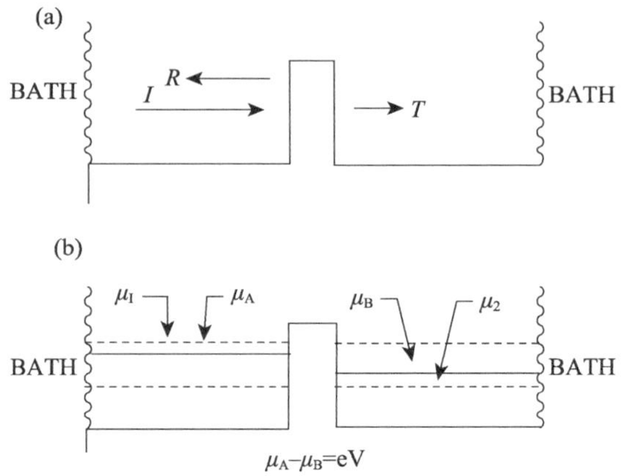

Figure 5.5 The scatterer is connected to two incoherent reservoirs by ideal 1D conductors.

quasi-Fermi levels is

$$I = \left(\frac{e}{\pi\hbar}\right)(\mu_1 - \mu_2). \tag{5.19}$$

When $T \neq 1$,

$$I = \left(\frac{e}{\pi\hbar}\right) T(\mu_1 - \mu_2). \tag{5.20}$$

Next, we have to determine the voltage across the scatterer. The carrier densities can be characterized by the chemical potentials μ_A and μ_B (see Fig. 5.5b). The respective levels μ_B, between μ_1 and μ_2, are determined such that the number of occupied states (electrons) above μ_B is equal to the number of empty states (holes) below μ_B. Since carriers have a transmission probability T, the number of occupied states is $T(\partial n/\partial E)(\mu_1 - \mu_B)$, and the number of unoccupied states is $(2 - T)(\partial n/\partial E)(\mu_B - \mu_2)$. Thus, the chemical potential μ_B is determined by

$$T\left(\frac{\partial n}{\partial E}\right)(\mu_1 - \mu_B) = (2 - T)\left(\frac{\partial n}{\partial E}\right)(\mu_B - \mu_2). \tag{5.21}$$

To the left of the barrier we have both incident carriers and reflected carriers. The number of occupied states is $(1 + R)(\partial n/\partial E)$ $(\mu_1 - \mu_A)$, and the number of unoccupied states is $[2 - (1 + R)]$ $(\partial n/\partial E)(\mu_A - \mu_2)$. Thus, the chemical potential μ_A is determined by

$$(1 + R)\left(\frac{\partial n}{\partial E}\right)(\mu_1 - \mu_A) = [2 - (1 + R)]\left(\frac{\partial n}{\partial E}\right)(\mu_A - \mu_2). \tag{5.22}$$

From Eqs. 5.21 and 5.22 we obtain

$$\mu_A - \mu_B = R(\mu_1 - \mu_2). \tag{5.23}$$

The charge neutrality requires that the separation between the chemical potential μ_A or μ_B, respectively, and the band bottom be the same as in equilibrium. The conduction-band bottoms are displaced against each other by a potential difference,

$$eV = \mu_A - \mu_B = R(\mu_1 - \mu_2). \tag{5.24}$$

From Eqs. 5.20 and 5.24 we obtain the conductance,

$$G = \frac{I}{V} = \frac{2e^2}{h} \cdot \frac{T}{R}. \tag{5.25}$$

Equation 5.25 is called the Landauer formula.

In the case of measurement of mesoscopic systems we run into three separate problems [1]. Firstly, mesoscopic probes are often invasive, that is, they change what we are trying to measure. With macroscopic conductors, the probes represent a minor perturbation. But for a small conductor, the probe can very well be the dominant source of scattering (and hence resistance). But this is not the main problem. It is likely that there will be more of less-invasive microscopic measurements as nanotechnology progresses, for example, using weakly coupled scanning tunneling probes.

Secondly, mesoscopic probes are seldom identical so the two voltage probes could very well couple differently to the $+k$ and $-k$ states. Only if the two-voltage probes couple identically to the $+k$ and $-k$ states, the measurement result of resistance will be given by Eq. 5.25. Otherwise the measured resistance will lie somewhere between the extremes

$$\frac{h}{2e^2 M}\frac{1}{T} \quad \text{and} \quad \frac{h}{2e^2 M}\frac{1-2T}{T}. \tag{5.26}$$

Finally, mesoscopic measurements are strongly affected by quantum interference effects unless the distance of the probes from the scatterer is much greater than the phase-relaxation length. We would expect that a probe to the left of the scatterer to measure a potential of approximately 1 (equal to that of the left reservoir). However, due to quantum interference it could measure any potential between 0 and 1 depending on its distance from the scatterer. The reason is that the probe may not "see" the electrons from the left reservoir due to destructive interference between the incident wave and the reflected wave. Therefore, in phase-coherent conductors the $+k$ and $-k$ states can be strongly correlated so distribution functions only tell part of the story.

The above discussion of Eq. 5.25 is under the assumption that $T = 0$. When $T \neq 0$ we should consider the electron Fermi distribution $f(E)$. The net current flowing from the left reservoir to the right reservoir is now given by

$$I = \frac{2e}{h}\left[\int dE\left(-\frac{\partial f}{\partial E}\right)T(E)\right](\mu_1 - \mu_2), \tag{5.27}$$

where

$$-\frac{\partial f}{\partial E} = [f(E-\mu_1) - f(E-\mu_2)]/(\mu_1 - \mu_2). \tag{5.28}$$

Similarly, to determine the chemical potential we have to multiply Eqs. 5.21 and 5.22 by $-\partial f/\partial E$ and integrate over the energy. This yields a voltage

$$eV = \frac{\int dE(-\partial f/\partial E)R(E)(\partial n/\partial E)}{\int dE(-\partial f/\partial E)(\partial n/\partial E)}(\mu_1 - \mu_2). \tag{5.29}$$

With Eqs. 5.27 and 5.29 we obtain the conductance

$$G = \frac{2e^2}{h}\left[\int dE\,(-\partial f/\partial E)\,T(E)\right] \times \frac{\int dE\,(-\partial f/\partial E)\,\upsilon^{-1}(E)}{\int dE\,(-\partial f/\partial E)\,R(E)\upsilon^{-1}(E)}, \tag{5.30}$$

where we have used $\partial n/\partial E = 1/\pi\hbar v(E)$. At zero temperature we have $-\partial f/\partial E = \delta(E - E_{\mathrm{F}})$ and Eq. 5.30 reduces to Eq. 5.25.

5.3 Many-Channel Case

If the ideal conductors include N independent conducting channels, at zero temperature all the channels have the same energy. The electrons in these channels can be transmitted or reflected each other, having the probabilities T_{ji} and R_{ji}, respectively, as shown in Fig. 5.6 [5]. T_{ji} represents the transmission probability of electrons from the left i-th channel to right j-th channel, and R_{ji} represents the reflection probability of electrons from the left i-th channel to

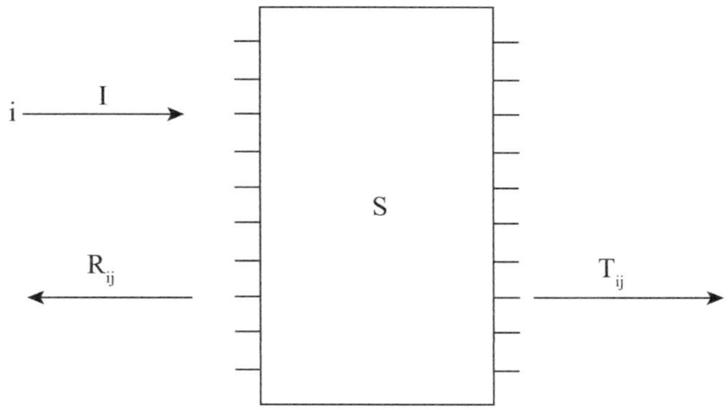

Figure 5.6 A multichannel system S. Indices i and j run from 1 to N.

left j-th channel. The scattering property can be described by a $2N \times 2N$ matrix S,

$$S = \begin{vmatrix} r & t' \\ t & r' \end{vmatrix}, \tag{5.31}$$

where r and t are the reflection and transmission coefficients of electrons from the left-hand side to right-hand side, respectively. The transmission probability $T_{ji} = |t_{ji}|^2$, and reflection probability $R_{ji} = |r_{ji}|^2$. r' and t' are those from the right-hand side to left-hand side.

The matrix S (Eq. 5.31) is unitary due to current conservation because T_{ij} and R_{ij} transform the lead currents. Furthermore, when time-reversal symmetry holds in the absence of a magnetic field,

$$SS^* = \mathbf{I} \quad \text{and} \quad S = \tilde{S}, \tag{5.32}$$

where the asterisk denotes complex conjugation, the tilde denotes the matrix transpose, and **I** is the unit matrix.

Assume that the interference effects between different input channels are neglected. The energy range is between μ_2 and μ_1, and carriers are injected into the perfect conductor only from the left-hand side reservoir. The current injected into the left channel j by the reservoir is $ev_j(\partial n_j/\partial E)(\mu_1 - \mu_2)$. The density of states (with a positive velocity) is given by $\partial n_j/\partial E = 1/\pi\hbar v_j$. Therefore, the current fed into the j-th channel is $(e/\pi\hbar)(\mu_1 - \mu_2)$, independent of the channel velocity. The current from the left j-th channel transmitted into the right i-th channel is $(e/\pi\hbar)T_{ij}(\mu_1 - \mu_2)$. The total current in the right i-th channel becomes

$$I_i = \frac{2e}{h}\left(\sum_{j=1}^{N} T_{ij}\right)(\mu_1 - \mu_2). \tag{5.33}$$

It is convenient to introduce a total transmission probability and reflection probability into the i-th channel,

$$T_i = \sum_j T_{ij}, \qquad R_i = \sum_j R_{ij}. \tag{5.34}$$

Thus the current in the right i-th channel is

$$I_i = \frac{2(\mu_1 - \mu_2)e}{h} T_i, \tag{5.35}$$

and the total current is given by

$$I_{\text{tot}} = \sum_i I_i = \frac{2(\mu_1 - \mu_2)e}{h} \sum_i T_i = \frac{2(\mu_1 - \mu_2)e}{h} \text{Tr}(tt^\dagger). \tag{5.36}$$

On the other hand, the current in the left i-th channel is

$$I_i = \frac{2(\mu_1 - \mu_2)e}{h}\left[1 - \sum_j R_{ij}\right] = \frac{2(\mu_1 - \mu_2)e}{h}(1 - R_i), \tag{5.37}$$

and the total current is

$$I_{\text{tot}} = \sum_i I_i = \frac{2(\mu_1 - \mu_2)e}{h} \sum_i (1 - R_i). \tag{5.38}$$

Comparing Eqs. 5.36 and 5.38, we find that current conservation implies

$$\sum_i T_i = \sum_i (1 - R_i). \tag{5.39}$$

Similar to the single-channel case (Eqs. 5.21 and 5.22), we determine the chemical potential μ_1 and μ_2. On the right side of the barrier the number of occupied states with energy higher than μ_B is

$$\sum_i T_i(\partial n_i/\partial E)(\mu_1 - \mu_\text{B}). \tag{5.40}$$

The number of unoccupied states with energy smaller than μ_B is

$$\sum_i (2 - T_i)(\partial n_i/\partial E)(\mu_\text{B} - \mu_2). \tag{5.41}$$

Again μ_B is determined such that the number of occupied states is equal to the number of unoccupied states,

$$\sum_i T_i \left(\frac{\partial n_i}{\partial E}\right)(\mu_1 - \mu_B) = \sum_i (2 - T_i)\left(\frac{\partial n_i}{\partial E}\right)(\mu_B - \mu_2). \quad (5.42)$$

Similarly, the chemical potential to the left is determined by

$$\sum_i (1 + R_i)\left(\frac{\partial n_i}{\partial E}\right)(\mu_1 - \mu_A) = \sum_i (1 - R_i)\left(\frac{\partial n_i}{\partial E}\right)(\mu_A - \mu_2). \quad (5.43)$$

Therefore, the voltage across the scatterer is given by

$$eV = \mu_A - \mu_B = \frac{\sum_i (1 + R_i + T_i)\upsilon_i^{-1}}{2\sum_i \upsilon_i^{-1}}(\mu_1 - \mu_2), \quad (5.44)$$

where we have used $\partial n_i/\partial E = 1/\pi\hbar v_i$. With the help of Eqs. 5.36 and 5.44 we obtain the conductance

$$G = \frac{2e^2}{h}\sum_i T_i \cdot \frac{2\sum_i \upsilon_i^{-1}}{\sum_i (1 + R_i - T_i)\upsilon_i^{-1}}. \quad (5.45)$$

5.4 Multi-Terminal Devices

Since 1985 many mesoscopic experiments have been conducted using miniature Hall bridges fabricated on both metallic and semi-conducting samples. However, because of the reasons mentioned above (see Section 5.2), for a while there was serious confusion about how such four-terminal measurements should be interpreted. Büttiker [6, 7] found a simple and elegant solution to this problem. He noted that since there is really no quantitative difference between the current and the voltage probes, one could treat all the probes on an equal footing and simply extend the two-terminal linear response formula to a multi-terminal response formula.

Consider the conductor shown in Fig. 5.7 [6]. The leads in Fig. 5.7 are connected to reservoirs that are at chemicals μ_1, μ_2, μ_3, and μ_4, and the center conductor is an Aharonov–Bohm ring with a uniform magnetic flux through the hole. Assume that these perfect leads are strictly one-dimensional (1D) quantum channels, that is, there are

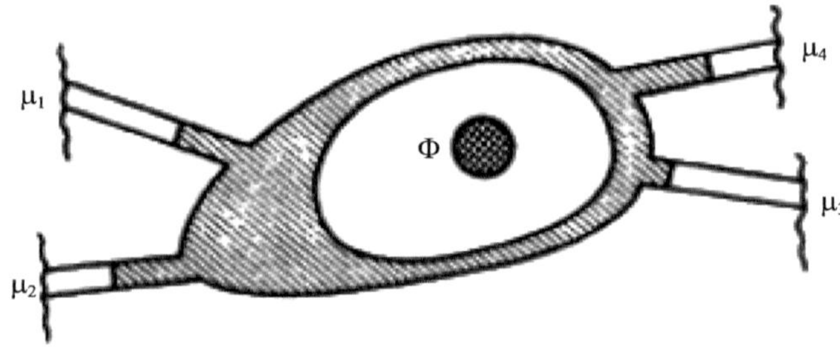

Figure 5.7 Conductor with four terminals connected via perfect leads to four reservoirs at chemical potentials μ_1, μ_2, μ_3, and μ_4. An Aharonov–Bohm flux Φ is applied through the hole of the sample.

only two states at the Fermi energy, one with positive velocity (taken to be the direction away from the reservoir) and one with negative velocity. Scattering in the sample is elastic. The elastic scattering properties of the sample are described by the probabilities $T_{ij}(\Phi)$ for carries incident in lead j to be transmitted into lead i and probabilities $R_{ii}(\Phi)$ for carriers to be reflected into lead i. Current conservation and time-reversal invariance in the presence of a flux imply

$$R_{ii}(\Phi) = R_{ii}(-\Phi), \quad T_{ij}(\Phi) = T_{ji}(-\Phi). \tag{5.46}$$

The potential μ_i are distributed in the range around the Fermi energy, which is so narrow that the energy dependence of the transmission and reflection probabilities in this range can be neglected. It is convenient to introduce a fifth chemical potential μ_0, which is smaller than the lowest of the four potentials μ_i. Below μ_0 the states with negative and positive velocities are filled and zero net current flows in each of the leads. The reservoir injects a current $e\mathbf{v}_i(dn_i/dE)\Delta\mu_i$ into lead i, $\Delta\mu_i = \mu_i - \mu_0$. Here v_i is the velocity at the Fermi energy in lead i and $dn_i/dE = 1/2\pi\hbar v_i$ is the density of states for carriers with a negative or a positive velocity at the Fermi energy. Thus the current injected by the reservoir i is $(e/h)\Delta\mu_i$. Consider the current in lead i. A current $(e/h)R_{ii}\Delta\mu_i$ is reflected back to the reservoir i. Carriers that are injected by the other reservoir j into lead i reduce the current in lead i by $-(e/h)T_{ij}\Delta\mu_j$. Collecting all results from lead j ($j \neq i$) we obtain the current in lead i,

$$I_i = \frac{e}{h}\left[(1 - R_{ii})\mu_i - \sum_{j \neq i} T_{ij}\mu_j\right]. \tag{5.47}$$

Note that these currents are independent of the reference potential μ_0 since the coefficients multiplying the potentials add to zero. Equation 5.47 is called the Büttiker formula [6].

Consider a four-probe configuration as shown in Fig. 5.7 where a current I_1 is led into lead 1 and leaves through lead 3, and a current I_2 is fed into lead 2 and leaves the sample through lead 4. Thus we have to solve Eq. 5.47 under the condition that $I_1 = -I_3$ and $I_2 = -I_4$. We obtain the currents I_1 and I_2 as a function of differences of voltages, $V_i = \mu_i/e$,

$$I_1 = \alpha_{11}(V_1 - V_3) - \alpha_{12}(V_2 - V_4),$$

and

$$I_2 = -\alpha_{21}(V_1 - V_3) + \alpha_{22}(V_2 - V_4), \tag{5.48}$$

where the conductance

$$\begin{aligned}
\alpha_{11} &= \left(\frac{e^2}{h}\right)[(-R_{11})S - (T_{14} + T_{12})(T_{41} + T_{21})]/S,\\
\alpha_{12} &= \left(\frac{e^2}{h}\right)(T_{12}T_{34} - T_{14}T_{32})/S,\\
\alpha_{21} &= \left(\frac{e^2}{h}\right)(T_{21}T_{43} - T_{23}T_{41})/S,\\
\alpha_{22} &= \left(\frac{e^2}{h}\right)[(1 - R_{22})S - (T_{21} + T_{23})(T_{32} + T_{12})]/S,
\end{aligned} \tag{5.49}$$

and

$$S = T_{12} + T_{14} + T_{32} + T_{34} = T_{21} + T_{41} + T_{23} + T_{43}. \tag{5.50}$$

Taking into account Eq. 5.46, we see that the diagonal elements are symmetric in the flux $\alpha_{11}(\Phi) = \alpha_{11}(-\Phi)$ and $\alpha_{22}(\Phi) = \alpha_{22}(-\Phi)$ and that the off-diagonal elements satisfy $\alpha_{12}(\Phi) = \alpha_{21}(-\Phi)$. Therefore, for the four-probe conductor the Onsager relations hold [6].

Suppose the current flows from lead 1 to lead 3. The potentials measured are $\mu_2 = eV_2$ and $\mu_4 = eV_4$ under the condition that the current in leads 2 and 4 is zero. Using Eq. 5.48 and taking

$I_2 = 0$, we obtain the current I_1 as a function of $V_2 - V_4$. Thus in this configuration the measured resistance is

$$R_{13,24} = \frac{V_2 - V_4}{I_1} = \frac{\alpha_{21}}{\alpha_{11}\alpha_{22} - \alpha_{12}\alpha_{21}}. \tag{5.51}$$

Since α_{21} is not symmetric the resistance $R_{13,24}$ is also not symmetric. Now we switch the current and the voltage leads but keep the flux fixed. This means that I_1 in Eq. 5.46 is zero. This yields a resistance

$$R_{24,13} = \frac{\alpha_{12}}{\alpha_{11}\alpha_{22} - \alpha_{12}\alpha_{21}}. \tag{5.52}$$

The sum of these two resistances is symmetric.

For a given flux there are generally six resistances,

$$R_{nm,kl} = \left(\frac{e^2}{h}\right)(T_{km}T_{ln} - T_{kn}T_{lm}) \Big/ D, \tag{5.53}$$

where $D = (\alpha_{11}\alpha_{22} - \alpha_{12}\alpha_{21})S$. The resistances given by Eq. 5.53 obey $R_{mn,kl} = -R_{mn,lk} = -R_{nm,kl}$ and the reciprocity relation $R_{mn,kl}(\Phi) = R_{kl,mn}(\Phi)$. The reciprocity relation states that the resistance measured in the presence of a flux Φ is equal to the resistance measured in the presence of a flux $-\Phi$ if the reversal of the flux is accompanied by an exchange in the role of the current and voltage leads.

If the lead has a definite width, there are many states at the Fermi energy. Motion in the transverse direction is quantized and characterized by a set of discrete energies, E_n. To this energy we have to add the kinetic energy for motion along the direction of the lead, $\hbar^2 k^2/2m$, such that $E_\mathrm{F} = \hbar^2 k^2/2m + E_n$. Each transverse state provides a channel. Suppose all the leads are identical and support N quantum channels. The probability of a carrier incident in channel n in lead i to be reflected into the same lead into channel m is denoted by $R_{ii,mn}$, and the probability of a carrier incident in lead j in channel n to be transmitted into lead i into channel m is $T_{ij,mn}$. Then the scattering matrix (Eq. 5.31) becomes a $4N \times 4N$ matrix.

Now we apply the Büttiker formula (Eq. 5.47) to the multichannel case. Assume that the reservoir feeds all channels equally up to the chemical potential μ_j. The current injected into each channel of lead j is $(e/h)\Delta\mu_j$ independent of the velocity and the density of states of this channel. The current in lead i due to the carriers injected in lead j is $I_{ij} = -(e/h)\Sigma_{mn} T_{ij,nm} \Delta\mu_j$. Therefore, if we introduce the trace $R_{ii} = \Sigma_{mn} R_{ii,mn}$, we can get $T_{ij} = \Sigma_{mn} T_{ij,mn}$, which has the symmetry properties given in Eq. 5.46. The conductance in the multichannel case is given by Eq. 5.47, except that $1 - R_{ii}$ in the single-channel case is replaced by $N - R_{ii}$ in the multichannel case. Thus the symmetry properties of the multichannel case are the same as those for the single-channel case.

5.5 Some Applications of the Büttiker Formula

5.5.1 Three-Probe Conductor

The three-probe conductor is shown in Fig. 5.8 [7], where probe 3 is used to measure the chemical potential μ_3. In general, the Büttiker formula can be written as a matrix form, in the case of a three-probe conductor,

$$\begin{pmatrix} I_1 \\ I_2 \\ I_3 \end{pmatrix} = \left(\frac{e^2}{h}\right) \begin{pmatrix} T_{12} + T_{13} & -T_{12} & -T_{13} \\ -T_{21} & T_{21} + T_{23} & -T_{23} \\ -T_{31} & -T_{32} & T_{31} + T_{32} \end{pmatrix} \begin{pmatrix} V_1 \\ V_2 \\ V_3 \end{pmatrix}. \tag{5.54}$$

Suppose the current always leaves the conductor through lead 2. V_2 is the lowest, and thus V_2 can be taken to be zero. Because of

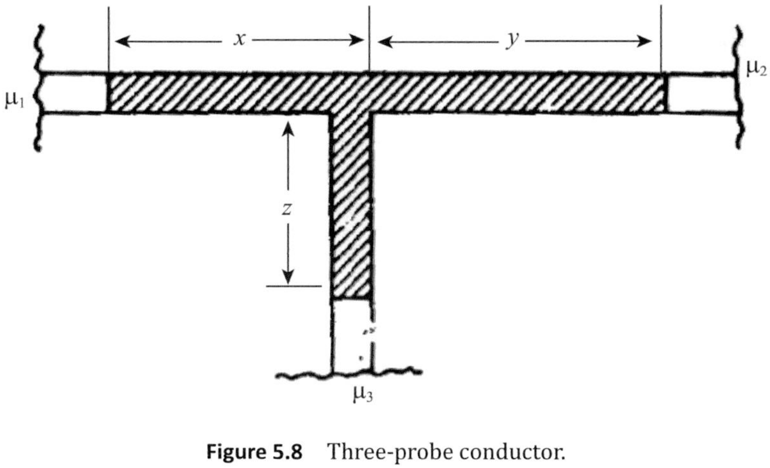

Figure 5.8 Three-probe conductor.

the current conservation, $I_1 + I_2 + I_3 = 0$, so the three equations in Eq. 5.54 are not independent. Equation 5.54 can be simplified to be 2×2 dimensional,

$$\begin{pmatrix} I_1 \\ I_3 \end{pmatrix} = \left(\frac{e^2}{h}\right) \begin{pmatrix} T_{12} + T_{13} & -T_{13} \\ -T_{31} & T_{31} + T_{32} \end{pmatrix} \begin{pmatrix} V_1 \\ V_3 \end{pmatrix}. \tag{5.55}$$

The inverse equation of Eq. 5.55 is

$$\begin{pmatrix} V_1 \\ V_3 \end{pmatrix} = \begin{pmatrix} R_{11} & R_{12} \\ R_{21} & R_{22} \end{pmatrix} \begin{pmatrix} I_1 \\ I_3 \end{pmatrix}, \tag{5.56}$$

where

$$(R) = \left(\frac{h}{e^2}\right) \begin{pmatrix} T_{12} + T_{13} & -T_{13} \\ -T_{31} & T_{31} + T_{32} \end{pmatrix}^{-1}. \tag{5.57}$$

Suppose that the currents from lead 1 to lead 2, the voltage is $V_1 - V_3$, thus $I_2 = -I_1$, $I_3 = 0$, and From Eq. (5.56) can calculate the resistance,

$$V_1 = R_{11} I_1, \quad V_3 = R_{21} I_1,$$

and

$$R_{12,13} = (V_1 - V_3)/I_1 = R_{11} - R_{21}. \tag{5.58}$$

Using Eq. 5.57 yields

$$R_{11} = \left(\frac{h}{e^2}\right) \frac{1}{D}(T_{31} + T_{32}), \quad R_{21} = \left(\frac{h}{e^2}\right) \frac{T_{31}}{D},$$

and

$$D = T_{12} T_{31} + T_{12} T_{32} + T_{13} T_{32}. \tag{5.59}$$

where D is the determinant of the matrix in Eq. 5.57. Inserting Eq. 5.59 into Eq. 5.58 yields

$$R_{12,13} = \left(\frac{h}{e^2}\right) \frac{T_{32}}{D}. \tag{5.60}$$

Similarly,

$$R_{12,32} = \frac{V_3}{I_1} = R_{21} = \left(\frac{h}{e^2}\right) \frac{T_{31}}{D}. \tag{5.61}$$

The determinant D is invariant under flux reversal, $D(\Phi) = D(-\Phi)$. The diagonal elements in the matrix of Eq. 5.57 are, respectively,

$$1 - R_{11} = T_{12} + T_{13} = T_{21} + T_{31}$$

and

$$1 - R_{33} = T_{31} + T_{32} = T_{13} + T_{23}. \tag{5.62}$$

Thus

$$D(-\Phi) = T_{21}T_{13} + T_{21}T_{23} + T_{31}T_{23}(\Phi). \tag{5.63}$$

The resistances $R_{12,13}$ and $R_{12,32}$ in Eqs. 5.60 and 5.61 are neither symmetric nor asymmetric under flux reversal due to T_{32} and T_{31}. However the combined resistance (the two-terminal resistance)

$$R_{12,12} = \frac{V_1 - V_2}{I} = \left(\frac{h}{e^2}\right)\frac{T_{31} + T_{32}}{D} \tag{5.64}$$

is symmetric. Thus, with regard to the two-terminal conductance, the fact that we have an additional lead does not change the symmetry.

Two-terminal conductance in the presence of an additional lead $G = (R_{12,12})^{-1}$ differs now from Eq. 5.25. It is given by

$$G = \left(\frac{e^2}{h}\right)(T_{\mathrm{el}} + T_{\mathrm{in}}), \tag{5.65}$$

where the elastic transmission probability describing the transmission of carriers that emanate from lead 1 and end up in lead 2 without entering reservoir 3 is given by

$$T_{\mathrm{el}} = T_{21}. \tag{5.66}$$

The inelastic transmission probability T_{in} describes carriers that emanate from lead 1 and reach reservoir 3 and from reservoir 3 reach reservoir 2. Comparing Eq. 5.65 with Eq. 5.64 yields

$$T_{\mathrm{in}} = \frac{T_{31}T_{23}}{T_{31} + T_{32}}. \tag{5.67}$$

Thus, the additional lead connected to an electron reservoir acts like an inelastic scatterer.

Equation 5.65 allows us to describe the continuous transition from completely coherent transmission through the conductor to completely incoherent or sequential transmission. In the coherent transmission case, carriers are not allowed to enter reservoir 3, Consequently, $T_{13} = T_{32} = 0$, and two-terminal conductance is given by Eq. 5.25, that is, $T_{\mathrm{el}} = T$ and $T_{\mathrm{in}} = 0$. In the completely incoherent case, $T_{21} = T_{\mathrm{el}} = 0$. In this case Eq. 5.64 yields

$$R_{12,12} = \left(\frac{h}{e^2}\right)\left(\frac{1}{T_{31}} + \frac{1}{T_{23}}\right). \tag{5.68}$$

Equation 5.68 is just the classical addition of series resistors.

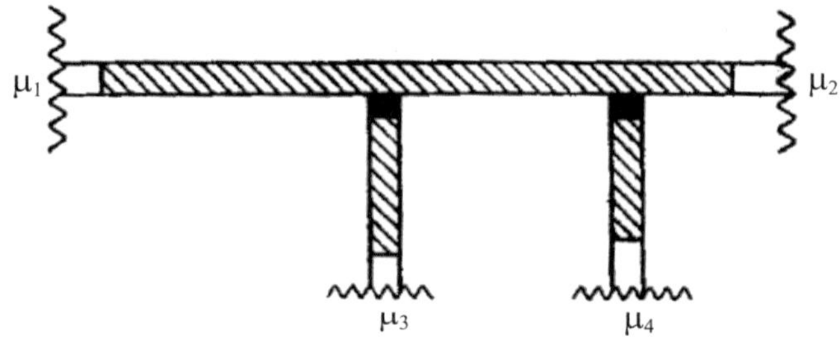

Figure 5.9 Four-probe conductor with tunneling barrier junctions (dark areas) for two of the probes.

5.5.2 Four-Probe Conductor

The four-probe conductor is shown in Fig. 5.9 [7].

Current is fed in at probe 1 and taken out at probe 2. Probes 3 and 4 serve to measure the voltage and are weakly coupled via tunneling barriers to the conductor, as shown in the dark parts in Fig. 5.9. The probabilities for transmission from probe 3 or 4 into the conductor and into reservoir 1 or 2 are small, denoted by ε. If the transmission probability T_{12} is zero order of ε; then T_{13}, T_{14}, T_{23}, and T_{24} are first order; and T_{34} is second order. Evaluation of Eq. 5.53 yields a resistance that reserves the lowest order of ε,

$$R_{12,34} = \frac{h}{e^2}\frac{1}{T_{12}}\frac{T_{31}T_{42} - T_{32}T_{41}}{(T_{31} + T_{32})(T_{41} + T_{42})}. \tag{5.69}$$

$T = T_{21} = T_{12}$ has the symmetry of the transmission probability, and the sums $T_{31} + T_{32}$ and $T_{41} + T_{42}$ are also symmetric with regard to flux reversal. Because $1 - R_{33} = T_{31} + T_{32} + T_{34}$, since T_{34} is zero to order ε, so $T_{31} + T_{32}$ is symmetric. Therefore, Eq. 5.69 has precisely the symmetry required by the reciprocity theorem,

$$R_{12,34}(\Phi) = R_{34,12}(-\Phi). \tag{5.70}$$

5.6 Experimental Results

5.6.1 Two-Terminal Conductor

Using molecular beam epitaxy, it is possible to grow AlGaAs-GaAs heterojunctions routinely with two-dimensional (2D) electron

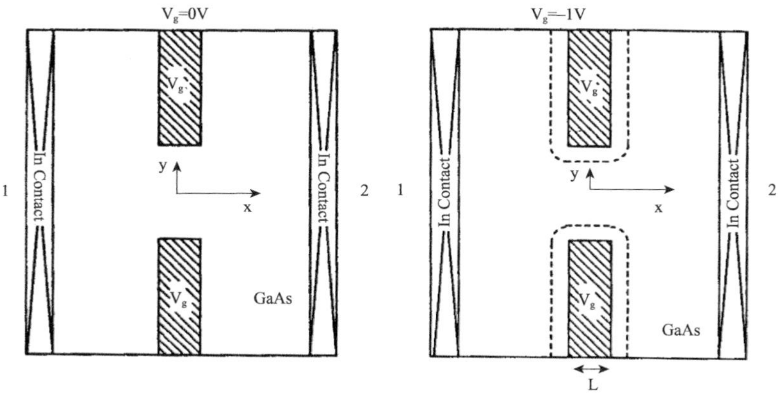

Figure 5.10 AlGaAs/GaAs-modulated doped heterojunctions with the electrodes over the 2DEG; (a) $V_g = 0$ V and (b) $V_g = -1$ V.

mobility $\mu \geq 10^6$ cm^2/Vs for a carrier density of $n \approx 3 \times 10^{11}$ cm^{-2}. The metallic electrodes are made over the 2DEG by using the electron beam lithography as shown in Fig. 5.10a. Timp [8] used the electrostatic potential provided by a split gate geometry to constrain laterally electron gas to the region within the gap between the gates to make a 1D constriction. Figures 5.10a and b schematically represent a typical device in which both gate electrodes are biased at $V_g = 0$ V and -1 V, respectively. In the case of a negative bias voltage the 2DEG at the AlGaAs/GaAs interface immediately beneath the gate electrodes is depleted and so the 2DEG is laterally constrained along the y axis and the electron gas becomes 1D.

Figure 5.11 shows the two-terminal resistance (conductance), $R_{12,12}$ ($G_{12,12}$), as a function of the applied gate voltage [8]; (a) and (b) correspond to the gate widths 200 nm and 600 nm. The series resistance, found at $V_g = 0$ V, was subtracted from the measured resistances. From Fig. 5.11 we see that the average conductance is approximately $2e^2N/h$, with N an integer ranging from 1 to about 10, and evidently is quantized in steps of $2e^2/h$ with about $1 - 5\%$ accuracy as an increasing negative gate voltage makes the constriction narrower. These results verify the theoretical results of Figs. 5.2 and 5.3. When the gate width increases from 200 nm to 600 nm (Fig. 5.11b) the quantized degree of the conductance becomes worse.

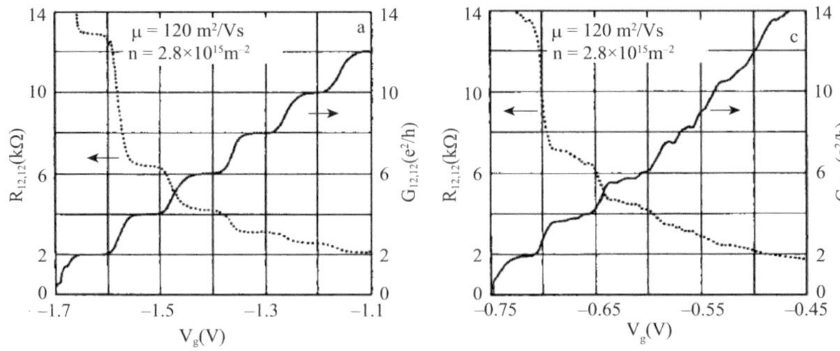

Figure 5.11 Two-terminal resistances (conductance), $R_{12,12}$ ($G_{12,12}$), as a function of the applied gate voltage; (a) and (b) correspond to the gate widths 200 nm and 600 nm.

5.6.2 Two-Terminal Device in the Magnetic Field

Consider an ideal 2D conductor without impurities or inhomogeneity of width w connecting two electron reservoirs as shown in Fig. 5.12 [9]. The chemical potentials of reservoirs are μ_1 and μ_2, respectively, the conductor is along the x direction, the transverse direction is the y direction, and the magnetic field is in the z direction. The electron Hamiltonian is

$$H = \frac{1}{2m^*}\left(\mathbf{p} - \frac{e\mathbf{A}}{\mathbf{c}}\right)^2 + V(y). \tag{5.71}$$

Take the vector potential

$$\mathbf{A} = (-By, 0, 0), \tag{5.72}$$

and let the electron wave function

$$\psi_{jk} = e^{ikx} f_{jk}(y). \tag{5.73}$$

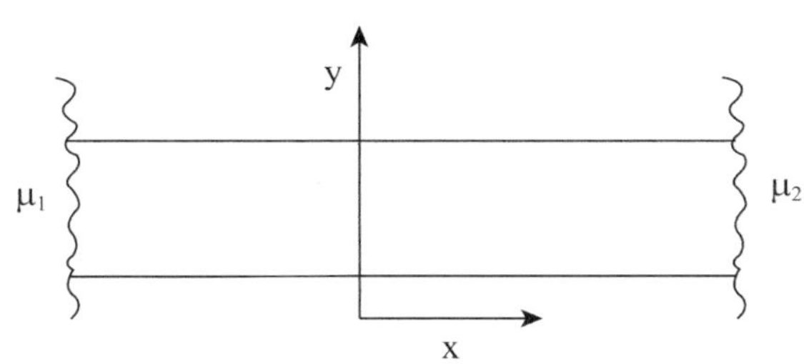

Figure 5.12 Perfect two-dimensional conductor connected to reservoirs.

We obtain the equation of the transverse wave function f,

$$\left[-\frac{\hbar^2}{2m^*}\frac{\partial^2}{\partial y^2}+\frac{1}{2}m^*\omega_c^2(y-y_0)^2+V(y)\right]f=Ef, \qquad (5.74)$$

where $\omega_c = eB/m^*c$, $y_0 = kl_B^2$, and $l_B = (\hbar c/eB)^{1/2}$ is the magnetic length. Therefore, the effect of the magnetic field is to add a parabolic potential to the original transverse potential; the origin of the parabolic potential is at y_0.

Suppose the transverse potential is a square wall with infinite barriers at the edges y_1 and y_2 of the conductor. Solving Eq. 5.74 yields the electron energy levels as a function of y_0 as shown in Fig. 5.13 [9]. When y_0 is near the center of the well, far from y_1 and y_2, the electron is less effected by the $V(y)$, and thus the energy levels are Landau levels,

$$E_{jk}=\hbar\omega_0\left(j+\frac{1}{2}\right). \qquad (5.75)$$

When y_0 is near the edges, the energy of a state depends on the y_0 through the distance $y_1 - y_0$ to the lower edge and $y_2 - y_0$ to the upper edge, as shown in Fig. 5.13. In general, the energy of a state is determined by

$$E_{jk}=E\left(j,\omega_c,y_0(k)\right). \qquad (5.76)$$

The carriers in an edge state acquire a longitudinal velocity,

$$v_{jk}=\frac{1}{\hbar}\frac{dE_{jk}}{\partial k}=\frac{1}{\hbar}\frac{dE_{jk}}{dy_0}\cdot\frac{dy_0}{dk}, \qquad (5.77)$$

which is proportional to the slope of the Landau level. dE/dy_0 is positive at the upper edge y_2 and negative at the lower edge (see Fig. 5.13). While $dy_0/dk = l_B^2$ is positive, therefore, the velocity is positive along the upper edge and negative along the lower edge. Note that it is only the edge states that contribute to carrier flow because the bulk Landau states have no velocity, as the region in Fig. 5.13, where E is independent of y_0.

Figure 5.14 depicts a single impurity near the edge of a sample. The quasi-classical skipping orbits are scattered by this impurity [9], but due to the magnetic field the scattered orbits are never further then a cyclotron radius away from the edge. After scattering by the impurity, the orbits return to the edge and continue to follow the

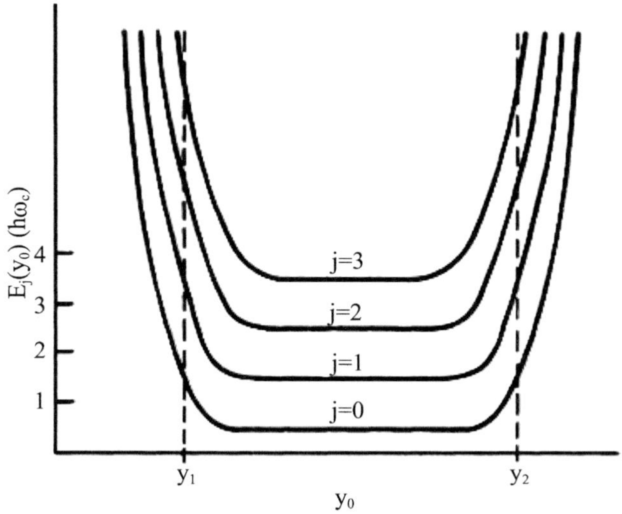

Figure 5.13 Energy spectrum of a perfect conductor in a high magnetic field for a rectangular confining potential (walls at y_1 and y_2).

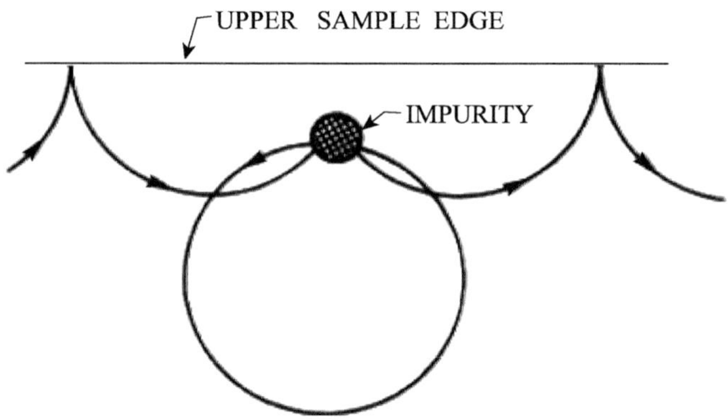

Figure 5.14 Quasi-classical skipping orbits along the upper edge of the sample in the presence of a localized impurity.

edge, and the velocity is positive. The key point is that an impurity cannot effectively reverse the direction of motion of a carrier. There is no scattering backward against the flow of carriers if the average distance between impurities is larger than the cyclotron radius.

To obtain the criterion for the strength of the field we can make use of the fact that the Landau states are a harmonic-oscillator wave function. The cyclotron radius of a state in the j-th Landau level is

$$\langle \Delta y^2 \rangle^{1/2} = l_B \left(j + \frac{1}{2} \right)^{1/2}. \tag{5.78}$$

From Eq. 5.78 we obtain the critical magnetic field B_{crit} for the onset of backscattering,

$$2\pi l_e^2 B_{crit} > \Phi_0 \left(N + \frac{1}{2} \right), \tag{5.79}$$

where $\Phi_0 = hc/e$ is magnetic flux quanta and N is the largest index of Landau state. l_e is the average distance between impurities. Therefore, Eq. (5.79) can be looked as the condition that the backscattering cannot be occurred.

In a strong magnetic field $B > B_{crit}$ the current fed into an edge state by a reservoir is the same as the current fed into a quantum channel in a zero-field perfect conductor. The resulting two-terminal resistance for a perfect conductor in a high magnetic field is thus

$$R = \frac{h}{e^2} \cdot \frac{1}{N}, \tag{5.80}$$

where N is the number of edge states (with a positive velocity).

If the conductor has a disordered section connected at its left and right ends to ideal perfect conductors, the disordered part of the conductor mixes the channels of the perfect conductors. Carriers incident in channel j (edge state j) from the left have a probability amplitude t_{ij} for transmission into channel i and a probability amplitude r_{ij} for reflection into channel i. Therefore, the transmitted current is

$$I_i = \frac{e}{h} \sum_{j=1}^{N} T_{ij} \Delta\mu, \tag{5.81}$$

where $T_{ij} = |t_{ij}|^2$. Summing over all incident channels gives a total transmitted current,

$$I = \frac{e}{h} \sum_{i,j=1}^{N,N} T_{ij} \Delta\mu. \tag{5.82}$$

The voltage drop is $eV = \Delta\mu$, and the Landauer resistance in this case is

$$R = \frac{h}{e^2} \cdot \frac{1}{T}, \quad T = \sum_{i,j=1}^{N,N} T_{ij}. \tag{5.83}$$

5.6.3 Quantum Hall Effect

We discuss how the quantum Hall effect is established in an open conductor with current probes and Hall bars. Figure 5.15 shows a conductor where the Hall bars are of the same width as the probes connected to the current source and sink [9]. In the conductor of Fig. 5.15, N edge states connect the four contacts in a cyclical fashion. Assume that all the contacts are ideal and same and there is no scattering from one edge to the other edge. Under these circumstances, the edge states provide perfect transmission channels for carriers and lead to transmission probabilities, $T_{41} = T_{34} = T_{23} = T_{12} = N$. All the other transmission probabilities are zero. The Hall resistance $R_{13,42}$ is determined by $T_{41}T_{23} - T_{43}T_{21}$, which is equal to N^2. Evaluation of the subdeterminant D in Eq. 5.53 yields $D = N^3$. All Hall resistance values of the conductor of Fig. 5.15 are quantized and given by $\pm(h/e^2)(1/N)$.

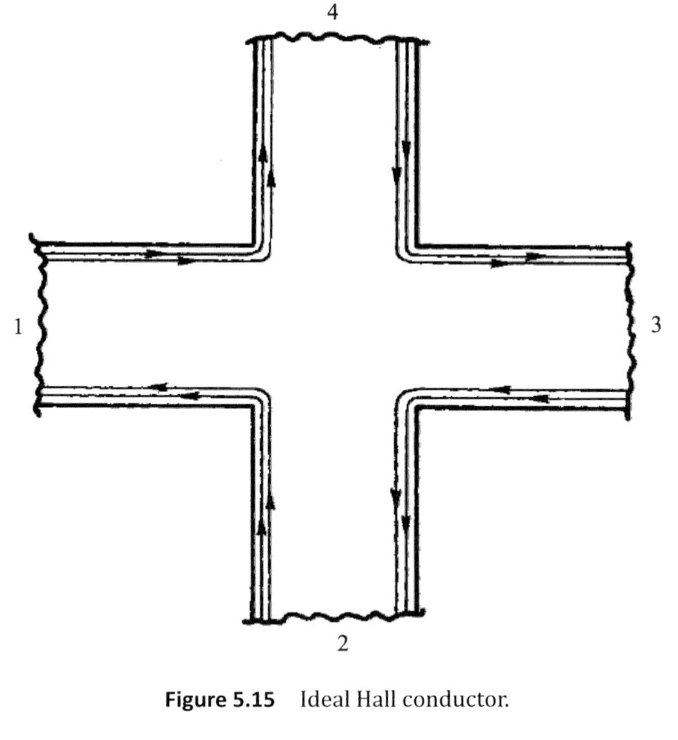

Figure 5.15 Ideal Hall conductor.

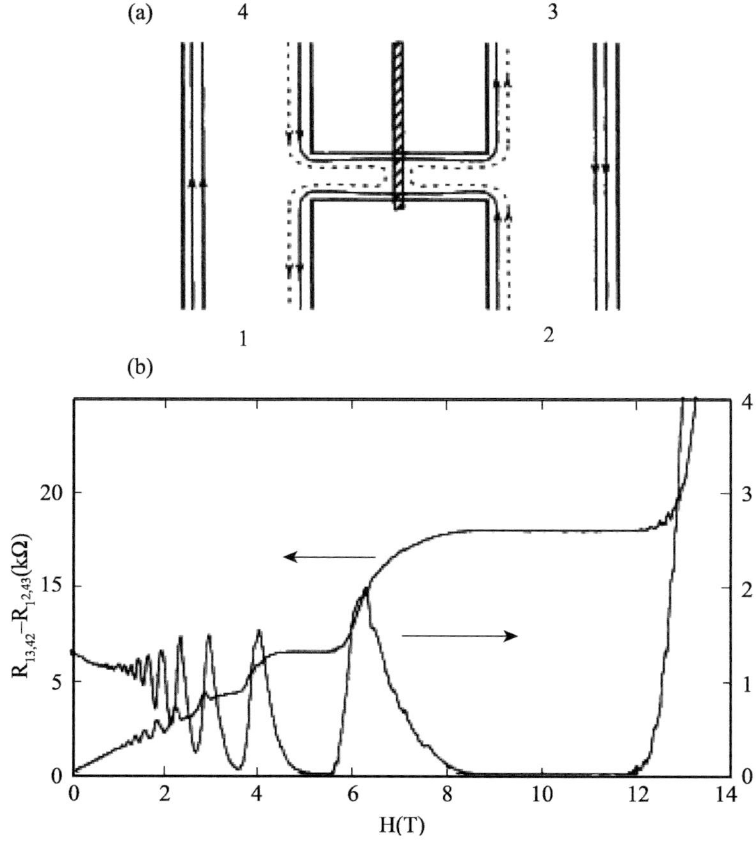

Figure 5.16 (a) A schematic view of the four-probe GaAs-Al$_x$Ga$_{1-x}$As wire with a narrow cross gate. (b) $R_{12,43}$ and $R_{13,42} - R_{12,43}$ at V_g as functions of magnetic field.

The longitudinal resistance, for example, $R_{12,43}$ are zero, since in the products forming the expression, $T_{41}T_{32} - T_{42}T_{31}$, at least one transmission probability always is zero.

The above conductor is an ideal conductor. Washburn et al. [10] added a potential barrier in this conductor, as shown in Fig. 5.16. They manufactured a narrow gate on the circuit center. Applying a negative bias voltage on the gate is equal to adding a potential barrier in the circuit; the barrier height is controlled by the bias voltage. Because the barrier is higher in the center region and lower in the edge region, the edge states near the edge (solid lines) can tunnel through the barrier. Those near the center region (dashed lines) are reflected into the opposite edge and move in the opposite direction, as shown in Fig. 5.16.

Assuming that there are K edge states that are reflected, we have $T_{41} = N$, $T_{34} = N - k$, $T_{14} = K$, $T_{23} = N$, $T_{12} = N - k$, and $T_{32} = K$ and all other T_{ij} equal zero. It can be proved from Eq. 5.53 that $D = N^2(N - k)$,

$$\begin{aligned} R_{13,42} &= \left(\frac{h}{e^2}\right)(T_{41}T_{23} - T_{43}T_{21})/D \\ &= \left(\frac{h}{e^2}\right)\frac{1}{N-K}, \\ R_{42,13} &= \left(\frac{h}{e^2}\right)(T_{14}T_{32} - T_{12}T_{34})/D \\ &= -\left(\frac{h}{e^2}\right)\frac{N-2K}{N(N-K)}, \end{aligned}$$

and

$$\begin{aligned} R_{12,43} &= \left(\frac{h}{e^2}\right)(T_{41}T_{32} - T_{42}T_{31})/D \\ &= \left(\frac{h}{e^2}\right)\frac{K}{N(N-K)}. \end{aligned} \tag{5.84}$$

Figure 5.16b shows $R_{12,43}$ and $R_{13,42} - R_{12,43}$ as functions of magnetic field H at $V_g = 0$ [10]. These combinations are the nearest correspondence to the classical resistance ρ_{xx} and ρ_{xy}, which are the parameters usually studied in large samples. From the figure we see that $R_{12,43}$ approaches zero over wide regions of field ($4.9 < H < 5.5$ T and $8.5 < H < 10$ T), and in the same ranges $R_{13,42}$ is quantized to $h/4e^2$ and $h/2e^2$, respectively.

Figure 5.17 shows $R_{13,42}$ as a function of gate voltage for several values of the magnetic field [10]. From the figure we see that when $V_g = 0$, $R_{13,42}$ changes from $h/(5e^2)$ and $h/(4e^2)$ to $h/(2e^2)$ with magnetic field increasing, indicating that the number of occupied Landau levels decreases. When V_g decreases, K changes from 0 to 1 or 2, and from Eq. 5.84 $R_{13,42}$ increases and raises to the next plateau, for example, from $h/(4e^2)$ to $h/(2e^2)$. From the above discussion we conclude that backscattering by the gate potential

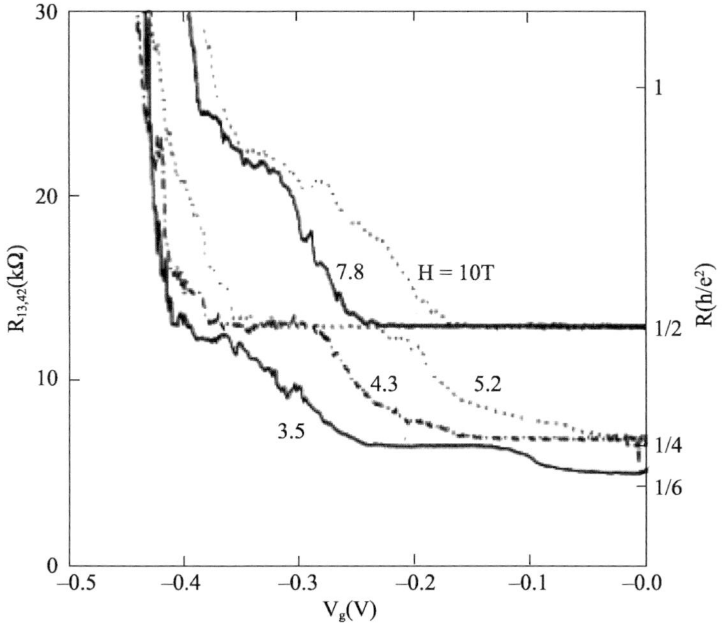

Figure 5.17 $R_{13,42}$ as a function of gate voltage for several values of the magnetic field.

does not destroy the quantized Hall effect. Instead, there is always a region in which the various four-probe resistances are quantized to h/e^2i.

References

1. Datta, S. (1995). *Transport in Mesoscopic Systems* (Cambridge, London).
2. Szafer, A., and Stone, A. D. (1989). Theory of quantum conduction through a constriction. *Phys. Rev. Lett.*, **62**, 300–303.
3. van Wees, B. J., van Houten, H., Beenakker, W. J., et al. (1988). Quantized conductance of point contacts in a two-dimensional electron gas. *Phys. Rev. Lett.*, **60**, 848–850.
4. Kirczenow, G. (1989). Resonant conduction in ballistic quantum channels. *Phys. Rev. B*, **39**, 10452–10455.

5. Büttiker, M., Imry, Y., Landauer, R., and Pinhas, S. (1985). Generalized many-channel conductance formula with application to small rings. *Phys. Rev. B*, **31**, 6207–6215.
6. Büttiker, M. (1986). Four-terminal phase-coherent conductance. *Phys. Rev. Lett.*, **57**, 1761–1764.
7. Büttiker, M. (1988). Symmetry of electrical conduction. *IBM J. Res. Dev.*, **32**, 317–334.
8. Timp, G. (1996). *Semiconductors and Semimetals*, Vol. 35, ed. Reed, M. (Academic Press, New York), p. 113.
9. Büttiker, M. (1988). Absence of backscattering in the quantum Hall effect in multiprobe conductors. *Phys. Rev. B*, **38**, 9375–9389.
10. Washburn, S., Fowler, A. B., Schmid, H., and Kern, D. (1988). Quantized Hall effect in the presence of backscattering. *Phys. Rev. Lett.*, **61**, 2801–2804.

Chapter 6

Transport in Quantum Dots

Since the single-electron effect in quantum dots and single-electron transistor were discovered, many new physical behaviors have been discovered, for example, core shell-like filling of electrons in quantum dots of regular shapes, electron levels of quantum dots in a perpendicular magnetic field, form of molecular state in two or three coupling quantum dots, coherent molecular states, covalent states, and ionic states. The high-order effects include resonant tunneling, Kondo effect caused by a strong coupling, mixed valence physics connected with the Anderson model, Fano resonance produced by the existence of resonance or non-resonance channels, etc. In the application aspects, many interesting proposals are suggested, for example, single-electron transistors (SETs) and quantum bite based on the electron spin used for quantum calculation. Some noticeable and useful devices have been invented, for example, scanning SET used to detect the local electric field and radio frequency (RF) SET used to detect electric charge at the level of $10^{-5}e/\sqrt{Hz}$.

Quantum Waveguide in Microcircuits (Second Edition)
Jian-Bai Xia, Duan-Yang Liu, and Wei-Dong Sheng

ISBN 978-981-4968-30-0 (Hardcover), 978-1-003-65146-8 (eBook)
www.jennystanford.com

6.1 Single-Electron Effect and Single-Electron Transistor

In the beginning of the twentieth century, physicist Millikan made the famous oil-drop experiment. He found that the electric charges brought by the tiny broken oil drops are always integral times of a basic electric quantity; then he deduced that the particle bringing the basic electric quantity is an electron and the basic electric quantity $e = 1.602 \times 10^{-19}$ C. But at that time Millikan could not obtain a single electron. Today the development of micro-lithography technology has made it possible to fabricate semiconductor micro-structures with a scale smaller than 100 nm. In this kind of structure one can operate electron motion one by one through a changing voltage. This is the single-electron effect.

In 1989 Scott-Thomas et al. created a dual-gate device with a 70 nm gap in the lower gate on narrow Si inversion layers. Figure 6.1 shows the schematic cross section (a) and the top view (b) of the slotted-gate device [1]. The inversion layer, formed by the positively biased upper gate, is confined by the lower gate. The width of the channel is controlled by the lower gate, and the electron density in

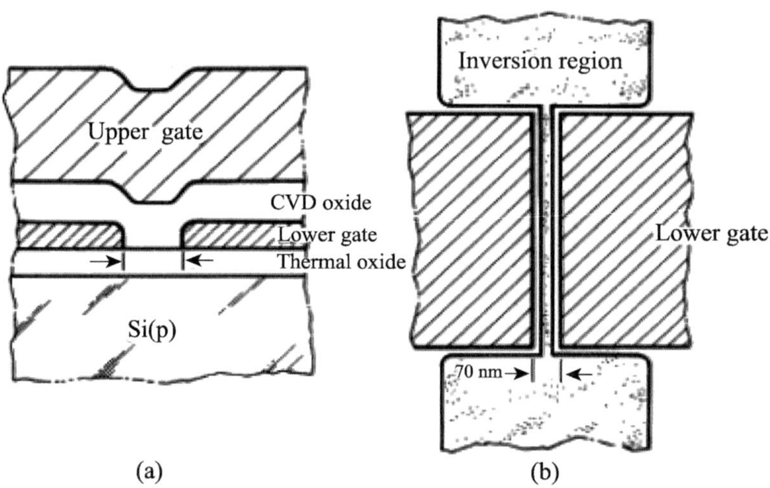

Figure 6.1 Schematic cross section (a) and top view (b) of the slotted-gate device.

the inversion layer is controlled by the upper gate voltage (positive). The width of the narrow channel is 20 nm.

Earlier, people also experimented with electrons through a narrow channel but did not obtained any regular results. Due to the limitation of the lithography technology the width of the channel is $3 \approx 5$ times of the present channel. Besides the electron mobility is lower. The experimental results of conductance from the device of Fig. 6.1 are shown in Fig. 6.2 [1]. The top panel shows G for a 10 μm long metal-oxide semiconductor field-effect transistor (MOSFET) as a function of the voltage on the upper gate V_G, which is proportional to the number of electrons per unit length. The periodic oscillations can be clearly identified, although they are accompanied by random, see [1]. That the oscillations are truly periodic is determined in the next three panels, which show the power spectral density, that is, the square of the modulus of the Fourier transform, for devices of 10, 2, and 1 μm lengths. Each device has a peak in its spectrum at a nonzero frequency. For a $L = 10$ μm device (second panel), the peak is at $1/\Delta V_G = 43\ \text{V}^{-1}$, that is, the oscillation period of the V_G is $\Delta V_G = 0.023$ V.

The electron charges in the channel equal the channel capacitance multiplied by V_G, C/L is the capacitance of unit length, and $e(N/L) = (C/L)V_G$. Therefore, in Fig. 6.2 G can be looked at as a function of number of electrons N. The oscillation of G with V_G is equivalent to that with N. The voltage difference ΔV_G for one period of oscillation is necessary for adding an electron. The experiments found that there is no correlation between ΔV_G and the channel length L, as shown in the next three panels of Fig. 6.2. The authors thought that the actual length of a channel is determined by the distance between two impurities in the channel, not the apparent length L as shown in Fig. 6.1b. The electrostatic potential produced by the impurities forms a barrier for electrons, forming a quantum dot. Thus, the period ΔV_G of G oscillation is decided by the distance between two impurities, not L. The bottom panel of Fig. 6.2 is the Fourier spectra for the 1 μm long channel in a magnetic field. Comparing the two lowest panels we found that the oscillation period of G is independent of the magnetic field.

The above impurity barrier model is only a surmise and has not been verified experimentally. To study the relation between

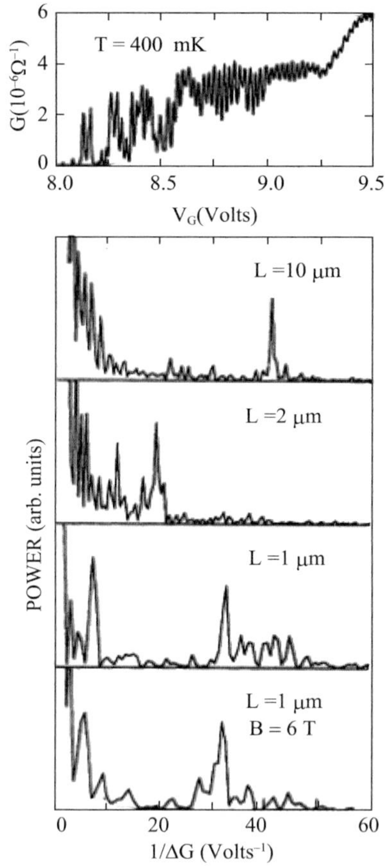

Figure 6.2 Top panel: G versus V_G for a 10 μm long inversion layer. Next three panels: Fourier power spectra of the data of the top panel and for 2 and 1 μm long channels. Bottom panel: Fourier spectra for the 1 μm long channel in a magnetic field.

the oscillation period of G, ΔV_G, and channel length Meirav et al. used a GaAs/AlGaAs-modulated doping heterojunction instead of a Si inversion layer and made the metal gate with two constrictions as shown in Fig. 6.3 [2]. Each constriction is about 100 nm long, defining a segment of length L_0 between them. When a negative V_b is applied, these constrictions induce saddle-shaped potential barriers for electrons moving along the channel.

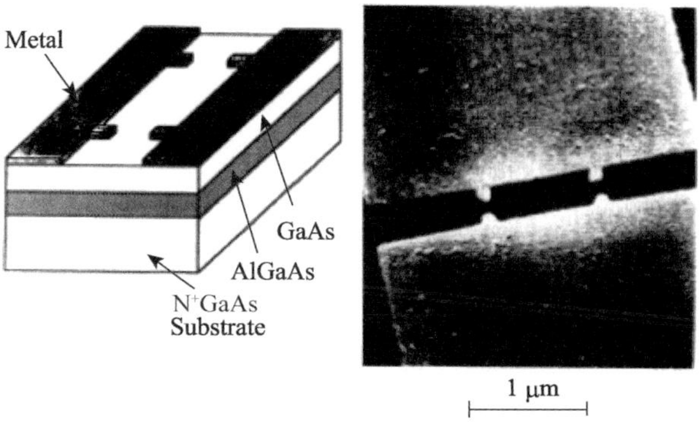

Figure 6.3 Schematic drawing of a device structure along with a scanning electron micrograph of one sample.

The measured G as a function of $V_g - V_t$ are shown in Fig. 6.4 [2], where V_t is a sample-dependent threshold and V_b was held fixed, typically at about -0.5 V. Figures 6.4a and 6.4b show data for two samples, which had the same geometry $L_0 = 1$ μm. The periodic oscillations are clearly seen, with the same period in both samples. The channels without constrictions had monotonic dependence or, occasionally, random fluctuation patterns, but no similar periodic behavior. Figures 6.4c and 6.4d show results for samples with progressively shorter lengths and correspondingly longer periods, consistent with the geometry.

The oscillations in Fig. 6.4 recurred with the exact same period even after samples were warmed to room temperature and cooled again, as shown in Fig. 6.5 [2]. In previous experiments, such thermal cycling invariably led to random changes or total disappearance of the periodicity, understood as the result of the redistribution of interface charges (impurities). Here no such randomness is seen. Thus, for the first time, we can control the period of the oscillation by the geometry of the devices. This is because the GaAs/AlGaAs heterostructure material is rather clean, making the incidence of charged impurities in the channel less common than in Si inversion layers.

There is still one problem: whether each oscillation corresponds to precisely one electron added to the system. We should estimate the increment in V_g required to add one electron to the segment

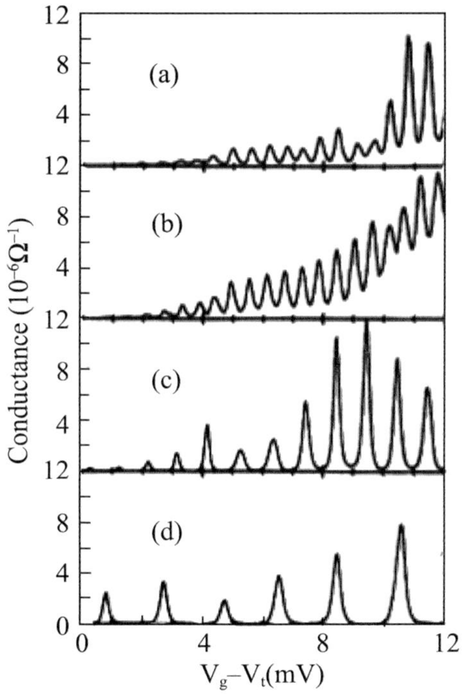

Figure 6.4 G as a function of $V_g - V_t$ for four samples of different L_0: (a) 1 μm, (b) 1 μm, (c) 0.8 μm, and (d) 0.6 μm.

of the channel between the two constrictions. Three-dimensional computer simulations of these structures gave estimates of the relation between V_g and the total charge in the channel. For all samples shown the estimated charge increase per ΔV_g is remarkably close to one electron, within 10% for most samples. The possibility of more than one electron per ΔV_g is outside the error margin.

Kouwenhoven et al. studied similar single charging effects in a lateral split-gate quantum dot defined by metal gates in the two-dimensional electron gas (2DEG) of a GaAs/AlGaAs heterostructure [3]. Figure 6.6 shows a scanning electron microscope (SEM) photograph of the gate geometry, which is fabricated on top of a GaAs/AlGaAs heterostructure containing 2DEG [3]. The gate F is denoted as the finger gate, gates 1 to 4 as quantum point contact (QPC) gates, and gate C as the center gate. A negative voltage −0.4 V

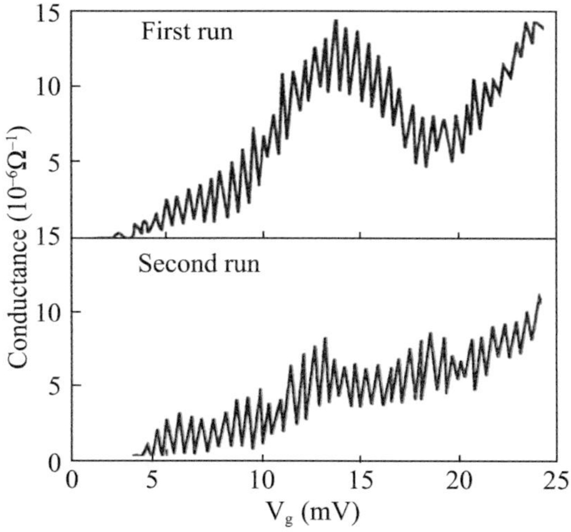

Figure 6.5 Conductance versus gate voltage for one sample taken in different runs, between which the sample was warmed to 300 K.

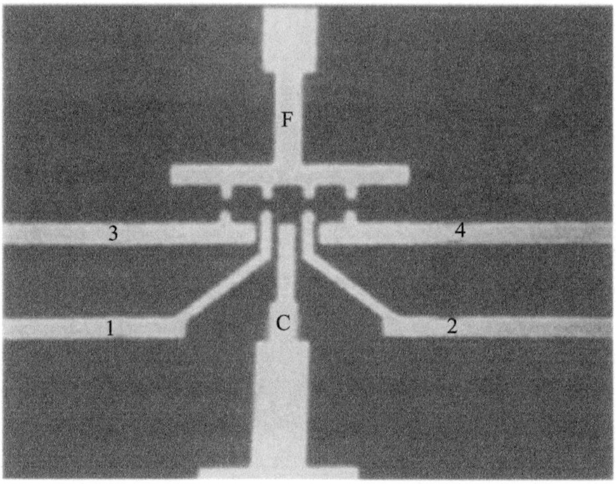

Figure 6.6 SEM photograph of the gate geometry, which is fabricated on top of a GaAs/AlGaAs heterostructure containing 2DEG.

depletes the 2DEG underneath the gates. The narrow channels between gates 3–1, 1–C, C–2, and 2–4 are completely pinched off at this gate voltage. Applying a negative voltage to the gates F, 1, 2, and C forms a dot in the 2DEG. The radius of the dot is estimated as 300 nm. QPC gates 1 and 2 are used to control the conductance of the tunneling barrier between the dot and the wide 2DEG regions, and the center gate is used to vary the number of electrons in the dot. This gate structure allows an independent control of the conductance of the two 2DEG leads and enables you to vary the number of electrons that are localized in the dot.

The gate voltages V_3 and V_4 are taken to be zero, so there is no quantum dot formed between gates 3, 4, and F. The gate voltages V_1 and V_2 are taken to be smaller than -1.4 V, so G_1 and $G_2 \ll 2e^2/h$. The measured conductance of the dot versus center gate voltage V_C is shown in Fig. 6.7 [3]. The oscillations appear as sharp peaks with an amplitude up to e^2/h. Experiments found that only when both QPC conductance values are below the quantized plateau value $2e^2/h$ does the amplitude of the Coulomb oscillations increase as the QPC conductance values decrease. From Fig. 6.7 the oscillation period $\Delta V_C = 8.3$ mV, a capacitance between the dot and the center

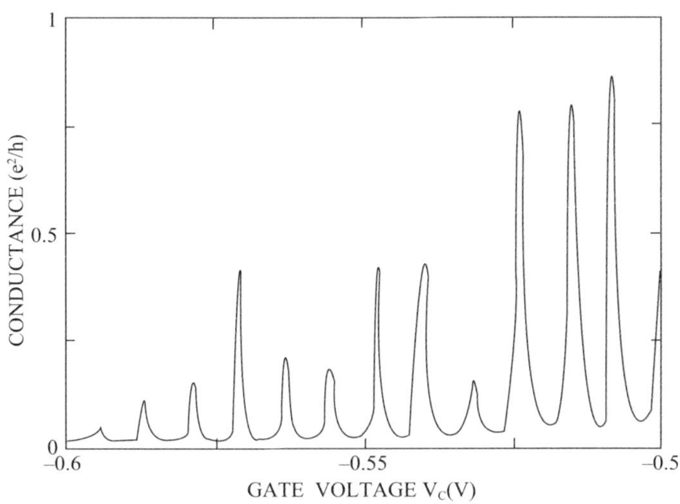

Figure 6.7 Conductance of the dot versus center gate voltage V_C for G_1 and $G_2 \ll 2e^2/h$.

gate is derived $C_C = e/\Delta V_C = 0.19 \times 10^{-16}$ F. To determine the total capacitance between the dot and the six gates of the sample, the authors measured the oscillations by varying the voltage on the different gates, while keeping the voltage on the remaining gates fixed. Then they obtained the capacitances between the dot and each gate, added these capacitances, and obtained the total capacitance $\Sigma C_g = 1.71 \times 10^{-16}$ F.

Using QPC gates 3 and 4, one can increase the dot size to two strongly coupled dots (i.e., one dot with a double size) and one dot with three times the size. Figure 6.8 shows the Coulomb oscillations versus the center gate voltage. The periods in the gate voltage are $\Delta V_{1\text{dot}} = 9.0$ mV, $\Delta V_{2\text{dot}} = 5.3$ mV, and $\Delta V_{3\text{dot}} = 3.6$ mV, yielding capacitances between the center gate and the dot $C_{\text{C,1dot}} = 1.8 \times 10^{-17}$ F, $C_{\text{C,2dot}} = 3.0 \times 10^{-17}$ F, and $C_{\text{C,3dot}} = 4.4 \times 10^{-17}$ F, which have the ratios 1.2:2:3. This result shows that the gate capacitance scales with the size of the dot.

The charging effect can be described in terms of the capacitance C between the dot and its environment. An estimate of C is obtained from the self-capacitance $C_0 = 8\varepsilon_r\varepsilon_0 R$ between a 2D dot with a disc shape and infinity, where R is the radius of the disc, ε_0 is the vacuum dielectric constant, and $\varepsilon_0 = 8.854 \times 10^{-12}$ F/m. A dot with $R =$ 300 nm and $\varepsilon_r = 13$ in GaAs gives $C_0 = 2.8 \times 10^{-16}$ F and a charging energy $e^2/C_0 = 0.6$ meV, which exceeds the thermal energy $k_B T$ for temperatures below 4 K. From the electron density of the ungated 2DEG and the dot area, the number N of electrons in the dot is estimated to be about 500. The average energy separation between the discrete states $2E_F/N$ is therefore $\sim$0.03 meV, where $E_F = 7$ meV is the 2DEG Fermi energy. This is less than one-tenth of the charging energy, so the charging effect is dominant in the electron transport process.

The transport of a quantum dot can be described schematically by Fig. 6.9 [3], where μ_l and μ_r denote the chemical potentials of left and right electron reservoirs, respectively, and $\mu_l > \mu_r$. The center part is the dot, separated from the reservoirs by the barriers, $e\varphi_n$ is the electrostatic energy of N electrons. At temperature $T = 0$ K, the ground state energy of N electrons is the sum of single electron

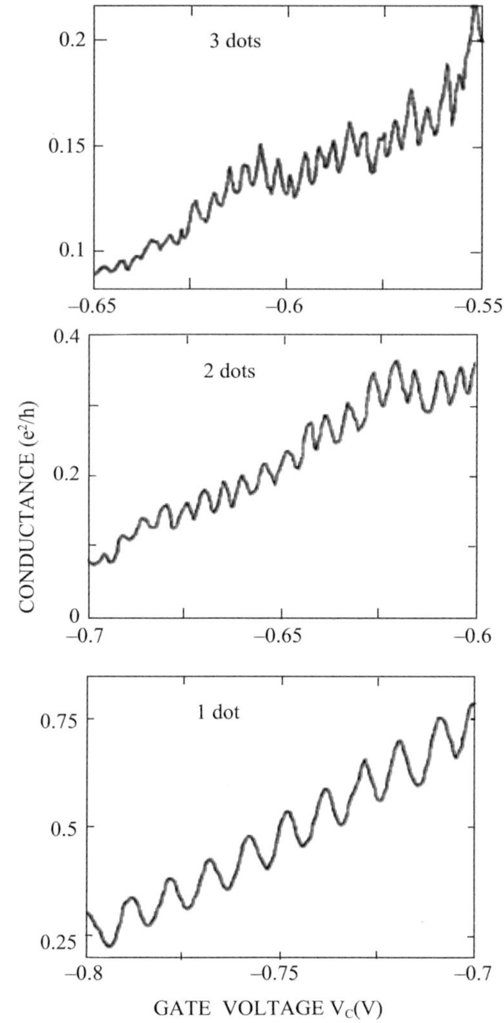

Figure 6.8 Conductance of the dot versus the center gate voltage V_C for different sizes of the dot.

energy E_p plus the electrostatic energy,

$$U(N) = \sum_{P=1}^{N} E_P + \frac{(-eN + C_g V_g)^2}{2C}, \tag{6.1}$$

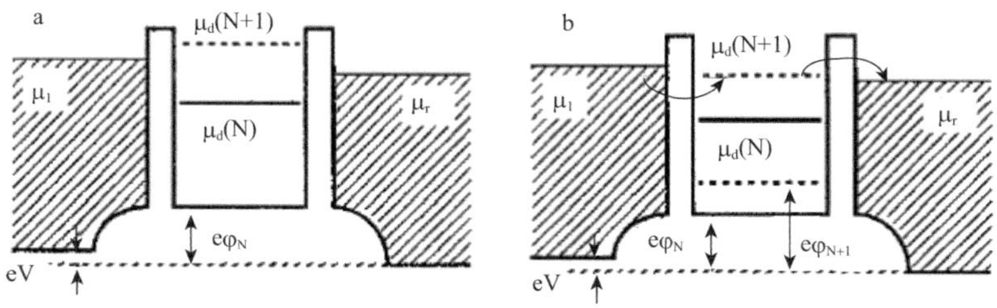

Figure 6.9 Two situations for different gate voltages. (a) Coulomb blockade of electron tunneling. (b) One-by-one electron tunneling at the $N \rightarrow N+1$ transition.

where C_g is the capacitance between the dot and one gate (e.g., the center gate) and C is the total capacitance between the dot and all gates.

The minimum energy necessary to add the N-th electron to the dot is

$$\mu_d(N) = U(N) - U(N-1) = E_N + \frac{(N-1/2)e^2}{C} - e\frac{C_g}{C}V_g. \qquad (6.2)$$

Thus, the difference of the chemical potentials of the $(N+1)$ and N electrons,

$$\mu_d(N+1) - \mu_d(N) = E_{N+1} - E_N + e^2/C. \qquad (6.3)$$

This energy gap leads to blockade for tunneling of an electron into and out of the dot, as shown schematically in Fig. 6.9a, where N electrons are localized in the dot. The $(N+1)$th electron cannot tunnel into the dot because the resulting electrochemical potential $\mu_d(N+1)$ is higher than the electrochemical potentials of the reservoirs. So for $\mu_d(N) < \mu_l$ and $\mu_l < \mu_d(N+1)$ the electron transport is blocked, which is known as the Coulomb blockade (CB). The CB can be eliminated by increasing the center gate voltage, so $\mu_d(N+1)$ is lined up between μ_l and μ_r [$\mu_l < \mu_d(N+1) < \mu_r$], as illustrated in Fig. 6.9b. In this case, the $(N+1)$th electron can tunnel into the dot and then tunnel out of the dot to the right 2DEG reservoir, causing the electrochemical potential to drop to $\mu_d(N)$. Now a new electron can tunnel into the dot and repeat the cycle. This process, where current is carried by successive discrete charging and discharging of the dot, is known as single-charge tunneling.

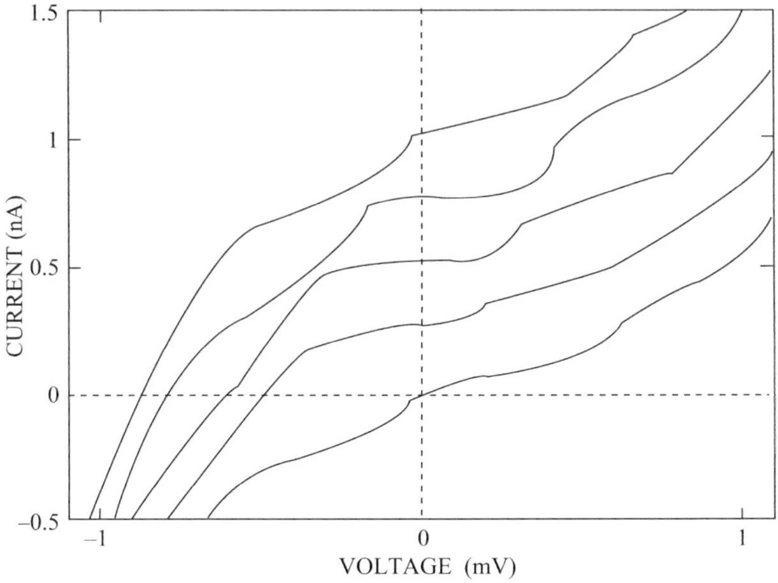

Figure 6.10 I–V characteristics for different center gate voltages, demonstrating the Coulomb staircase. The curves correspond to different values of V_C (in steps of 1 mV) and are offset for clarity ($I = 0$ occurs at $V = 0$).

In the case of Fig. 6.9b the potential difference of left and right reservoirs $eV = \mu_l - \mu_r$ is small, so only one electron is allowed to move through the dot, that is, the dot only opens a one-electron channel. If the bias voltage eV increases continuously, so the chemical potential difference $\mu_l - \mu_r$ can include two electron states, then the dot opens a second electron channel. Therefore, in the I–V characteristic curve, when V increases, the current will step-likely increase, as shown in Fig. 6.10 [3], where each step corresponds to one electron transport. This I–V curve is called the Coulomb staircase. The current steps $\Delta I \approx 0.2$ nA occur at voltage intervals $\Delta V \approx 0.67$ mV. This voltage difference is a direct measure of the charging energy in the sample $\Delta V = e/C$, which yields a total capacitance $C = 2.4 \times 10^{-16}$ F. Suppose that the total tunnel conductance is G. One extra elementary charge e is transported through the dot in a typical time C/G, yielding $\Delta I \approx eG/C$. This gives $G \approx (4\ \mathrm{M}\Omega)^{-1}$ and the tunnel time $C/G \approx 10^{-9}$ s.

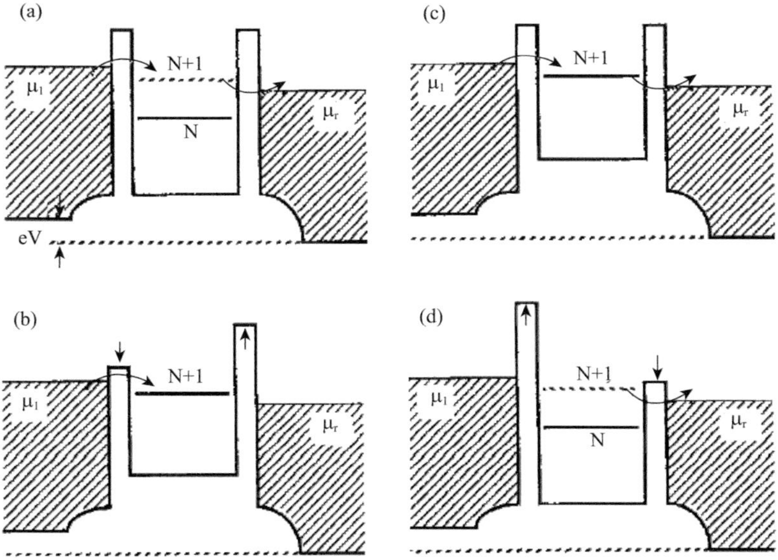

Figure 6.11 Schematic potential landscape through the quantum dot. (a) to (d) represent 4 stages of potential barrier height variation in a RF cycle.

When the two gates (gates 1 and 2 in Fig. 6.6) are applied alternating voltages with phase difference π (frequency $f = 10$ MHz), in one period the heights of two barriers beside the dot vary such as shown in Fig. 6.11a–d. The solid arrow denotes allowed electron motion, and the dashed arrow denotes forbidden electron motion. Thus, in one period of voltage variation just one electron tunnels through the dot. The AC frequency is f, and the current $I = ef$. If the bias voltage V is increased, so that the number of charge states in the interval between μ_l and μ_r increases to n, then n electrons can tunnel through the dot per radio-frequency (RF) cycle, yielding a quantized current $I = nef$.

The measured I–V curves for RF signals with a phase difference of π applied to gates 1 and 2 are shown in Fig. 6.12 [4]. The I–V curves have current plateaus at multiple values of ef, demonstrating that a discrete number of electrons are transferred through the quantum dot for each RF cycle. The curves correspond to different center gate voltages V_C (in steps of 1 mV) and are offset for clarity

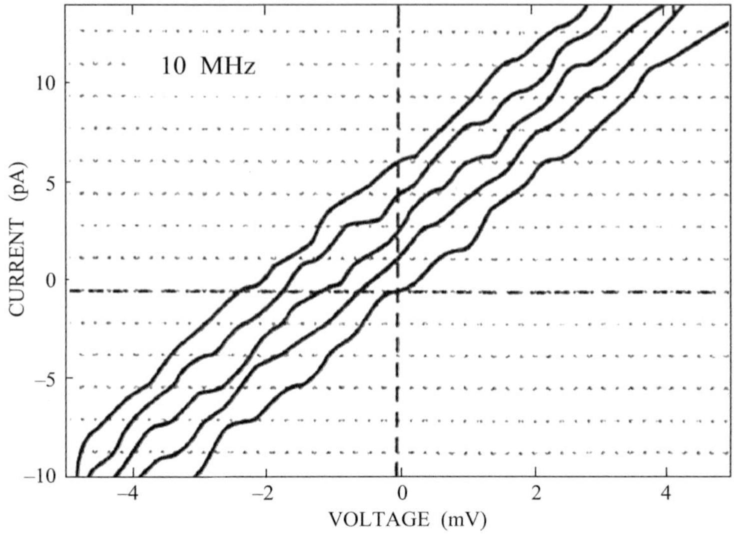

Figure 6.12 I–V characteristics when two phase shifted RF signals are applied with a frequency $f = 10$ MHz. The curves correspond to different center gate voltages V_C (in steps of 1 mV) and are offset for clarity by an integer times *ef*. Dotted lines show the current plateaus at integer values of *ef*.

by an integer times *ef* ($I = 0$ occurs at $V = 0$). The height of each plateau is $e \times 10$ MHz $= 1.6$ pA.

Figure 6.13 shows I–V characteristics for different frequencies $f = 5$, 10, and 20 MHz. The different curves in each group correspond to different V_C [4]. The current is proportional to f, varying in the interval between *nef* and $(n + 1)ef$. This behavior of transport in the RF field is called a turnstile effect, which is like the turnstile in the hotel allowing one or several people through it in each cycle. Comparing Figs. 6.12 and 6.10, we found that the current scales are pA and nA for the turnstile effect and the Coulomb staircase, respectively.

The most hopeful application of the single-electron effect is the single-electron transistor (SET), which can be used to fabricate memory of a large capacity. Now in the 4G DRAM, each bite accumulates about 500,000 electrons, dissipating a lot of energy. If we would fabricate 16 T memory, we should reduce largely

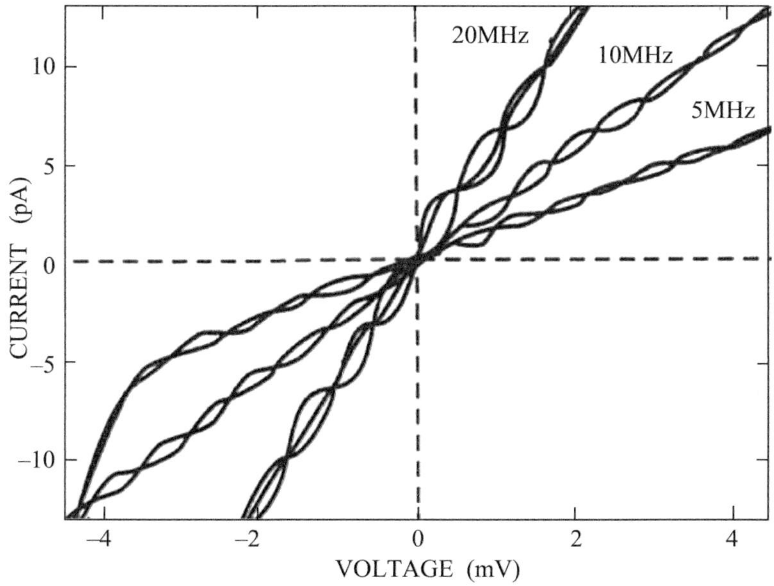

Figure 6.13 I–V characteristics for different frequencies $f = 5$, 10, and 20 MHz.

the number of electrons in each bite, for example, to less than 100. Therefore, the SET is the best choice of future large-capacity memory. But now it is far from the actual application of the SET, its volume is not small enough, the capacitance is too large and so the charging energy e^2/C is too small, and the device should operate at an ultra-low temperature so $k_B T \ll e^2/C$. The goal is to reduce the size of the quantum dot, reduce the capacitance and the number of electrons in the dot, and raise the operating temperature.

6.2 Transport of a Quantum Dot in a Magnetic Field

Experiments found that the conductance oscillation with the gate voltage exists for a quantum dot even in the perpendicular magnetic field. The geometry of the device is shown schematically in Fig. 6.14a [5]. It is a modulation doping GaAs/AlGaAs heterostructure, in which

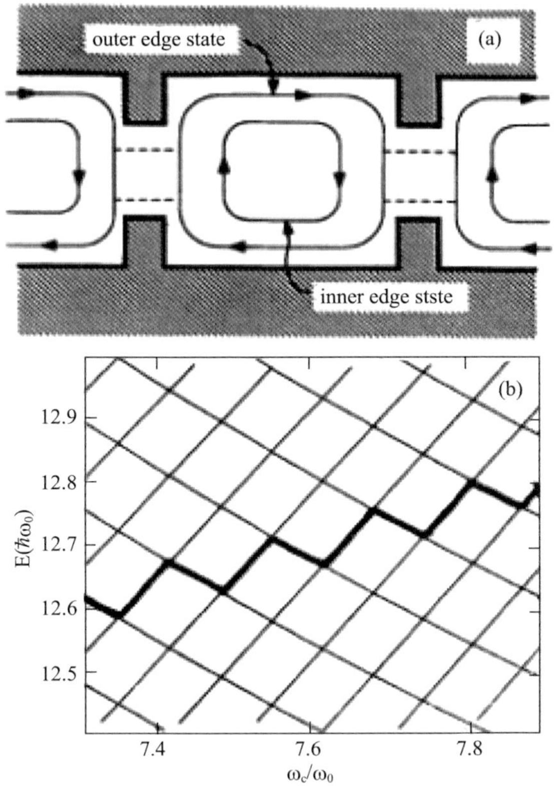

Figure 6.14 Schematic top view of the device, showing the path of the edge states associated with the lowest two Landau levels. The lithographic dimension is 500 nm by 700 nm. (b) Energy levels of a dot with a parabolic confining potential as a function of $\omega_c = eB/m^*$.

the upper gate (a negative bias) is used to define the dot and the lower gate (a positive bias) is to adjust the electron density. The magnetic field is perpendicular to the 2DEG, which forms the edge states in the dot. Figure 6.14b shows the calculated magnetic energy levels as functions of magnetic field ω_c/ω_0 in the parabolic confining potential $(1/2)m^*\omega_0^2 r^2$. The heavy line represents the energy of the single-particle state that is 78th lowest in energy.

The conductance G versus the gate voltage V_g applied to the lower gate is shown in the inset of Fig. 6.15 [5]. The conductance

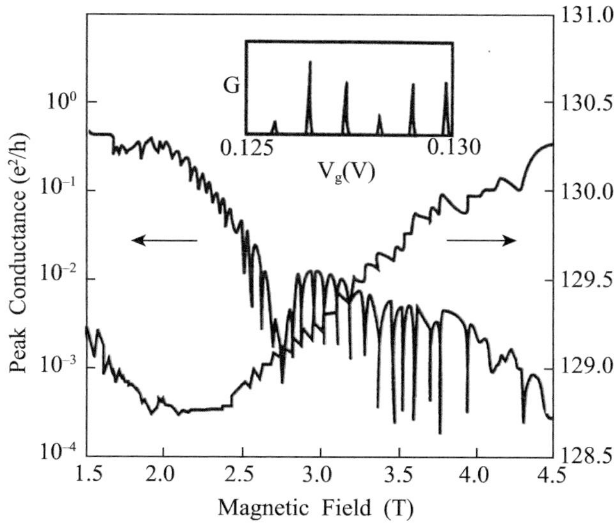

Figure 6.15 Amplitude and position of a conductance peak as a function of magnetic field at $T = 0.1$ K. Inset: Conductance versus V_g for the device at $\mathbf{B} = 3$ T.

consists of a periodic series of sharp peaks, but the amplitudes are much smaller than those in the absence of a magnetic field. The full scale of the ordinate is 0.03 e^2/h. The amplitude and position of a particular conductance peak as a function of magnetic field for $\mathbf{B} = 1.5 - 4.5$ T is shown in Fig. 6.15. The amplitude oscillates periodically with the magnetic field and drops by as much as orders of magnitude. Commensurate with these dips, oscillations are observed in the position of the peak. This structure washes out rapidly with an increasing temperature and is almost entirely destroyed by $T \approx 0.3$ K.

The conductance oscillation in the inset of Fig. 6.15 can be explained by the standard CB model (see Fig. 6.9). The gate voltage corresponding to a conductance peak is given from Eq. 6.2,

$$V_g(N) = \frac{\mu_d(N) - \mu}{ae} = \left(\frac{1}{ae}\right)\left[\left(N - \frac{1}{2}\right)U + (E_N - \mu)\right] + \text{const}, \tag{6.4}$$

where $U = e^2/C$ is the electron Coulomb energy, μ is the chemical potential in the leads, and a is a dimensionless constant, which can be determined from the temperature dependence of the width of a conductance peak and is found to be 0.4 for this device. Therefore, the position of a conductance peak is determined mainly by the Coulomb term $(N - 1/2)U$ and the single-particle term $E_N - \mu$. The Coulomb energy does not vary with the magnetic field, so the variation of the position of the peak shown in Fig. 6.15 results from variation in E_N.

The single-particle energy levels as a function of magnetic field is shown in Fig. 6.14b, which consists of two Landau levels (LLs). Due to the confinement potential of the dot the LLs are composed of discrete nondegenerate states. States in the first LL fall in energy with increasing B, while those in the second LL rise. The electron alternately occupies a state in the first LL and a state in the second LL as the magnetic field is increased, as shown by the dark line in Fig. 6.14b. Consequently, the position of the N-th peak oscillates.

At a particular B, if the N-th single-particle state is in the first LL (the outer-edge state shown in Fig. 6.14a), it couples well to the leads and transport can occur by resonant tunneling through this state. If the N-th state is in the second LL (the inner-edge state), the coupling to the lead is weak and the peak amplitude is suppressed. A dip in amplitude is thus expected whenever the N-th state is in the second LL, that is, when the position of the peak is moving up in energy. The dip in conductance disappears when $k_B T$ becomes comparable to the single-particle level spacing in the first LL, since transport can then occur by thermal activation to the nearest energy state in the first LL.

The above discussion has not taken into account electron spin. Experiments found that the difference between an even- and odd-numbered finite fermion system, known as the even-odd parity effect, is a distinct feature reflecting the unique behavior of fermions in the presence of both orbital and spin degrees of freedom. The peak spacing fluctuation in CB peaks provides unique information about single-particle energy level, many-body interaction effect, and the parity of electron numbers. To study the parity effect of the CB peaks, Chang et al. fabricated a quantum dot of a small diameter (lithographic diameter 160 nm), as shown in Fig. 6.16a [6], which contains the number of electrons $N \approx 10$, made from a crystal of

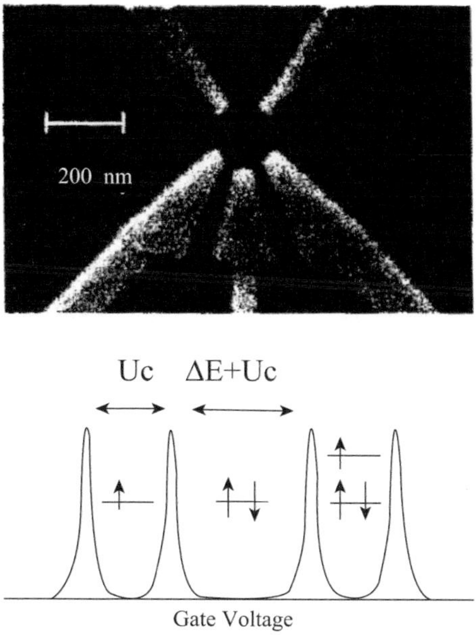

Figure 6.16 Scanning electron micrograph of a small-scale device (top). Schematic diagram of the peak positions as a function of gate voltage (bottom).

density $n = 3.5 \times 10^{11}$ cm^{-2}. According to the interaction model, the spacing between conductance peaks should be

$$\Delta V_{\mathrm{g}} = \frac{C}{eC_{\mathrm{g}}}\left(\Delta E + \frac{e^2}{C}\right), \tag{6.5}$$

where ΔE is the interaction energy of an even number of particles. The theoretical conductance versus the V_{g} curve will be shown as Fig. 6.16 (bottom).

Figure 6.17 shows the fluctuation of peak spacing in device 2 for a strong coupling case (a) and a weak coupling case (b) at $T = 75$ mK [6]. The lithographic diameter of device 2 is 230 nm, which contains about 50 electrons, with the electron density 3.5 $\times$ 10^{11} cm^{-2} ($r_{\mathrm{s}} = 0.93$). For the strong coupling case (Fig. 6.17a) the peak pairing is observable obviously. The spacing of 2, 4, 6, 8, ... is larger than that of 1, 3, 5, 7, ... Moreover, peak height pairing is also visible over several peaks. On the other hand, for

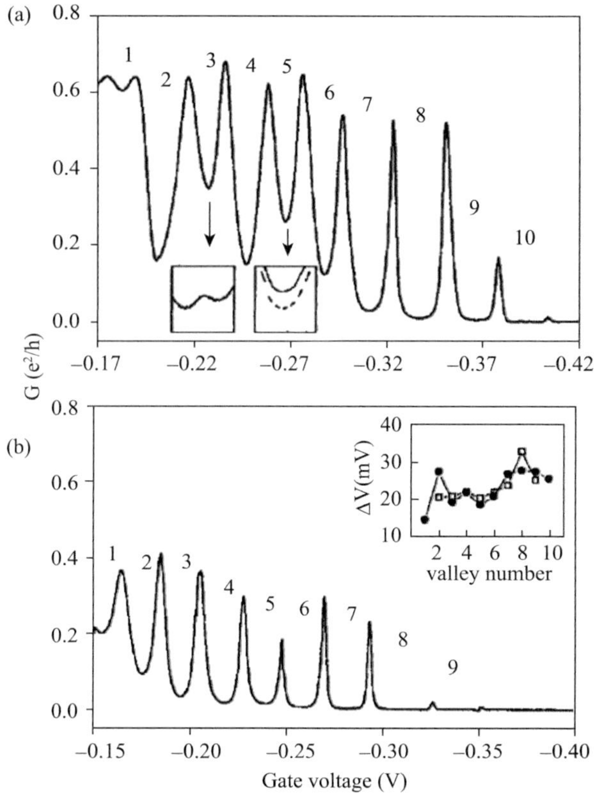

Figure 6.17 Fluctuation of peak spacing in device 2 in the case of strong coupling to the leads (a) and in the case of weak coupling (b) at $T = 75$ mK. Inset in (b): Spacing of conductance peaks as a function of order number of valleys; solid circles for the case of strong coupling and empty squares for weak coupling.

the weak case (Fig. 6.17b) the peak pairing behavior is not as obvious as that for device 2. The difference in behavior is believed to be a consequence of the modulation in the electron–electron interaction due to the difference of coupling strengths in two cases and suggests that strong Coulomb interaction plays an important role in deciding the peak spacing fluctuation. The inset in Fig. 6.17a shows several features of the Kondo effect. The Kondo valley shows the characteristic of unpaired electrons.

6.3 Kondo Effect in Quantum Dot Transport

In the preceding section it was found that the spacing of conductance peaks for even and odd numbers of electrons is non-equal (see Fig. 6.17). This behavior can be explained by Eq. 6.5. Experiments found further that when the temperature descends to very low the conductance valley of the odd number of electrons rises with a decreasing temperature, approaching the limiting value $2e^2/h$. This behavior cannot be explained by the single-electron energy and the Coulomb energy (Eq. 6.5).

In 1964 Kondo explained the anomalous temperature dependence of the resistance of metal. When the temperature is decreased, the resistance of a pure metal decreases, and at a certain temperature it is saturated. The temperature dependence changes considerably when a small concentration of magnetic atoms, such as cobalt, is added to the metal. Rather than getting saturated at a low temperature, the resistance increases as the temperature is lowered further, as shown in Fig. 6.18, where the solid line is for the pure metal and the dotted line is for the doped metal. Because in 1964 Kondo provided a theoretical explanation, since

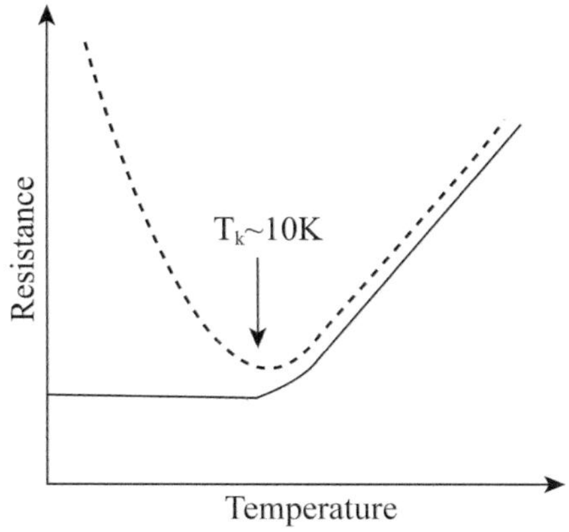

Figure 6.18 Schematic temperature dependence of the resistance of a pure metal (solid line) and metal with a small concentration of magnetic impurity atoms (dashed line). The latter curve shows a minimum around the Kondo temperature.

then the phenomenon has been referred to as the Kondo effect, and the turning temperature (~10 K) in Fig. 6.18 is called the Kondo temperature.

When the temperature is lower than T_K, the mobile electrons in the host metal tend to screen the non-zero total spin of the electrons in the magnetic impurity atom. In the simplest Anderson model of a magnetic impurity, there is only one electron level with energy ε_0 and the impurity spin is 1/2. The exchange processes can effectively flip the impurity spin while simultaneously creating a spin excitation in the Fermi sea. When many such processes are added coherently, a new state—the Kondo resonance—is generated with the same energy as the Fermi level. Such resonance is very effective at scattering electrons with energies close to the Fermi level. The strong scattering contributes greatly to the resistance. The whole system, that is, the magnetic impurity atom plus its surrounding electrons, forms a spin singlet. The energy scale for this singlet state is the Kondo temperature.

A quantum dot connected to source and drain leads can nicely mimic the above situation of a localized spin impurity in a Fermi sea. Therefore, the Kondo effect was expected to occur in quantum dot systems as well. However, there is an important difference between metal and quantum dot systems. In a metal, electrons are described by plane wave functions. Scattering from impurities mixed electron waves with different momentums. The momentum transfer increases the resistance. In a quantum dot, on the contrary, all the electrons have to travel through the device as there is no electrical path around it. In this case, the Kondo resonance makes it easier for states belonging to a bulk metal. Figure 6.19 shows the Anderson model of a magnetic impurity, applied to a single-level quantum dot connected to source and drain leads [7]. The level has an energy ε_0 below the Fermi energy of the leads and is initially occupied by one spin-up electron (leftmost diagram of Fig. 6.19). Adding another electron is prohibited by the charging energy E_C. By virtue of quantum uncertainty, the spin-up electron can momentarily tunnel out of the dot, leaving the dot-lead system in a classically forbidden virtual state (middle diagram of Fig. 6.19). If the spin-up electron is replaced by a spin-down electron from the leads (rightmost diagram of Fig. 6.19), the dot spin has effectively

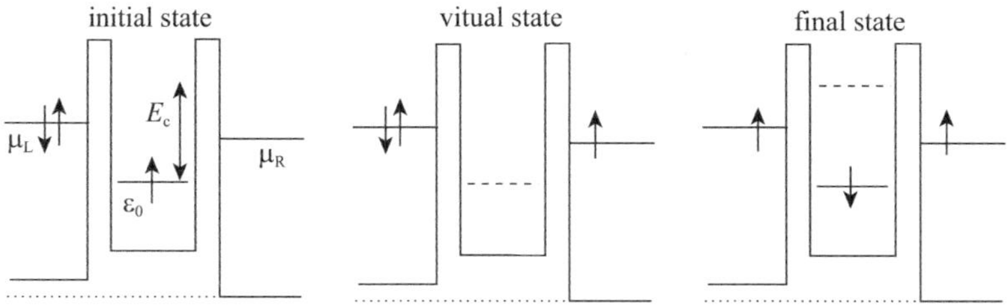

Figure 6.19 The Anderson model of a magnetic impurity, applied to a single-level quantum dot connected to source and drain leads.

been flipped. Many such events combine to produce the Kondo effect in quantum dots, which leads to the formation of an extra resonance at the Fermi level of the leads.

When the energy level ε_0 is occupied by two electrons, the total spin is zero, the virtual state does not exist, and thus there is no spin exchange interaction, that is, no resonant tunneling caused by the Kondo state. Only when the E_C is lowered to the Fermi level of the lead can conductance occur. The conductance as a function of the gate voltage V_g is shown in Fig. 6.20 [7], where the solid curve is for $T \ll T_K$, the dotted curve for $T \approx T_K$, and the dashed curve for $T < T_K$. Although the dot has two tunnel barriers and the charging energy tends to block electrons from tunneling into or out of it, the Kondo effect ensures that electrons are transmitted perfectly through the dot. The Kondo effect only occurs for odd electron numbers, resulting in the odd-even asymmetry between the different Coulomb valleys. Figure 6.20b shows that in the Kondo valleys conductance increases logarithmically with a lowering temperature, getting saturated at $2e^2/h$. Figure 6.20c shows that the Kondo resonance leads to a zero-bias resonance in the differential conductance, dI/dV, versus the bias voltage V.

Haldane [8] derived the Kondo temperature using the parameters of the Anderson model,

$$T_K = \frac{\sqrt{\Gamma E_C}}{2} e^{\pi \varepsilon_0 (\varepsilon_0 + E_C)/\Gamma E_C}, \tag{6.6}$$

where Γ is the width of the dot level, which is broadened by electrons tunneling to and from it. So the parameters that characterize the single-level dot system—E_C, ε_0, and Γ—can be replaced by a single energy scale, T_K. The advantage of quantum dots

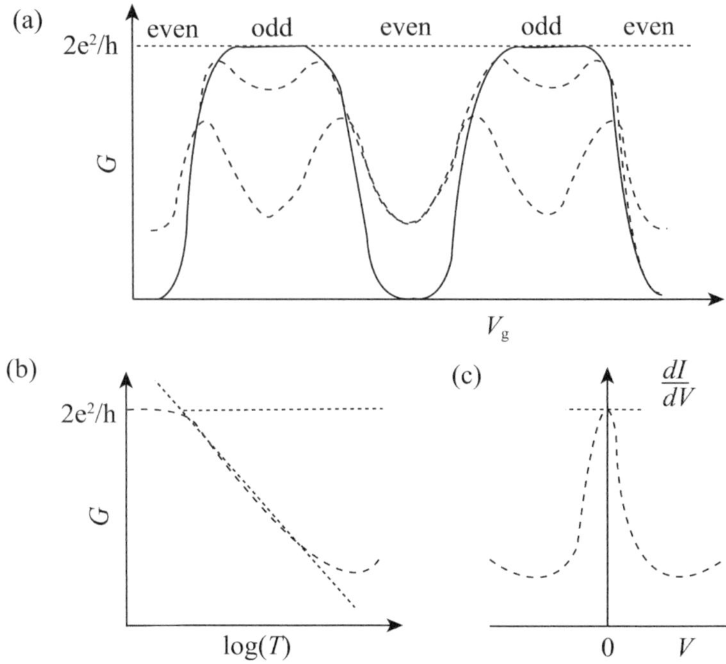

Figure 6.20 (a) Conductance as a function of the gate voltage V_g, where the solid curve is for $T \ll T_K$, the dotted curve for $T \approx T_K$, and the dashed curve for $T < T_K$. (b) In the Kondo valley G as a function of $\log(T)$. (c) Differential conductance dI/dV versus bias voltage V.

in studying the Kondo effect is that the parameters that determine the Kondo temperature can be easily changed by adjusting the voltages on the gates.

As shown in Fig. 6.20b in the Kondo valleys conductance increases logarithmically with a lowering temperature, getting saturated at $2e^2/h$, which is called "unitary limit" of conductance. Van der Wiel et al. [9] first observed experimentally the Kondo effect of the quantum dot in the unitary limit, that is, the CB for electron tunneling is overcome completely by the Kondo effect.

In Fig. 6.21a [9] an AB ring is defined in a 2DEG; in both arms of the ring a quantum dot can be defined by applying negative voltages to the gate electrodes. A quantum dot of size $\sim$200 $\times$ 200 nm, containing $\sim$100 electrons, is formed in the lower arm using gate

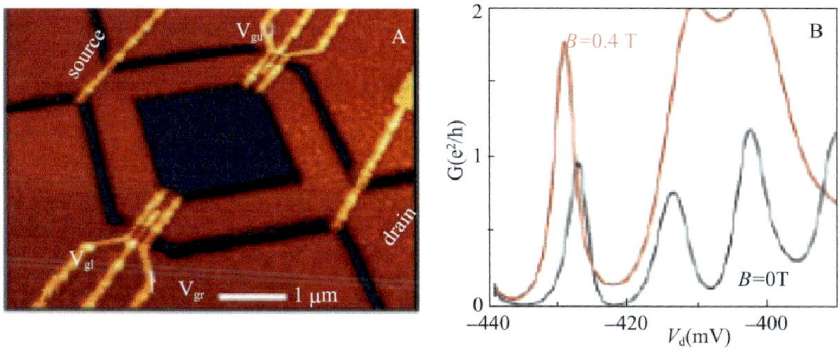

Figure 6.21 (a) Device structure diagram. (b) G as a function of V_{gl} for $B = 0$ T and 0.4 T at $T = 15$ mK.

voltages V_{gl} and V_{gr}. The gate voltage of the upper arm V_{gu} is kept at zero. The energy-level spacing of a single electron in the dot is about 100 μeV. Figure 6.21b shows G as a function of the left gate voltage. The right gate voltage is fixed at $V_{gr} = -488$ meV. At $B = 0$ regular Coulomb oscillations are observed with low valley conductance, while in the magnetic field the valley conductance increases considerably and can even reach $2e^2/h$. The pronounced peak reflects the Kondo resonance at the Fermi energy at the unitary limit.

More detailed results are shown in Fig. 6.22 [9]. Figure 6.22a shows Coulomb oscillations for different temperatures. At the base temperature, the valley $V_{gl} = -413$ and -372 mV reach the maximum possible conductance value of $2e^2/h$. When the temperature is increased, two separate Coulomb peaks develop with growing peak spacing. The conductance in the center of the valley has a logarithmic T dependence with a saturation at $2e^2/h$ for a low T. Figure 6.22b shows the differential conductance for different T values in the middle of the Kondo plateau at $2e^2/h$. The pronounced peak around $V_{SD} = 0$ reflects the Kondo resonance at the Fermi energy. The peak height has the same T dependence as shown in Fig. 6.22a, right inset. The width of the peak increases linearly with temperature (Fig. 6.22b, inset).

Last we return to the effect of a magnetic field. Experiments found near $\mathbf{B} = 0$, regular Coulomb oscillations are observed, but the Kondo effect typically changes the valley conductance by only

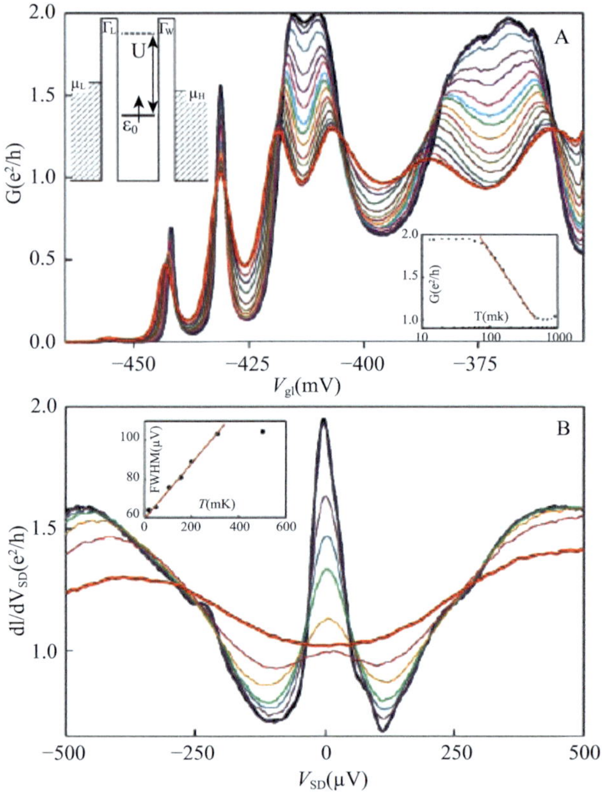

Figure 6.22 (a) Coulomb oscillations in G versus V_{gl} at $B = 0.4$ T for different temperatures. T ranges from 15 mK (thick black trace) to 800 mK (thick red trace). V_{gr} is fixed at -448 mV. The right inset shows the logarithmic T dependence between ~90 and ~500 mK for $V_{gl} = -413$ mV. (b) Differential conductance dI/dV_{SD} versus dc bias voltage between source and drain contacts V_{SD} for T ranging from 15 mK up to 900 mK. The inset shows that the width of the zero-bias peak measured from the FWHM increases linearly with T.

~20%. A big change occurs at $B \approx 0.1$ T, reflecting the onset of a different transport regime, an observation that seems common for half-open quantum dots. The magnetic scale corresponds to adding a flux quantum to the area of the dot, implying that time-reversal symmetry is broken.

6.4 Single-Electron Transport in Vertical Quantum Dots

Apart from the quantum dot defined on the 2DEG by the gate voltage, there is another kind of quantum dot—the three-dimensional (3D) quantum dot fabricated by the lithography technology, which has a well-defined confinement potential. Such a quantum dot can be regarded as an artificial atom. Associated with the rotational symmetry, as well as the parabolicity in the in-plane 2D harmonic confinement potential, atom-like properties such as shell-filling and abidance of Hund's first rule are all observed. Furthermore, this allows us to discuss the way of spin-filling for a given characteristic of the quantum state. When the single-particle states in the dot are separated by a large energy ΔE, an antiparallel spin-filling, forming a singlet state, is favored. For a small ΔE, parallel filling, forming a triplet state, is observed, which is in line with Hund's first rule. The magnetic field B can be used to tune $\Delta E(B)$; this allow us to alter the way of spin-filling.

6.4.1 Quantum Dot and Single-Electron Energy Levels

The schematic diagram of the device is shown in Fig. 6.23 [10]. The vertical quantum dot is fabricated by a double-barrier structure (DBS), which consists of an undoped 12 nm $In_{0.05}Ga_{0.95}As$ well and two undoped $Al_{0.22}Ga_{0.78}As$ barriers of thicknesses 9.0 and 7.5 nm, respectively. This is processed to form a circular mesa with a nominal top-contact diameter D of 0.54 μm. Figure 6.23b shows typical scanning electron micrographs of a circular mesa and a rectangular mesa. The dot is strongly confined in the vertical direction by the heterostructures, whereas it is softly confined in the lateral direction by the Schottky-gate-induced depletion region. This depletion region is well approximately by a harmonic potential, and the characteristic energy $\hbar\omega_0$ is evaluated from measurements of the magnetic field dependence of the Coulomb oscillations. The values of $\hbar\omega_0$ obtained for various numbers of electrons in the dot are shown in Fig. 6.23c. The effective lateral confinement becomes weak as the number of electrons in the dot increases, due to the effect of Coulomb screening.

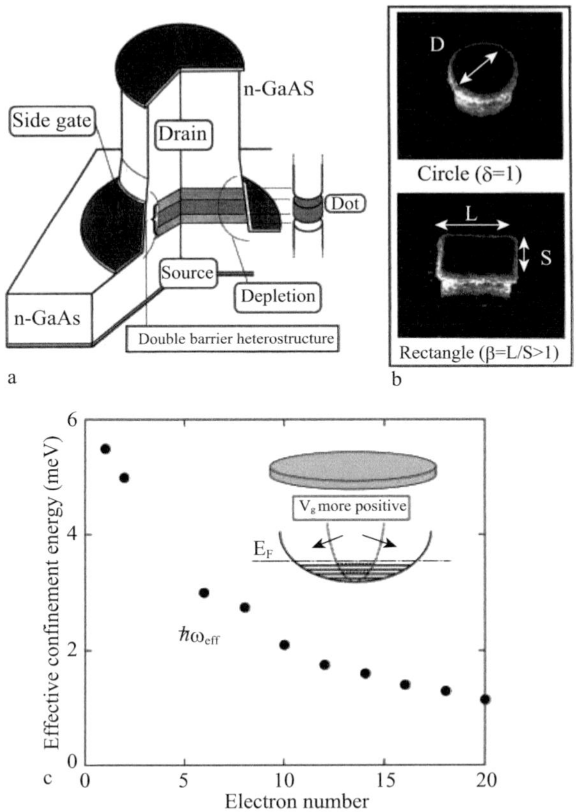

Figure 6.23 (a) Schematic diagram of the device. (b) Scanning electron micrographs of a circular mesa and a rectangular mesa. (c) Values of $\hbar\omega_0$ as a function of number of electrons in the dot.

The eigenstates for a 2D harmonic quantum dot in the perpendicular magnetic field are the Fock–Darwin states, and their energies are expressed by

$$E_{n,l} = -\frac{l}{2}\hbar\omega_0 + \left(n + \frac{1}{2} + \frac{1}{2}|l|\right)\hbar\sqrt{4\omega_0^2 + \omega_c^2}, \qquad (6.7)$$

where n and l are the radial quantum number and angular momentum quantum number, respectively, and $\hbar\omega_c = eB/m^*$ is the cyclotron energy.

6.4.2 Shell Filling and Hund's First Rule

The Coulomb oscillations in the current versus gate voltage at $B = 0$ T for a $D = 0.5$ μm dot is shown in Fig. 6.24a, and Fig. 6.24b shows the addition energy versus electron number for two different dots with $D = 0.5$ and 0.44 μm [11]. This series of current peaks is caused by electrons filling in the dot from $N = 1$, but the spacing between peaks is not equal, differing from that on the 2DEG.

From Eq. 6.7, when $\omega_c = 0$,

$$E_{n,l} = (2n + 1 + |l|)\,\hbar\omega_0. \tag{6.8}$$

Thus, according to the energy the electron energy level forms various shells: $(n, l) = (0, 0)$, $\hbar\omega_0$; $(0, \pm 1)$, $2\hbar\omega_0$; $(1, 0)$, $(0, \pm 2)$, $3\hbar\omega_0 \ldots$. Each orbital state fills two electrons, then the first three shells fill 2, 4, and 6 electrons. The current peak spacing in Fig. 6.24a is caused by Coulomb energy and also by the energy-level spacing $\hbar\omega_0$. Thus the spacing of current peaks after 2, 6, and 12 electrons is larger than others. This is also reflected in Fig. 6.24b, which shows the addition energy as a function of electron number for two dots.

The electronic states are expected to be significantly modified by a magnetic field B perpendicular to the dot. The B dependence on the position of the current oscillations is shown in Fig. 6.25 for B increasing from 0 to 3.5 T in steps of 0.05 T [11]. The positions of first three peaks depend monotonously on B, whereas the other peaks oscillate back and forth a number of times. The number of wiggles increases with N. A close inspection of the figure reveals that the current peaks generally shift in pairs with B. This even-odd effect persists up to $N = 40$.

For the simplest explanation of the magic number and the B dependence we ignore, for the moment, the Coulomb interaction. The magnetic energy spectrum is given in Eq. 6.7. At $B = 0$, E_{nl} has degenerate sets of states, which are separated by $\hbar\omega_0$ from each other and are completely filled by $N = 2$, 6, 12, 20, etc., as shown in Fig. 6.24. When B increases the original degenerate states split into a series of singlet states. The rearrangement of the pairing can be understood in terms of Hund's rule. Hund's rule favors the filling of electrons with spin-up and spin-down in a singlet state, decreasing the interaction energy. The detailed discussion can be seen in Refs. [10] and [11].

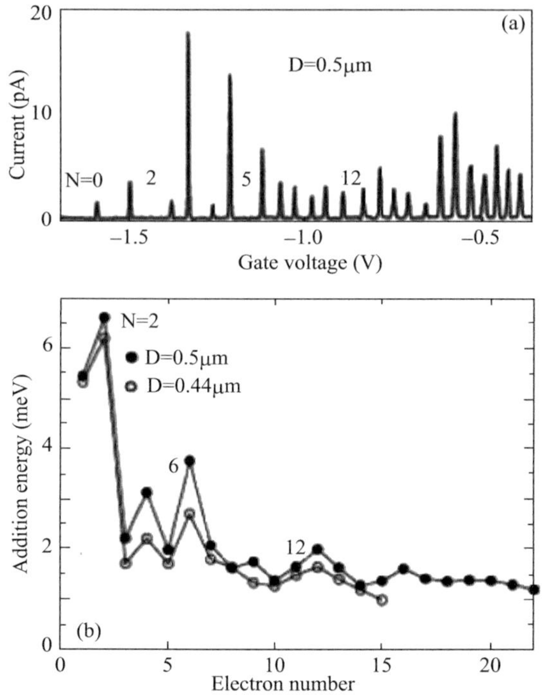

Figure 6.24 (a) Coulomb oscillations in the current versus gate voltage at $\mathbf{B} = 0$ T for a $D = 0.5$ μm dot. (b) Addition energy versus electron number for two different dots with $D = 0.5$ and 0.44 μm.

6.4.3 Single-Electron Tunneling Spectrum in the Magnetic Field

Below we introduce the elastic tunneling between two vertical coupled quantum dots [12]. The vertical double quantum dot device is like that shown in Fig. 6.23, but is made from a triple barrier heterostructure (TBS) and is located in a cylinder mesa whose diameter is 0.6 μm.

By appropriately tuning both V_G and the source-drain voltage V_S one can obtain configurations as shown in Fig. 6.26 [12]. The lowest $(n,l) = (0,0)$ orbital state of the left dot is aligned with the Fermi level of the left reservoir (source), and the lowest orbital state of the right dot is always located below the lowest orbital state of the left dot, and above the Fermi level of the right reservoir (drain). In

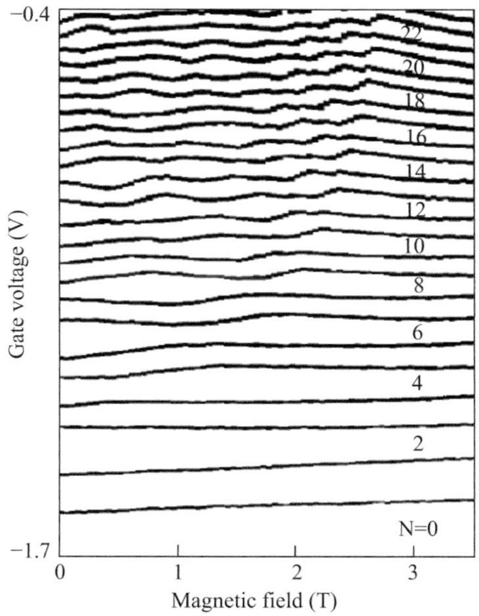

Figure 6.25 Gate voltage positions of the current oscillations versus magnetic field for a dot with $D = 0.5$ μm.

such a configuration, due to the CB in each dot, as well as inter-dot electrostatic interactions, the left dot acts as a "turnstile" so that the electron number in the right dot is always zero or one, no matter how large the source-drain voltages are. When the source-drain voltage is turned in such a way that a higher orbital (n,l) state in the right dot is aligned with the left (0,0) state in the left dot, elastic tunneling gives rise to a peak of current. This "inter-dot tunneling spectroscopy" enables us to demonstrate the single-particle nature of electronic states in the right dot.

The positions of the current peaks changing with magnetic field B are plotted in Fig. 6.27a [12]. The positions agree well with the dependence of the calculated Fock–Darwin states $E_{n,l}$ after subtracting $E_{0,0}$, that is, the B dependence of the orbital states of the right dot. Thus we can label (n,l) for each current peak by comparison with the Fock–Darwin diagram. The peak heights measured 30 mV above in V_G from the threshold line are plotted in Fig. 6.27c. It is clear that the current peak is significantly smaller

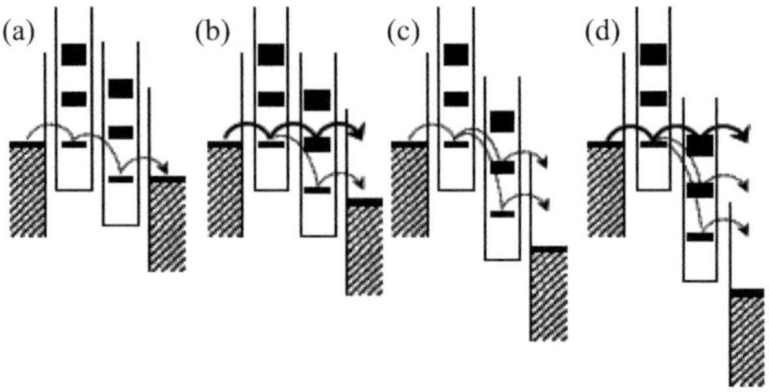

Figure 6.26 Potential landscape for the double quantum dot structure at various source-drain voltages. The horizontal lines in each dot represents single-particle energy levels at zero magnetic field. Solid (dotted) arrows indicate elastic (inelastic) three-step sequential tunneling processes via two 0D states.

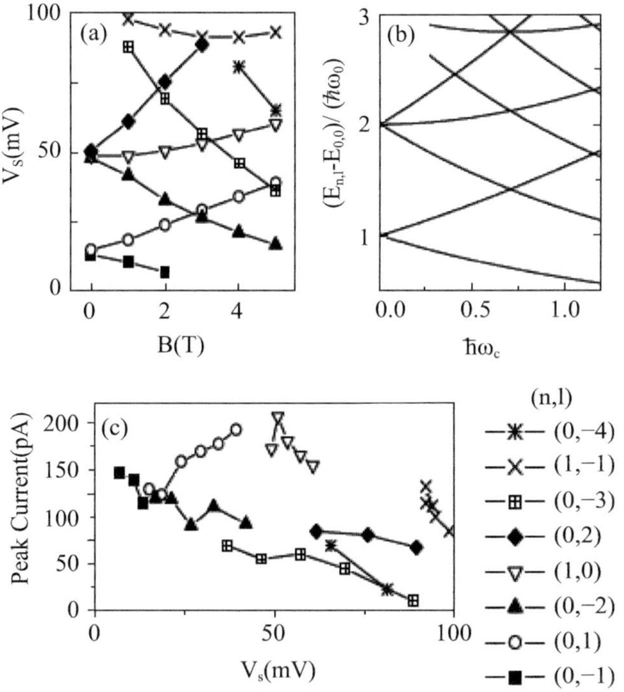

Figure 6.27 (a) Experimental data showing the positions in V_S on the varies current peaks. (b) Theoretically calculated Fock–Darwin states. (c) Heights of the current peaks shown in (a).

when the peak is associated with a higher angular momentum state. This demonstrates the existence of angular momentum selectivity for the electron inter-dot tunneling between the (0,0) state of the left dot and the higher orbital state of the right dot. The tunneling probability between the (0,0) state and the higher orbital states is small but not zero. For explanation of this, the authors proposed an incoherent sequential tunneling model; the tunneling probability is proportional to the product of the "local density of state,"

$$T \propto \int dxdy|\psi_{0,0}(x,y)|^2|\psi_{n,l}(x,y)|^2. \tag{6.9}$$

The transmission becomes small (but are non-zero) when the angular momentum of $\psi_{n,l}$ is large.

6.4.4 Spin Blockade Effect

Except the CB effect, one also observed the spin blockade effect caused by the Pauli exclusion principle, which prevents two electrons of parallel spin from occupying a single spatial orbital. Ono et al. [13] found that the Pauli effect, in combination with the CB, can be used to block current altogether in one direction while permitting it to flow in the opposite direction, thereby realizing a fully controllable spin-Coulomb rectifier.

The schematic of the double-dot device is shown in Fig. 6.28b, and the electron transport in the forward bias (upper) and backward bias (down) is shown in Fig. 6.28a [13]. Suppose there are two sites, site 1 and site 2, weakly coupled together and one electron is permanently localized on site 2. Now consider the transport of a second electron through the system between two contact leads. The number of electrons on site 1 (or site 2), $N1$ ($N2$), varies between 0 and 1 (1 and 2). Crucially, the spin effect also markedly influences electron transport. Because the tunnel coupling between the two sites is sufficiently weak, the $(N1,N2) = (1,1)$ spin singlet and spin triplet are practically degenerate. For $(N1,N2) = (0,2)$, only a spin singlet is permitted because of Pauli exclusion. Therefore, electron transport is only allowed for a channel made from the (1,1) and (0,2) singlet states. For a forward bias, either the (1,1) singlet or triplet can be populated with the same probability by injection of an electron onto site 1 from the left lead. If the (1,1) singlet is populated,

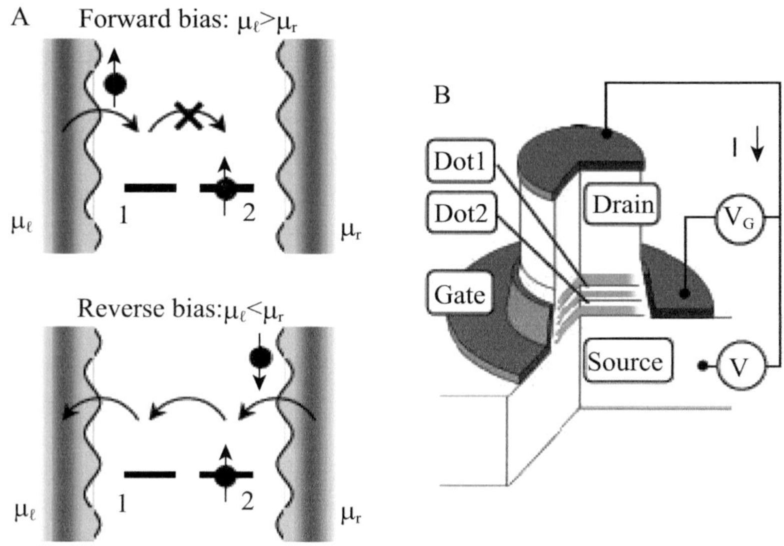

Figure 6.28 (b) Schematic of the double-dot device. (a) Electron transport in the forward bias (upper) and backward bias (down).

a single-electron tunneling current can flow through the singlet state. Once the triplet is populated, however, subsequent electron transfer from site 1 to 2 is blockaded by Pauli exclusion. Note that an electron arriving on site 1 usually cannot go back to the left lead because of the fast relaxation of the hole state left behind the lead. Thus the (1,1) triplet will sooner or later be occupied on a time scale sufficiently longer than the electron tunneling time between the leads and this should lead to clear current suppression. Because this blockade is due to spin and not charge, we hereafter refer to this process as a "spin blockade." On the other hand, for a reverse bias an antiparallel spin electron can always be injected onto site 2 from the right lead.

The differential conductance dI/dV and current I, flowing vertically through the two dots as a function of V_G and V, is shown in Fig. 6.29 [13]. V_G and V are used to change the electrostatic potential of the two dots together (number of electrons) and the potential offset between the two dots, respectively. In the lower-right inset

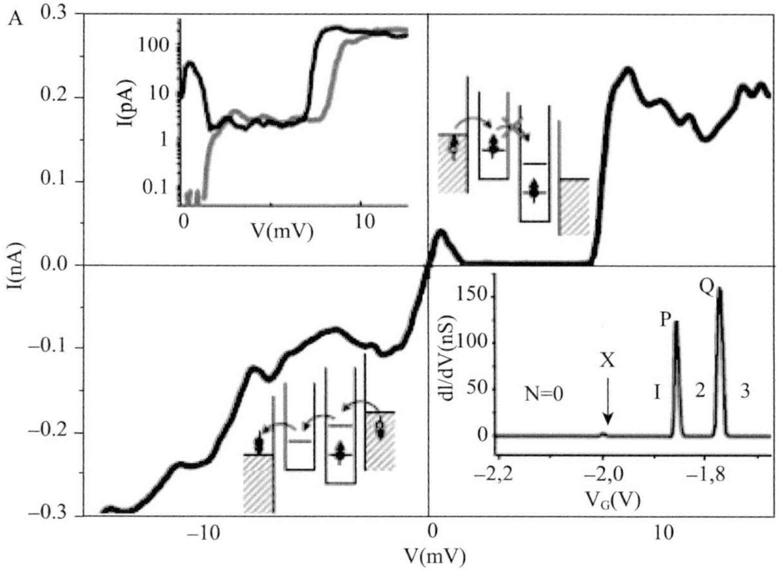

Figure 6.29 The differential conductance dI/dV and current I flowing vertically through the two dots as a function of V_G and V. Lower-right inset: $dI/dV \approx V_G$ measured for a small V ($\sim$0 V). Upper-left inset: magnified plots of I versus V for V_G fixed at peak P (black curve) and in the middle between the peaks P and Q (red curve) on a logarithmic scale.

$dI/dV \approx V_G$ measured for a small V ($\sim$0 V) shows clear Coulomb oscillation peaks. The second peak (P) and the third peak (Q) are much large. This implies that tunneling is elastic between the source and drain leads for $N = 1 \to 2$ and $2 \to 3$. For $|V| \geq 1$ mV, transport is nonlinear (see potential diagrams in the main diagram). The current is clearly suppressed in the forward bias because of a spin blockade, whereas a large current still flows in the reverse bias because of inelastic tunneling via the singlet states. The upper-left inset shows magnified plots of I versus V for V_G fixed at peak P (black curve) and in the middle between the peaks P and Q (red curve) on a logarithmic scale. Both curves show strong current suppression ($I \approx 2$ pA) due to the spin blockade. In the vicinity of $V = 0$ V, the red curve shows strong current suppression due to a CB.

6.4.5 Single-Electron Tunneling in Coupled Quantum Dots

Using the coupled quantum dots structure (e.g., Fig. 6.28b) the electron filling and the CB effects can also be studied by changing the gate voltage V_G. Figure 6.30a shows calculated addition energy spectra, $\Delta_2(N) = U(N+1) - 2U(N) + U(N-1)$ for different spacings between two dots b [14]. $U(N)$ is the total energy of the N-electron system, and $\Delta_2(N)$ can reveal a wealth of information about the energy required to place an extra electron into a quantum dot system. When b is small the coupled quantum dot is rather similar to a single quantum dot. At intermediate dot separation, the spectra pattern becomes more complex. However, at a larger inter-dot distance, a simple picture emerges that the coupled quantum dot is about to dissociate. For example, at $b = 7.2$ nm a strong peak at $N = 2$, 4, and 12 and a weaker peak at $N = 8$ appear that can be easily interpreted from the peaks appearing in the single quantum dot spectrum. The peaks at $N = 4$ and 12 are a consequence of symmetric dissociation into two closed shells $N = 2$ and 6 of quantum dots, respectively, whereas the peak at $N = 8$ corresponds to the dissociation into two identical stable quantum dots holding four electrons each, filled according to Hund's rule to give maximal spin. The peak at $N = 2$ is related to the localization of one electron on each constituent dot, the two-electron state being a spin-singlet configuration.

Due to the difficulty in fabricating two perfectly identical constituent quantum dots in the coupled quantum dot, the experimental spectra (Fig. 6.30b) are different from those of theoretical spectra. (i) The spectrum for the most strongly coupled quantum dot ($b = 2.5$ nm) resembles that of the quantum dot up to the third shell ($N = 12$). (ii) For intermediate coupling ($b = 3.2$ to 4.7 nm), the spectra are quite different from the quantum dot spectrum, and a fairly noticeable peak appears at $N = 8$. (iii) For weaker coupling ($b = 6.0$ and 7.5 nm) the spectra are different again, with prominent peaks at $N = 1$ and 3.

In calculation the authors took the well width $w = 12$ nm, the well depths $V_0 \pm d$, $V_0 = 225$ meV, $d = 0$ or a realistic

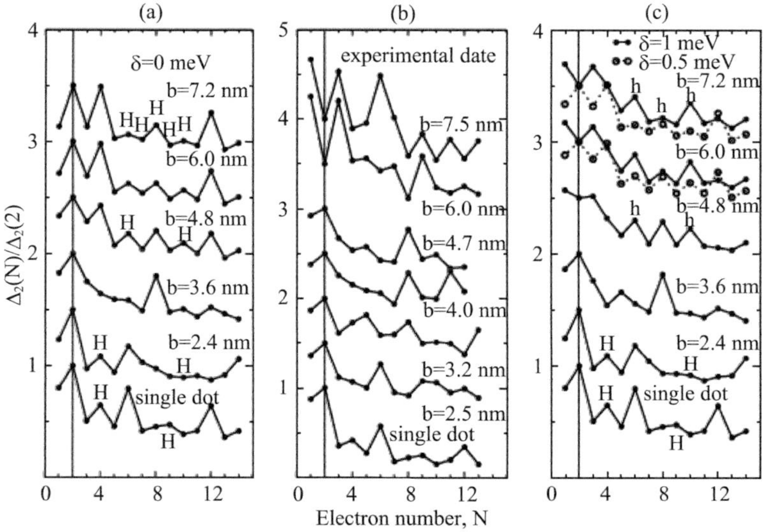

Figure 6.30 (a) Calculated $\Delta_2(N)/\Delta_2(2)$ for coupled QDs with different inter-dot distances b. Also shown is the calculated reference spectrum for a single QD. (b) Experimental $\Delta_2(N)/\Delta_2(2)$ for several inter-dot distances between 2.5 and 7.5 nm. (c) Same as panel (a), but for coupled QDs obtained using $\delta = 1$ or 0.5 meV.

value 0.5 or 1 meV. Taking into account this mismatch of well depth the authors confirmed the spectra character of the coupled quantum dot by performing the theoretical calculations with $\delta =$ 1 or 0.5 meV. The results are shown in Fig. 6.30c. The overall agreement between theory and experiment of the general spectral shape is quite good, indicating the crucial role played by a mismatch. In particular, the appearance of the spectra in the weak coupling limit for small N values is now correctly given, as well as the evolution with b of the peak appearing at $N = 8$ for intermediate coupling. A comparison between Figs. 6.30b and 6.30c reveals that for small values of b ($\leq$4.8 nm), for a reasonable choice of parameters (w, δ), a mismatch does not produce sizable effects.

References

1. Scott-Thomas, J. H. F., Field, S. B., Kastner, M. A., Smith, H. I., and Antoniadis, D. A. (1989). Conductance oscillations periodic in the density of a one-dimensional electron gas. *Phys. Rev. Lett.*, **62**, 583–586.
2. Meirav, U., Kastner, M. A., and Wind, S. J. (1990). Single-electron Charging and periodic conductance resonances in GaAs nanostructures. *Phys. Rev. Lett.*, **65**, 771–774.
3. Kouwenhoven, L. P., van der Varrt, N. C., Johnson, A. T., et al. (1991). Single electron charging effects in semiconductor quantum dots. *Z. Phys. B*, **85**, 367–373.
4. Kouwenhoven, L. P., Johnson, A. T., van der Vaart, N. C., et al. (1991). Quantized current in a quantum dot turnstile. *Z. Phys. B*, **85**, 381–388.
5. McEuen, P. L., Foxman, E. B., Meirav, U., Kastner, M. A., Meir, Y., and Wingreen, N. S. (1991). Transport spectroscopy of a Coulomb island in the quantum Hall regime. *Phys. Rev. Lett.*, **66**, 1926–1929.
6. Chang, A. M., Jeong, H., and Melloch, M. R. (2003). *Electron Transport in Quantum Dots*, ed. Bird, J. P., "Novel phenomena in small individual and coupled quantum dots" (Kluwer Academic Publishers, Boston/Dordrecht/London), pp. 123–157.
7. Tarucha, S., Ono, K., Fujisawa, T., van der Wiel, W. G., and Kouwenhoven, L. P. (2003). *Electron Transport in Quantum Dots*, ed. Bird, J. P., "Interactions, spins and the Kondo effect in quantum-dot systems" (Kluwer Academic Publishers, Boston/Dordrecht/London), pp. 1–42.
8. Haldane, F. D. M. (1978). Scaling theory of the asymmetric Anderson model. *Phys. Rev. Lett.*, **40**, 416–419.
9. van der Wiel, W. G., De Franceschi, S., Fujisawa, T., et al. (2000). The Kondo effect in the unitary limit. *Science*, **289**, 2105–2108.
10. Tarucha, S., Austing, D. G., Sasaki, S., Tokura, Y., van der Wiel, W., and Kouwenhoven, L. P. (2000). Effect of Coulomb interactions on spin states in vertical semiconductor quantum dots. *Appl. Phys. A*, **71**, 367–378.
11. Tarucha, S., Austing, D. G., Honda, T., van der Hage, R. J., and Kouwenhoven, L. P. (1996). Shell filling and spin effect in a few electron quantum dot. *Phys. Rev. Lett.*, **77**, 3613–3616.
12. Ono, K., Austing, D. G., Tokura, Y., and Tarucha, S. (2002). Angular momentum selectivity in tunneling between two quantum dots. *Physica B*, **314**, 450–454.

13. Ono, K., Austing, D. G., Tokura, Y., and Tarucha, S. (2002). Current rectification by Pauli exclusion in a weakly coupled double quantum dot system. *Science*, **297**, 1313–1317.

14. Pi, M., Emperador, A., Barranco, M., Garcias, F., et al. (2001). Dissociation of vertical semiconductor diatomic artificial molecules. *Phys. Rev. Lett.*, **87**, 066801(1–4).

Chapter 7

Silicon Single-Electron Transistor

The operation principle of the single-electron transistor (SET) is to control one electron through a very small quantum dot. The performance of the SET is enhanced with the decreasing scale of the quantum dot. Therefore, the quantum dot is generally smaller than 10 nm, very suitable to be high-density integrated. Meanwhile, the property of a single electron permits it to operate at ultra-low power because the power dissipated in the circuit is proportional to the number of participating electrons. SETs have a peculiar I–V characteristic, which is absent for a general MOSFET. All these properties make it possible for SETs to become high-performance circuits in a very small area.

At the early stage of single-electron transistor research the used materials are metals or compound semiconductors, mainly studying quantum dot physics. But considering the application of the SET, the most suitable material is silicon. Because Si SETs are compatible with general MOSFETs, it is easier to integrate a Si SET into the present very large scale integration (VLSI). Another convenience is that one can use the present mature microprocessing technology to fabricate a SET. Now the Si SET has been considered as one of the most suitable candidates of the unit device in future ultra-high density and ultra-low power VLSI.

Quantum Waveguide in Microcircuits (Second Edition)
Jian-Bai Xia, Duan-Yang Liu, and Wei-Dong Sheng

ISBN 978-981-4968-30-0 (Hardcover), 978-1-003-65146-8 (eBook)
www.jennystanford.com

7.1 Principle of a Single-Electron Transistor

Figure 7.1a is a schematic diagram of the equivalent circuit of a single-electron quantum dot [1]. The two leads of the dot are connected with the source and gate, and C_t and C_g are the tunneling capacitance and gate capacitance, respectively.

When an electron tunnels into the dot, the electrostatic energy (Coulomb energy) changes,

$$\Delta U = \frac{e^2}{2C_{dot}}, \tag{7.1}$$

where C_{dot} is the total capacity of the dot. If the dot is small enough, ΔU can be comparable with the thermal energy, even larger than it, thus only when the gate provides the extra energy can the electron tunnel into the dot. This is the Coulomb blockade (CB) effect, which is the foundation of the single-electron transistor (SET). A more detailed description of a SET is that the electrostatic energy of the system

$$U = \frac{Q_{dot}^2}{2C_{dot}} + \left(\frac{C_g}{C_{dot}}\right) Q_{dot} V_g + \text{const.}, \tag{7.2}$$

where $Q_{dot} = -N_{dot}e$ is the charge in the dot and C_g is the gate-dot capacitance. U can be rewritten as

$$U = \frac{\left(N_{dot}e - C_g V_g\right)^2}{2C_{dot}}. \tag{7.3}$$

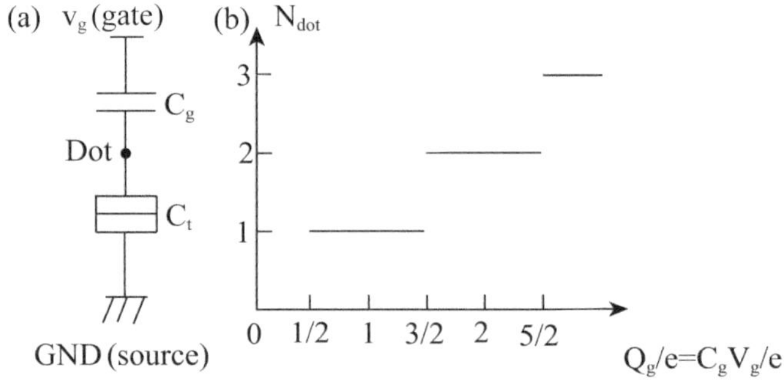

Figure 7.1 (a) Schematic diagram of the equivalent circuit of a single-electron quantum dot. (b) N_{dot} as a function of $Q_s = C_g V_g$ at $T = 0$ K.

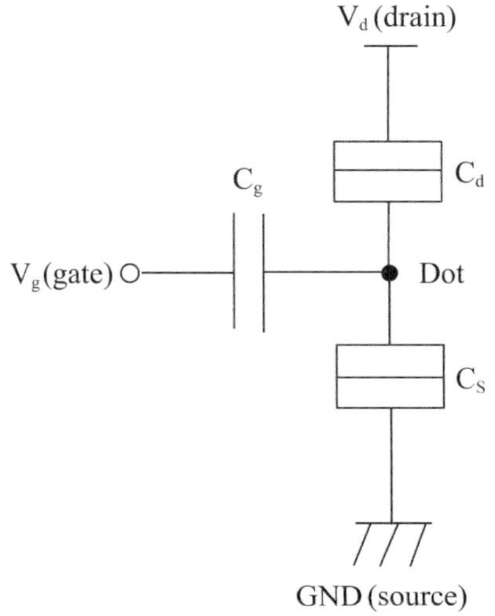

Figure 7.2 Equivalent circuit of the SET.

At $T = 0$ K, taking U as the minimum, one obtains N_{dot} as a function of V_g, as shown in Fig. 7.1b. When V_g increases by each e/C_g, N_{dot} increases one step. At a finite temperature, this step character becomes indistinct due to thermal fluctuation.

Figure 7.2 is the equivalent circuit diagram of a SET [1], where the dot is connected with the source and drain by the tunneling barriers C_s and C_d, respectively. Meantime, it is connected with the gate by the capacitively coupling, so it becomes a three-terminal device similar to the metal-oxide semiconductor field-effect transistor (MOSFET). Different from the single-electron quantum dot shown in Fig. 7.1a, in a SET there is current flowing from the source to the drain under the control of the gate.

The total energy of the N electrons in the dot is

$$E(N) = U(N) + \sum_{k=1}^{N} \varepsilon_k = \frac{\left(Ne - C_g V_g - C_d V_{ds}\right)^2}{2C_{\text{dot}}} + \sum_{k=1}^{N} \varepsilon_k, \quad (7.4)$$

where C_d is the source-dot capacitance, V_{ds} is the source-drain voltage, and ε_k is the k-th quantum level of the quantum dot. The

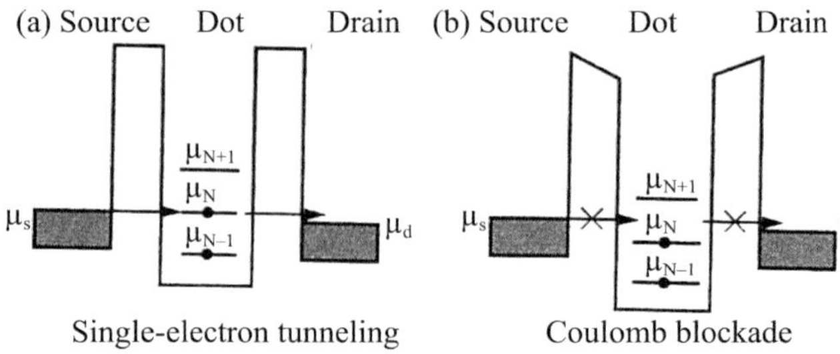

Figure 7.3 Conduction band potential energy diagram of the SET. (a) Condition for single-electron tunneling. (b) Condition for a Coulomb blockade.

chemical potential of the N-th electron in the quantum dot is

$$\mu_N = E(N) - E(N-1) = \frac{(2N-1)e^2 - 2eC_gV_g - 2eC_dV_{ds}}{2C_{dot}} + \varepsilon_N. \qquad (7.5)$$

Figure 7.3 is the conduction band potential energy diagram of the SET [1]. When the chemical potential of the electron in the quantum dot satisfies

$$\mu_{N+1} > \mu_s \geq \mu_N \geq \mu_d > \mu_{N-1}, \qquad (7.6)$$

the electron can enter the quantum dot from the source and then move out to the drain, as shown in Fig. 7.3a, where μ_s and μ_d are the chemical potential of the source and drain, respectively. Because only one electron can enter into and move out of the quantum dot, the transport is called single-electron tunneling. When V_g increases, so μ_N is lower than μ_d,

$$\mu_{N+1} > \mu_s \geq \mu_d > \mu_N, \qquad (7.7)$$

the electron cannot tunnel into the dot and this case is called a CB. Thus the drain current oscillates with the gate voltage, as shown in Fig. 7.4 [4], where each peak represents the flowing of one electron.

Define the energy needed to increase an electron in the quantum dot as the single electron additional energy E_a,

$$E_a = \mu_N - \mu_{N-1} = \frac{e^2}{C_{dot}} + (\varepsilon_N - \varepsilon_{N-1}) = E_C + \Delta E, \qquad (7.8)$$

where E_C is the Coulomb charge energy of a single electron and ΔE is the spacing of the quantum levels. When the potential energy

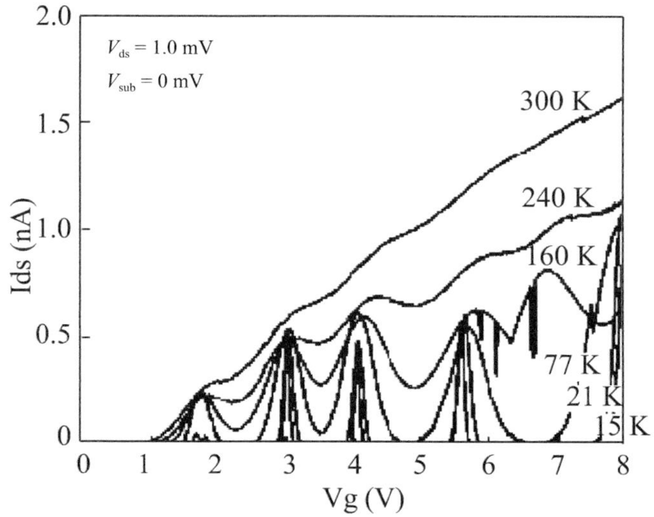

Figure 7.4 $I_d - V_g$ curves for $V_{ds} = 1.0$ mV; the temperature decreases from room temperature to 15 K.

of the quantum dot increases to E_a, it satisfies the single-electron-tunneling condition in Eq. (7.6). In the I–V curve (Fig. 7.4) the oscillation period is

$$\Delta V_g = \frac{E_a}{e\alpha} = \frac{C_{dot}}{eC_g}E_a = \frac{e}{C_g} + \frac{C_{dot}}{eC_g}\Delta E, \tag{7.9}$$

where $\alpha = C_g/C_{dot}$ is a gain modulation factor, which changes the dot potential into V_g. When the dot is large enough, ΔE can be neglected compared with E_C and ΔV_g is approximately a constant; this case is called the Coulomb oscillation.

Figure 7.4 shows the $I_d - V_g$ curves for $V_{ds} = 1.0$ mV [4]. When V_g is fixed at the peak value of the Coulomb oscillation, I_d increases monotonously with an increasing V_{ds}, as shown in Fig. 7.5. When V_g is fixed at the CB regime, I_d keeps a low value as V_{ds} increases until V_{ds} is larger than a definite value so that μ_d is lower than μ_N; then I_d increases continuously with V_{ds}, as shown in Fig. 7.5.

Figure 7.6 shows the contour plot of I_d in a SET as a function of V_g and V_{ds} [1], which gives the whole picture of the CB characteristic. At a low V_{ds} along the V_g axis, there are a series of rhombus current blockade regimes (deep gray), called the CB regime or the Coulomb

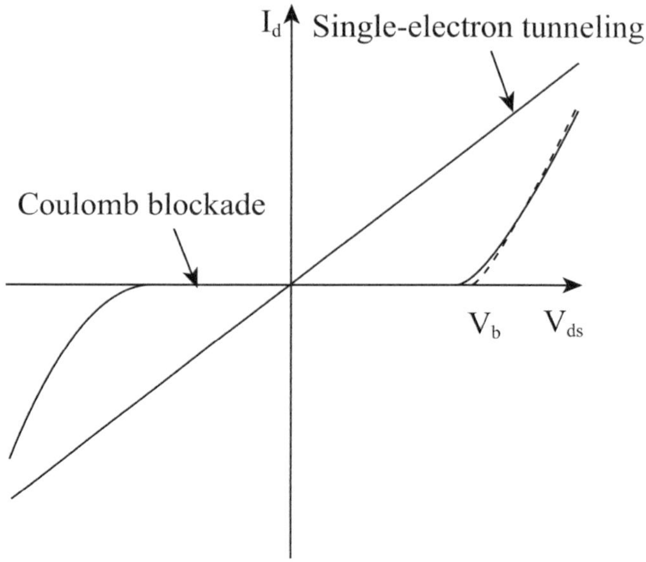

Figure 7.5 $I_d - V_{ds}$ characteristic curve of a SET under the single-electron-tunneling condition and the Coulomb blockade condition.

diamond. In each CB regime the number of electrons keeps an integer number. Between two neighboring CB regimes there exist light-gray regimes called single-electron-tunneling regimes. In each single-electron-tunneling regime the number of electrons jumps between $N - 1$ and N and the electron flows from the source to the drain, producing a tunneling current.

The slope of the boundary between the CB regime and the single-electron-tunneling regime reflects the difference of the influence of V_g and V_{ds} to the quantum dot potential; thus they are determined by the ratio of capacitances of the SET. The slope of the descending boundary of the diamond is C_g/C_d, and the slope of the raising boundary is $C_g/(C_g + C_s)$. The slope of the descending boundary corresponds to the voltage gain of the SET, which has an important influence on the performance of the SET circuit.

The operation of the SET has to meet two conditions; the first is that the Coulomb charging energy is much larger than the thermal energy,

$$E_a > 20 - 100\, k_B T. \tag{7.10}$$

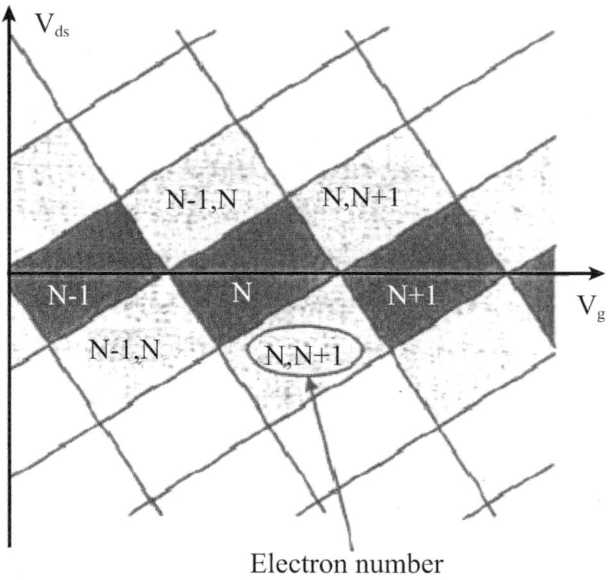

Figure 7.6 Contour plot of I_d in a SET as a function of V_g and V_{ds}.

If the temperature is not low enough, even in the case that the condition (7.6) does not satisfy, the thermally exited electron will tunnel through the quantum dot, resulting in closing of the neighboring CB peaks and increase of the valley current of the CB oscillation. Theoretically, the full width at half maximum (FWHM) of the CB peak is proportional to $k_B T/(\alpha e)$. While the spacing between gate voltages of peaks $\Delta V_g = E_a/(\alpha e)$, the ratio of ΔV_g and FWHM demonstrates the ratio of peak to valley of a CB oscillation $E_a/k_B T$. To increase the ratio of peak to valley the condition (Eq. 7.10) should be satisfied.

Figure 7.7 shows calculated E_a, E_C, and ΔE of a Si SET as a function of quantum dot size. The Si SET is a spherical Si quantum dot buried in the SiO^2 base [1]. From the figure we see that the room operation of the SET demands that the size of the SET be smaller than 2 or 3 nm. The capacitance of this ultra-small quantum dot is smaller than 1 aF. In this condition the proportion of the ΔE in the total energy will increase rapidly and there will appear the quantum effect in the SET characteristics.

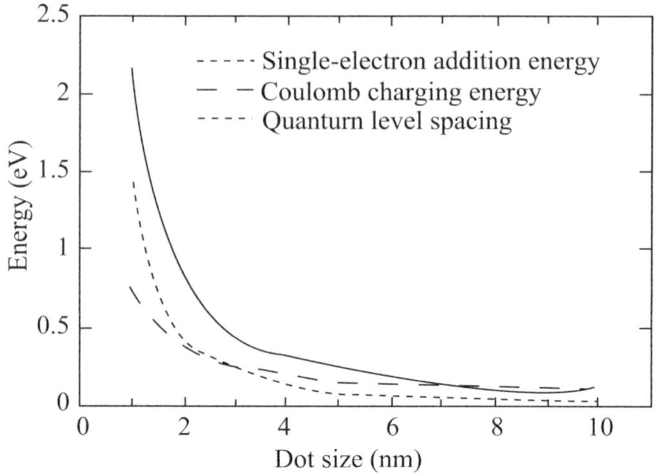

Figure 7.7 Calculated E_a, E_C, and ΔE of a Si SET as a function of QD scale.

The second condition comes from the uncertainty relation. To avoid the fluctuation or uncertainty of the number of electrons in a quantum dot, the electron should be localized well in the quantum dot, separated from the source and drain, which gives the condition

$$R_t \gg h/e^2 = 25.8\ \text{k}\Omega, \tag{7.11}$$

where R_t is the resistance of the tunneling barrier. $R_q = h/e^2$ is called the quantum resistance. A too-high tunneling resistance will result in very low drain current and low drivability, which is the main shortcoming of the SET.

7.2 Early Works of Set Operating at Room Temperature

Takahashi et al. [2] fabricated a Si SET by converting a one-dimensional (1D) Si wire substrate into a small Si island with a tunneling barrier at each end by means of pattern-dependent oxidation of a very thin Si layer on SiO_2. With this structure, the total capacitance was reduced to ~2 aF, which enabled conductance oscillation of the SET at room temperature.

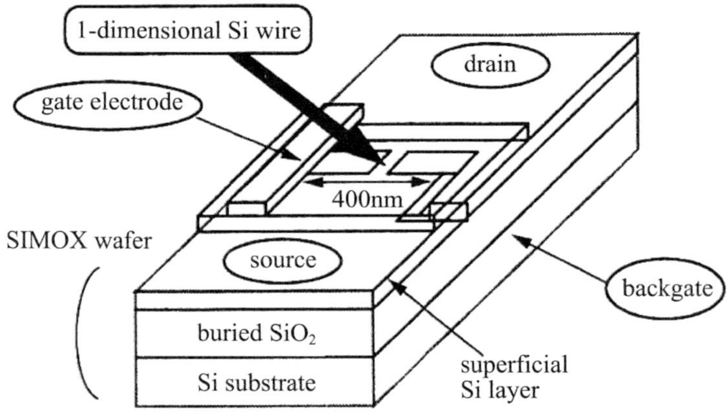

Figure 7.8 Schematic diagram of the Si SET.

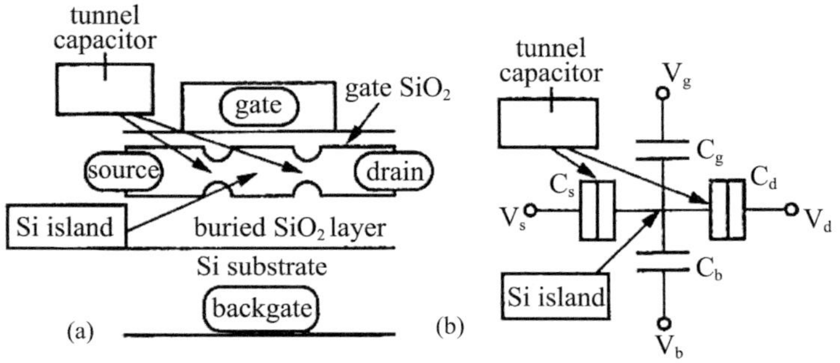

Figure 7.9 Cross-sectional view of a SET (a) and an equivalent circuit (b).

Figure 7.8 shows the schematic diagram of the Si SET [2]. Figure 7.9 shows the cross-sectional view of a SET (a) and an equivalent circuit (b) [2].

The conductance as a function of gate voltage in different temperatures is shown in Fig. 7.10 [2]. Here, the drain voltage was 1 mV and the source and back-gate voltages were fixed at 0 V. Conductance oscillations are observed even at room temperature. These oscillations are due to the CB by tunneling barriers with a very small capacitance. The capacitance can be estimated from the conductance characteristics. From the oscillation period ΔV_g in the gate voltage, ~500 mV, the gate capacitance C_g is calculated to be

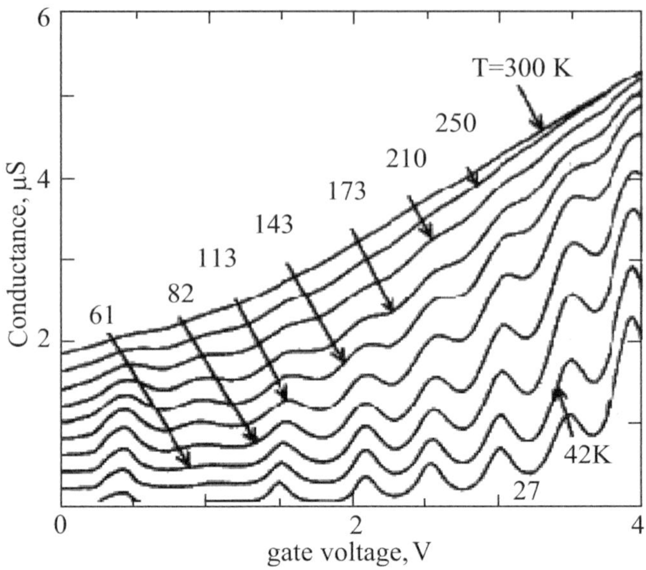

Figure 7.10 Conductance as a function of gate voltage in different temperatures.

~0.3 aF according to $C_g = e/\Delta V_g$. This value is consistent with the capacitance estimated from the dimensions of the structure. Hence, the back-gate capacitance C_b is determined to be ~0.01 aF because the cycle of the conductance oscillation as a function of the back-gate voltage is 30 times that for the front-gate voltage.

In Fig. 7.11 the source-drain current as a function of drain voltage V_d at the gate voltage corresponding to the conductance valley is shown [2]. The figure is similar to Fig. 7.5 in the CB condition. The gate voltage V_g and back-gate voltage are fixed at 0.7 and 0 V, respectively. The drain current is blocked over a V_d range of 160 mV. The drain capacitance C_d is found to be ~1 aF on the basis of the offset in the drain voltage of 160 mV. Taking into account that the source capacitance C_s does not exceed C_d in the present structure, one obtains the total capacitance $C_t = C_s + C_d + C_g + C_b < 2.3$ aF. This small value corresponds to a very large charging energy (e^2/C_t) of 70 meV, ensuring conductance oscillation at room temperature.

Ono et al. [3] developed a new oxidation method related to the original pattern-dependent oxidation (PADOX) method,

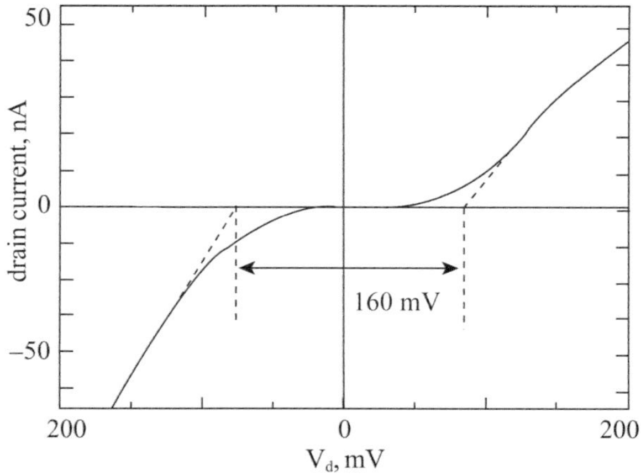

Figure 7.11 Source-drain current as a function of drain voltage V_d at the gate voltage corresponding to the conductance valley.

the vertical pattern-dependent oxidation (V-PADOX). The method applies thermal oxidation to a Si wire with a fine trench across it on a silicon-on-insulator substrate. During the oxidation, the Si wire with the fine trench is converted, in a self-organized manner, into a twin SET structure with two single-electron islands, one along each edge of the trench, due to position-dependent oxidation rate modulation caused by stress accumulation. Figure 7.12 shows the Si wire with a fine trench across it [3]. The key aspect of the method, thermal oxidation, is carried out just after the wire with a fine

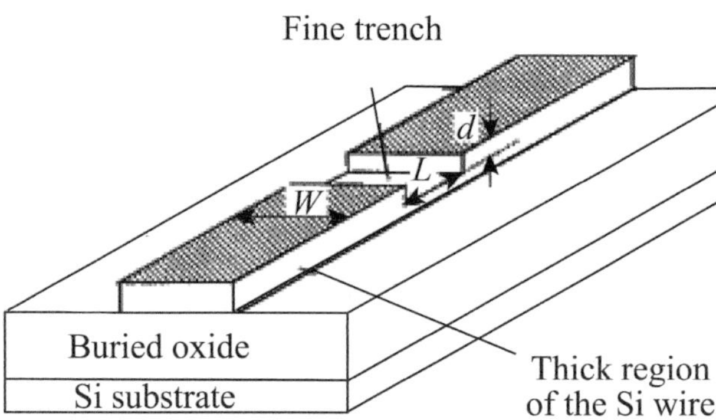

Figure 7.12 Si wire with a fine trench across it.

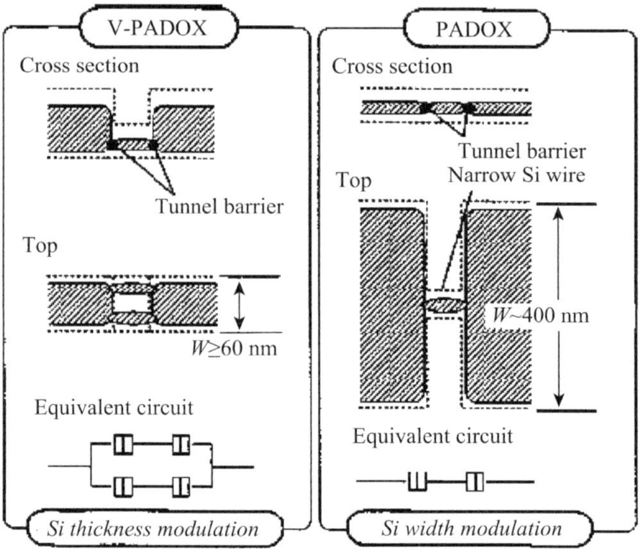

Figure 7.13 Resulting SET structure and the corresponding equivalent circuit. Those for the original PADOX are also shown for comparison.

trench is defined. During the oxidation the region below the trench is turned into twin Si islands, one along each edge of the trench. The resulting SET structure and the corresponding equivalent circuit are schematically shown in Fig. 7.13 [3]. Those for the original PADOX are also shown for comparison. In this figure, the broken lines represent preoxidation Si patterns and the hatched regions represent islands and leads after oxidation. The width of the island is 60 nm, and the length is 30 nm.

Figure 7.14 shows the conductance G, measured at 40 K with a drain voltage of 10 mV, as a function of the top-gate voltage V_g, the side-gate voltage $V_A = -4$ V, and $V_B = 0$ V [3]. Three peaks are observed, peaks A1, B1, and B2. Changing the side-gate voltages V_A and V_B, one can demonstrate that peak A1 originates from island A, which is closer to side gate A than island B is, and peaks B1 and B2 originate from island B. This confirms that a SET is formed along each edge of the fine trench and both SETs can be individually controlled. But in the present device the independent control effect of the side gate is not obvious.

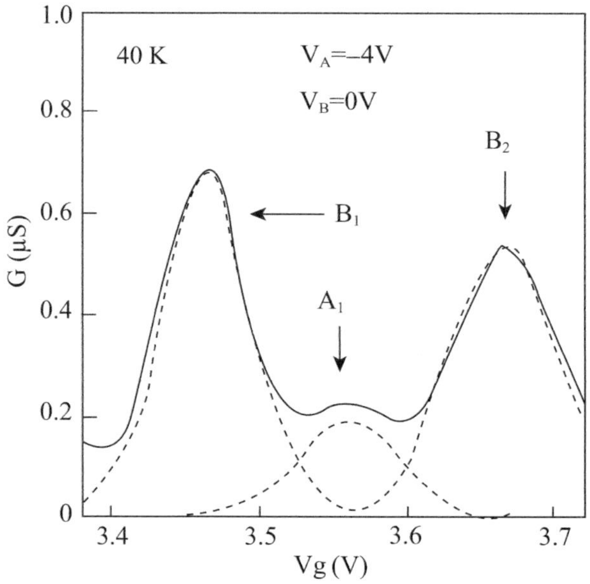

Figure 7.14 Conductance G, measured at 40 K with a drain voltage of 10 mV, as a function of the top-gate voltage V_g.

7.3 Si Set Operating at Room Temperature

In Section 7.1 we discussed the condition of operation of a SET at room temperature. As in Eq. 7.10 and Fig. 7.7, the size of the quantum dot should be smaller than 2 or 3 nm. In this case the spacing between single-electron energy levels ΔE will be much larger than the thermal energy and there will appear new quantum effects. Ishikuro et al. [4] fabricated the SET in the form of point-contact MOSFETs with various channel widths using electron beam lithography and the anisotropic etching technique on a silicon-on-insulator substrate. The device with an extremely narrow channel shows CB oscillations at room temperature. At a low temperature, negative differential conductance and fine structures are superposed on the device characteristics, which are attributed to the quantum mechanical effects in the quantum dot.

The schematic view of the point-contact MOSFET is shown in Fig. 7.15 [4]. It is found that the device with a 30 nm width shows

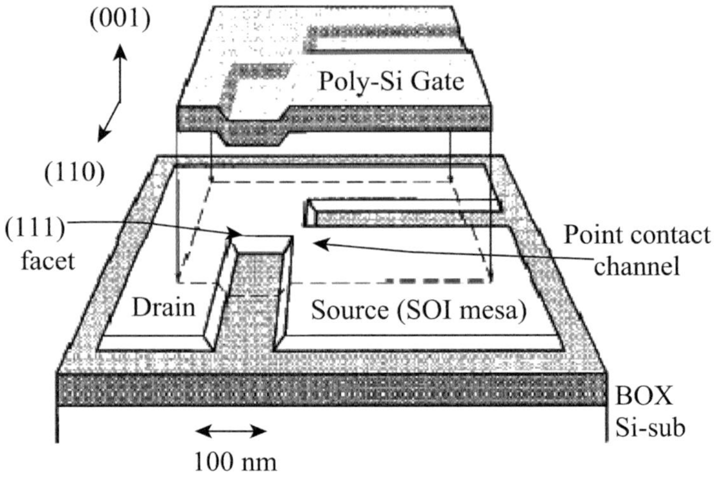

Figure 7.15 Schematic view of the point-contact MOSFET.

normal MOSFET operation even at 4.2 K. As the width decreases, the drain current decreases and, when the channel is extremely narrow, the device shows clear CB oscillations, which indicates that the channel is separated by tunnel barriers.

Figure 7.4 shows the drain current as a function of V_g. The temperature is changed from room temperature down to 21 K. The CB oscillation can be observed even at room temperature. Different from the CB characteristics of the semi-classical model, (i) the intervals between the peaks are not constant and (ii) negative differential conductance (NDC) and the fine structures are superposed on the $I_{ds} - V_{ds}$ characteristics. From the experimental results one can estimate that the single-electron charging energy $E_C = 58$ meV and the spacing between quantum levels $\Delta E =$ 30 meV, which are all much larger than the thermal energy at room temperature. Knowing E_C, one obtains $C_g = 0.085$, $C_d = 0.50$, $C_s =$ 0.80, and $C_{sub} = 0.009$ aF. Suppose the quantum dot is spherical and its diameter is estimated to be 6 nm.

Figure 7.16 shows the three-dimensional (3D) plot of the I_{ds} as a function of the V_g and V_{ds} at 4.2 K [4]. Clear rhombus shapes caused by CB oscillations are observed. NDCs and fine structures appear in parallel with the sides of the rhombus shapes.

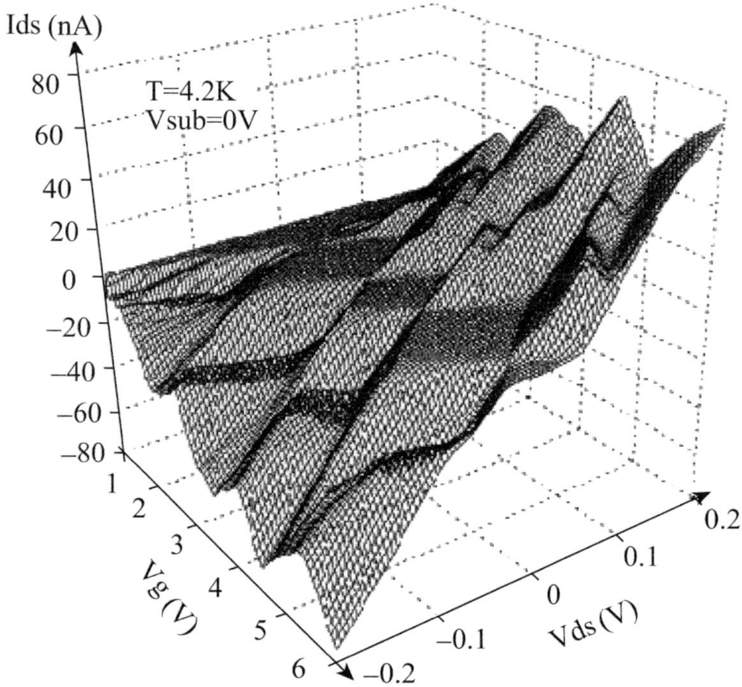

Figure 7.16 Three-dimensional plot of the I_{ds} as a function of the V_g and V_{ds} at 4.2 K.

Saitoh et al. [5] reported room temperature (RT) observation of NDC in a silicon single-dot single-hole transistor (SHT). They reduced the dot size to as small as 2 nm by improving the fabrication technique. In the single-dot SHT, both large CB oscillation and clear NDC due to large quantum-level spacing ΔE are observed at RT. The NDC appears as clearly as CB oscillations in the present SHT so the merged current blockade region (extended CB region) is observed.

The schematic of the point-contact MOSFET SHT is shown in Fig. 7.17a [5]. Two rectangles A and A' form the source and drain, and at the opposite angles a SHT (a single dot sandwiched by two tunneling barriers) is self-formed in the point-contact channel. To reduce the channel thickness as well as the channel width, they removed the buried oxide (BOX) just under the point-contact channel by hydrogen fluoride (HF). As a result, the etching of

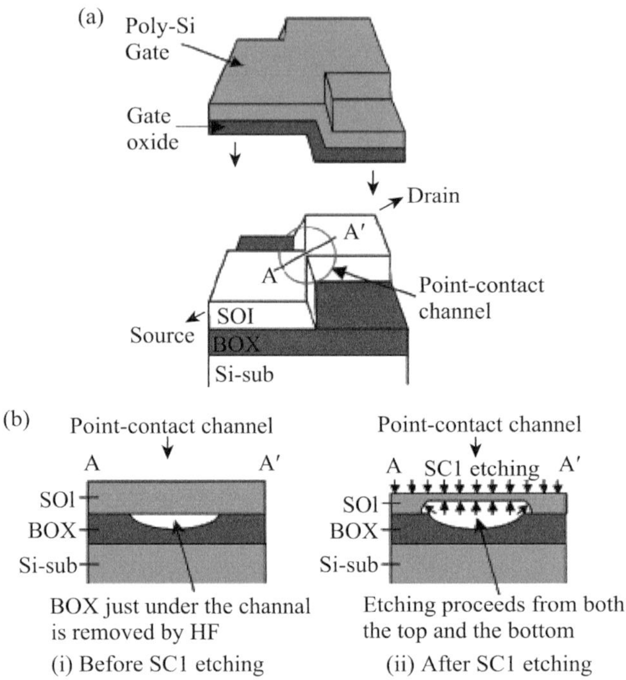

Figure 7.17 (a) Schematic of the point-contact MOSFET SHT. (b) The point-contact channel before and after SCI ($NH_4OH/H_2O_2/H_2O$) etching.

the channel proceeds from both the top and the bottom and the channel thickness is effectively reduced while the source/drain silicon on insulator (SOI) regions are kept relatively thick, as shown in Figs. 7.17b and 7.17c. The height and the width of the final channel are estimated to be less than 5 nm.

The drain current I_d as a function of the gate voltage V_g at RT is shown in Fig. 7.18a [5]. A large CB oscillation with one current peak is observed. The peak-valley current ratio (PVCR) is as high as 40.4. An enormously high PVCR of the observed CB oscillation indicates an ultra-small dot is successfully formed in the extremely constricted channel. Figure 7.18b shows I_d as a function of the drain voltage V_{ds} at RT. In the negative V_{ds} region, clear NDC with the PVCR of 11.8 is observed.

The most probable origin of clear NDC in the single-dot SETs/SHTs is the resonant tunneling at the source-side barrier due

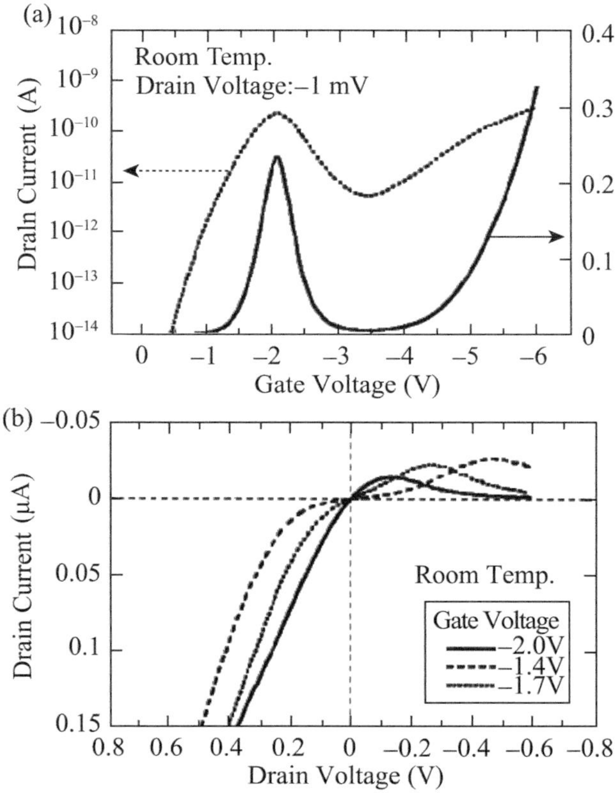

Figure 7.18 (a) Drain current I_d as a function of the gate voltage V_g at RT. (b) I_d as a function of the drain voltage V_{ds} at RT.

to large ΔE in the ultra-small dot. The schematics of the conduction band profile in a SET with an ultra-small dot are shown in Fig. 7.19a [5]. For simplicity, electron transport in SETs is considered instead of hole transport in SHTs. Since ΔE is much larger than the thermal energy at RT in the present device, electrons in the source cannot tunnel to the excited states in the dot and only ground state is accessible. At low V_{ds}, the tunneling rate of electrons from the source to the dot is relatively high because electrons tunnel through the thin tunneling barrier close to the Fermi level in the source. As V_{ds} increases, the energy levels in the dot fall down and electrons must tunnel through the thicker barrier close to the

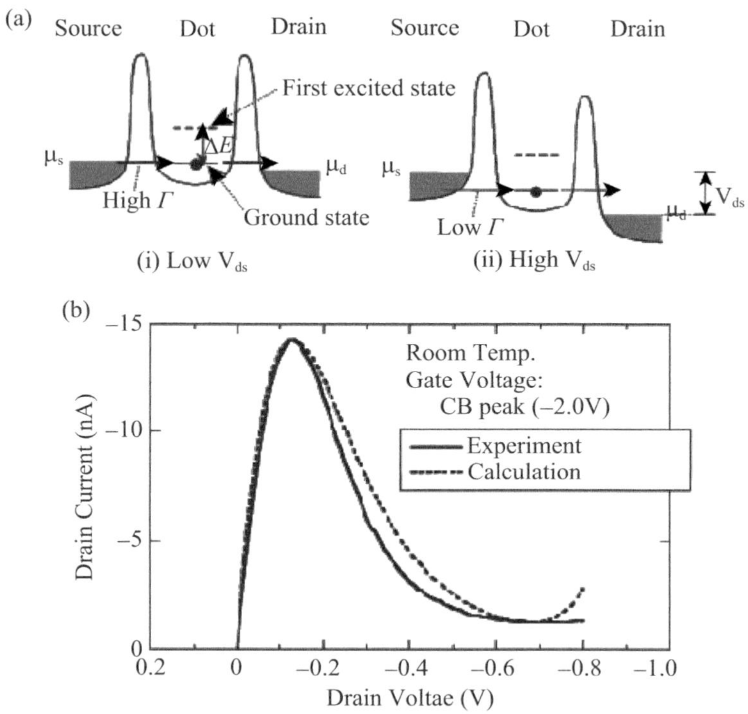

Figure 7.19 (a) Schematics of the conduction band profile in a SET with an ultra-small dot. (b) Comparison of the calculated $I_d - V_{ds}$ characteristics of the SHT with a 2 nm spherical silicon dot and the experimental data.

band edge in the source. Consequently, the tunneling rate decreases and NDC appears. Figure 7.18b compares the calculated $I_d - V_{ds}$ characteristics of the SHT with a 2 nm spherical silicon dot and the experimental data; the agreement is good.

The contour plots of the differential conductance $\partial I_d/\partial V_{ds}$ and I_d as a function of V_g and V_{ds} are shown in Figs. 7.20a and 7.20b, respectively [5]. In Fig. 7.20a two rhombus-shaped CB regions and a large NDC region are observed. In Fig. 7.20b the CB region is extended from low V_{ds} to negative and high V_{ds} as a result of the existence of the large NDC region. Such extended CB region and the NDC region have so far been observed only at a much low temperature. The NDC in a single-dot SHT is very simple and more suitable for circuit application.

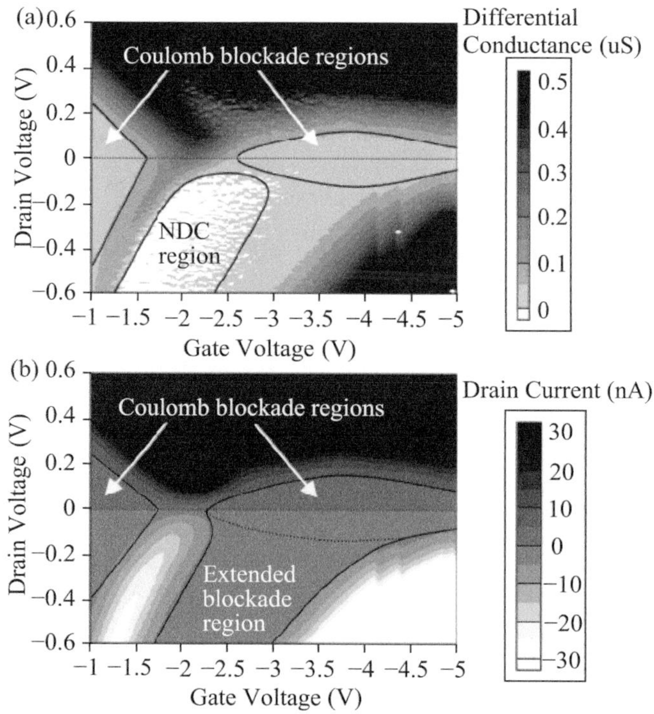

Figure 7.20 Contour plot of the differential conductance $\partial I_d/\partial V_{ds}$ (a) and I_d (b) as a function of V_g and V_{ds}, respectively.

7.4 Si Set Used as a Logic Circuit

The ultimate aim of the Si SET is to fabricate VLSI with ultra-high device density and ultra-low power loss. The single-electron logic circuit can be divided into two kinds: one is the CMOS-like SET logic circuit, in which the SET is used as an on/off switch device, similar to the general MOSFET in a CMOS logic circuit; and the other is the charging state logic circuit, where one bite is represented by one electron.

Ono et al. [6] fabricated a complementary single-electron inverter in which two identical SETs are packed on a silicon-on-insulator substrate (see Fig. 7.12). The resulting circuit occupies a very small area: 100 × 100 nm for each SET. For complementary

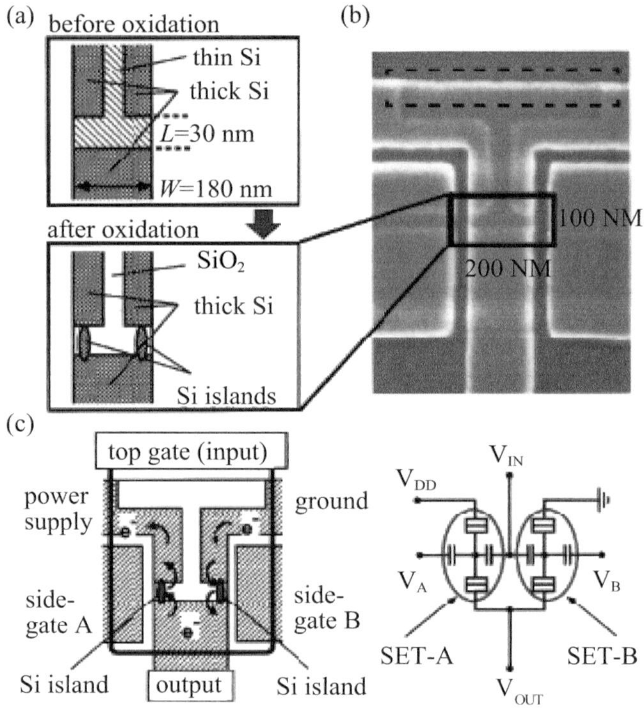

Figure 7.21 Structure of the complementary single-electron inverter. (a) Basic pattern before (top) and after (bottom) oxidation. (b) SEM image of the inverter. (c) Its equivalent circuit.

operation, the electric characteristics of one of the SETs are shifted using a side gate situated near the SET. The structure of the complementary single-electron inverter is shown in Fig. 7.21a [6], the equivalent circuit is shown in Fig. 7.21c, and the scanning electron microscope (SEM) image of the inverter is shown in Fig. 7.21b. The flow of electrons is indicated by the arrows. In the equivalent circuit, V_{IN} and V_{OUT} represent the input and output voltages. V_{DD}, V_A, and V_B are the voltages applied to the power-supply terminal and the two side gates, A and B, respectively.

Figure 7.22a shows input-output transfer characteristics of the inverter for a power-supply voltage V_{DD} of 20 mV [6]. For this measurement, a negative constant voltage ($V_B = -7$ V) was applied

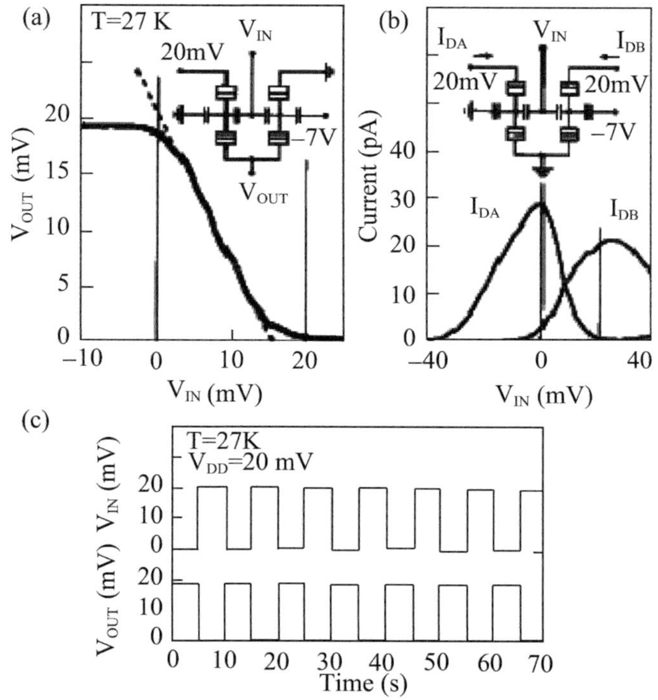

Figure 7.22 (a) Input-output transfer characteristics of the inverter for a power-supply voltage V_{DD} of 20 mV. (b) Drain-current characteristics of the two SETs for the drain voltage of 20 mV with the same side-gate voltage $V_A = 0$ V and $V_B = -7$ V. (c) Output voltage for a square-wave input with an amplitude of 20 mV.

to the side gate B to shift the current curve of SET B in the positive V_{IN} direction while gate A was grounded. Thus SET A and SET B operated as p- and n-type transistors, respectively. The drain-current characteristics of the two SETs for the drain voltage of 20 mV with the same side-gate voltage $V_A = 0$ V and $V_B = -7$ V are shown in Fig. 7.22b. The output voltage for a square-wave input with an amplitude of 20 mV is shown in Fig. 7.22c. The amplitude of the output is nearly the same as that of the input and the power supply voltage. But the switching speed is low, and it is caused by the slow response of the external circuit due to a large capacitance in the measurement system.

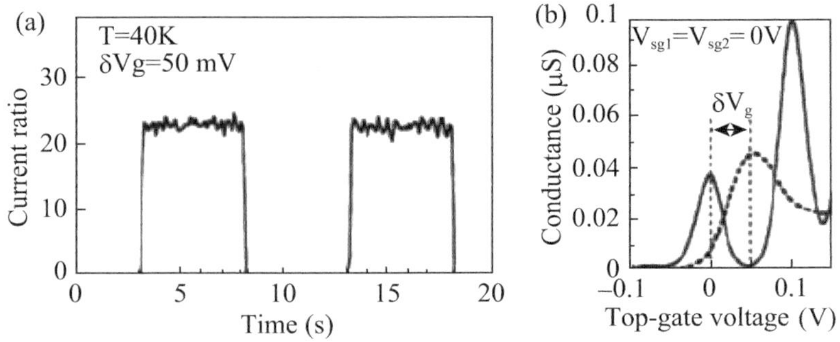

Figure 7.23 (a) Switching operation for a square-wave input with an amplitude of 50 mV. (b) Corresponding conductance characteristics of SET 1 (solid curve) and SET 2 (dotted curve).

Ono et al. [7] fabricated a Si SET current switching device on the basis of the Si SET inverter. The structure of the switching device is similar to that of the inverter. The switching operation for a square-wave input with an amplitude of 50 mV is shown in Fig. 7.23a [7], and the corresponding conductance characteristics of SET 1 (solid curve) and SET 2 (dotted curve) are shown in Fig. 7.23b. δV_g represents the amplitude of the square-wave input. Similarly, the switching speed is low.

The main shortcoming of the Si SET used as a device is that the speed is too slow. Apart from the tunneling resistance R_t, there is also the parasitic resistance of the external circuit. Due to the uncertainty relation R_t should be larger than the quantum resistance $h/e^2 \approx 25.8$ kΩ (see Eq. 7.11). The parasitic resistance comes from the undoped ultra-thin SOI region between the channel and the gate, which is serially connected with the intrinsic SET, resulting in large parasitic resistance. Although the upper limit of the peak conductance of SETs is determined by the quantum conductance ($= e^2/h \approx 38.8$ μS), further degradation of the peak conductance due to parasitic series resistance is a crucial issue.

Saitoh et al. [8] improved the silicon point-contact SET (see Fig. 7.17). By narrowing only the point-contact region and suppressing the parasitic series resistance, a peak conductance as large as 8.8 μS and single-electron addition energy as large as 128 meV are simultaneously obtained. Figure 7.24 shows the

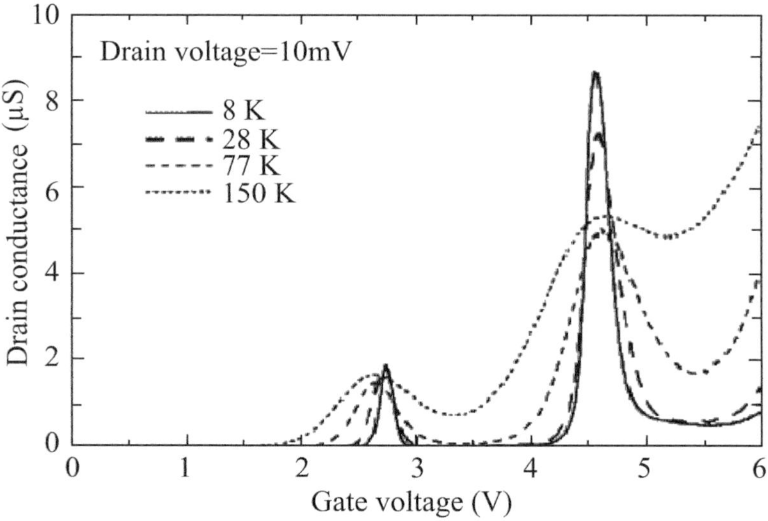

Figure 7.24 $I_d - V_g$ characteristics of the fabricated SET at a fixed $V_{ds} =$ 10 mV at several temperatures.

$I_d - V_g$ characteristics of the fabricated SET at a fixed $V_{ds} =$ 10 mV at several temperatures [8]. In Fig. 7.24 the conductance of the second peak at 8 K is 8.8 μS, which is much larger than the reported values (usually about 0.1 μS). This indicates that the relatively thick SOI layer successfully suppresses the parasitic series resistance. In the meantime, the height of the second peak decreases with an increasing temperature from 8 to 77 K. This temperature dependence is quite different from previously reported Si SETs, where the peak current increases with an increasing temperature. Since in this SET the quantum-level spacing ΔE is sufficiently large, only one quantum level, closest to the Fermi level, contributes significantly to one current peak when $k_B T \ll \Delta E$; the integrated area of one peak is independent of temperature. Therefore, the peak conductance decreases as the temperature increases and the peak width broadens. However, the first current peak keeps almost constant with an increasing temperature. This is because the conductance of the first peak is easily affected by the tunneling through the next level with higher coupling to reservoirs at a relatively high temperature.

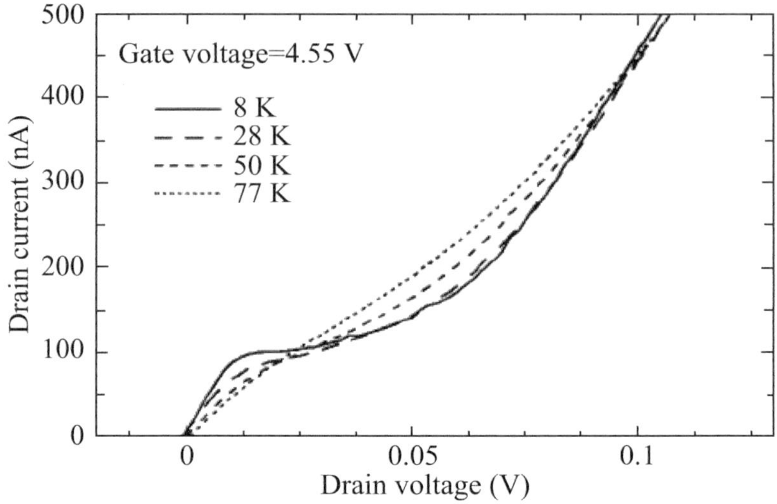

Figure 7.25 $I_d - V_{ds}$ characteristics at a fixed $V_g = 4.55$ V, where the $I_d - V_g$ curve has its peak, at several temperatures.

Figure 7.25 shows the $I_d - V_{ds}$ characteristics at a fixed $V_g = 4.55$ V, where the $I_d - V_g$ curve has its peak, at several temperatures [8]. A current staircase is observed at 8 K, and the fine structure (nonlinear behavior) persists up to 77 K. In this case, the staircase feature arises from the discreteness of quantum levels. When $k_B T \ll \Delta E$, only the ground state is accessible for an electron for a low drain voltage and the current is kept constant until a drain voltage at which the first excited state becomes aligned with the Fermi level in the reservoirs and begins to contribute to conduction. Thus, each plateau of the staircase corresponds to a different quantum state in the dot.

The energy spacing between the ground state and the first excited state ΔE can be derived from the drain voltage at which the drain current starts to increase from the first plateau, because the first excited state starts to contribute to electron transport at this drain voltage. Because the onset drain voltage is about 0.05 V in Fig. 7.25, the quantum-level spacing is estimated to be about 19 meV. The coulomb charging energy is roughly estimated to be 109 meV as the difference between the single-electron addition energy and the quantum-level spacing.

References

1. Saitoh, M., and Hiramoto, T. (2006). *Handbook of Semiconductor Nanostructures and Nanodevices*, Vol. 3, eds. Baladin, A. A., and Wang, K. L. (American Scientific Publishes, Los Angeles), p. 279.
2. Takahashi, Y., Nagase, M., Namatsu, H., et al. (1995). Fabrication technique for Si single-electron transistor operating at room temperature. *Electron. Lett.*, **31**, 136–137.
3. Ono, Y., Takahashi, Y. Yamazaki, K., et al. (2000). Fabrication method for IC-oriented Si single-electron transistors. *IEEE Trans. Electron Devices*, **47**, 147–153.
4. Ishikuro, H., and Hiramoto, T. (1997). Quantum mechanical effects in the silicon quantum dot in a single-electron transistor. *Appl. Phys. Lett.*, **71**, 3691–3693.
5. Saitoh, M., and Hiramoto, T. (2004). Extension of Coulomb blockade region by quantum confinement in the ultrasmall silicon dot in a single-hole transistor at room temperature. *Appl. Phys. Lett.*, **84**, 3172–3174.
6. Ono, Y., Takahashi, Y., Yamazaki, K., Nagase, M., and Namatsu, H. (2000). Si complementary single-electron inverter with voltage gain. *Appl. Phys. Lett.*, **76**, 3121–3123.
7. Ono, Y., Takahashi, Y., Yamazaki, K., et al. (2000). Single-electron transistor and current-switching device fabricated by vertical pattern-dependent oxidation. *Jpn. J. Appl. Phys.*, **39**, 2325–2328.
8. Saitoh, M., and Hiramoto, T. (2002). Observation of current staircase due to large quantum level spacing in a silicon single-electron transistor with low parasitic series resistance. *J. Appl. Phys.*, **91**, 6725–6728.

Chapter 8

Silicon Single-Electron Memory

Memory is the first and foremost application of silicon single-electron devices (SEDs) and has been extensively researched. One reason is that the SEDs can operate in smaller dimensions and this directly leads to the increased memory density. Another is that memory usually has a periodic and simple structure and this is expected to make the implementation of the emerging devices straightforward. The scaling limit in the current silicon memories, both flash memory and dynamic random access memory (DRAM), also adds to the demand for single-electron memories.

The memory unit with a discrete trap-like memory node has attracted much attention because it operates using fewer electrons and can possibly operate at a high density, a high speed, and low power dissipation. In these memories, the memory node includes floating-node type or natural nitride trap and semiconductor nanocrystals. The electrons are injected into the memory node, causing the shift of the critical voltage, whose principle is the same as that of general erasable programmable read-only memories (EPROMs) or electrically erasable and programmable read-only memories (FEPROMs), but the total number of electrons involved in the memory node is far less than the latter. This kind of memory is called a few-carriers memory.

Quantum Waveguide in Microcircuits (Second Edition)
Jian-Bai Xia, Duan-Yang Liu, and Wei-Dong Sheng

ISBN 978-981-4968-30-0 (Hardcover), 978-1-003-65146-8 (eBook)
www.jennystanford.com

8.1 Memory of Floating-Gate-Node Type

Most of the proposed single-electron memories are floating-node types, in which a limited number of electrons are stored in the floating memory node and the presence of the charge is detected by a charge-sensing device. SEDs can be used in either electron storage or charge sensing schemes. Both schemes are shown in Figs. 8.1 and 8.2 [1], respectively. Possible combinations of them are numerous.

A distinctive device in electron storage schemes is the electron trap, which consists of one, two, or more serial tunnel junctions and a capacitor (Fig. 8.1a–c). Another electron storage scheme is based on a serially connected transistor and a storage capacitor (Figs. 8.1d and 8.1e). Since transistors are three-terminal devices, this scheme provides flexibility in memory operations. Figure 8.2 shows various kinds of charge-sensing devices used in single-electron memories. Their drain current is modulated by the charge in the memory node. Thus the memory charge can be detected by the current of the sensing line. Apart from the single-electron transistor (SET)

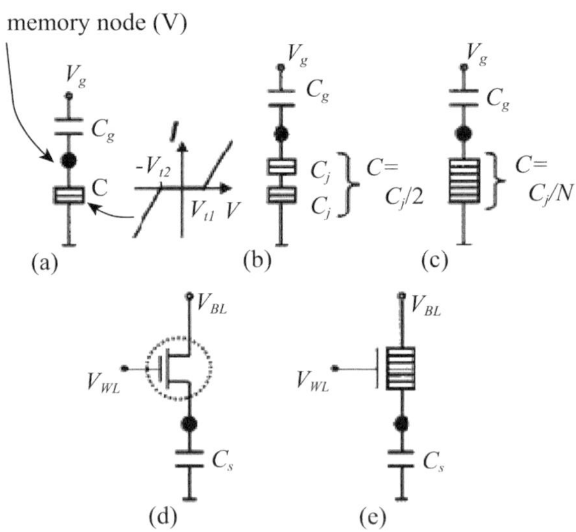

Figure 8.1 Various electron storage schemes used in single-electron memories.

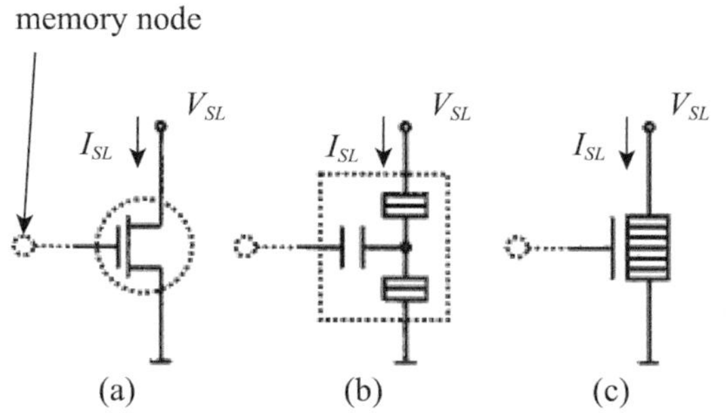

Figure 8.2 Various kinds of charge-sensing devices used in single-electron memories.

and multiple-tunnel-junction (MTJ) SET, the minimum metal-oxide semiconductor field-effect transistor (MOSFET) is also sensitive enough to detect a single electron.

In the relationship between the potential of the memory node V and the gate voltage V_g, Fig. 8.3 [1], tunneling of the electron into and out of the memory node is prohibited by the Coulomb blockade (CB) in the region $-Q_c/C < V < Q_c/C$. The critical charge Q_c is given by

$$Q_c = \frac{eC}{C_\Sigma}\left(\frac{1+\Delta}{2}\right), \tag{8.1}$$

where C_Σ is the total capacitance $C + C_g$, C is the capacitance of the MTJ, and C_g is the gate capacitance. Δ determines a multiple value

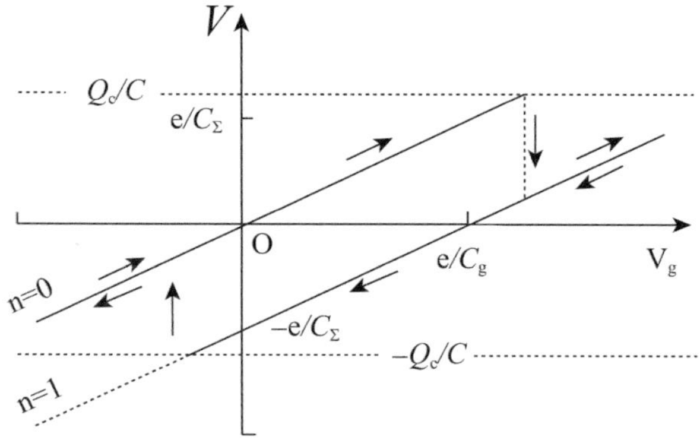

Figure 8.3 Potential of the memory node V as a function of gate voltage V_g for double- and multiple-tunnel-junction traps.

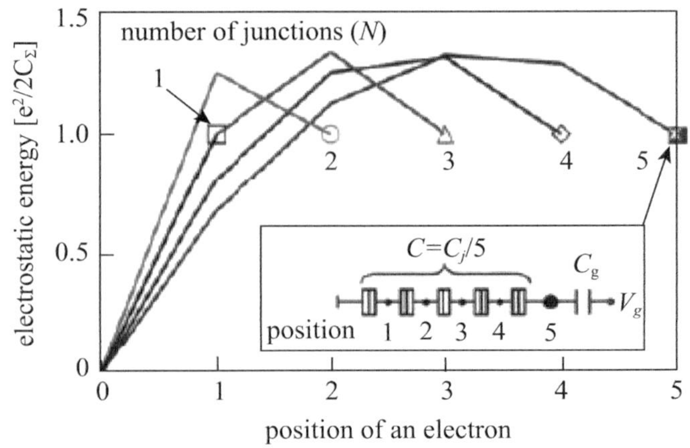

Figure 8.4 Electrostatic energy of the MTJ trap system with an extra electron as a function of its position when $V_g = 0$ and $C_g/C = 3$ [1].

condition, given by

$$\Delta = \left(1 - \frac{1}{N}\right)\frac{C_g}{C}, \tag{8.2}$$

where N is the number of serially connected tunnel junctions.

The operation principle of the memory is shown in Fig. 8.3 [1]. V increases with a slope of C_g/C_Σ until the boundary of the CB regime ($V = Q_c/C$) is reached, where an electron tunnels into the memory node and the V drops abruptly by e/C_Σ. Even if the V_g sweep is reversed, the electron is trapped until the sweep reaches the other boundary ($V = -Q_c/C$), where the electron tunnels out of the memory node. Thus a hysteresis loop appears, and bistability is attained at $V_g = 0$ for $1 < \Delta < 3$.

The stability of a trapped electron can be assessed by calculating the electrostatic energy of the system as a function of the position of an extra electron, see Fig. 8.4 [1]. For the number of tunnel junctions more than one ($N > 1$), the curves become convex and an energy barrier appears between states with and without an extra electron in the memory node. This results in the stable storage of an electron in the node. If there is only one tunnel node ($N = 1$) the stable storage is impossible, only if the tunnel resistance is nonlinear, as shown in Fig. 8.1a.

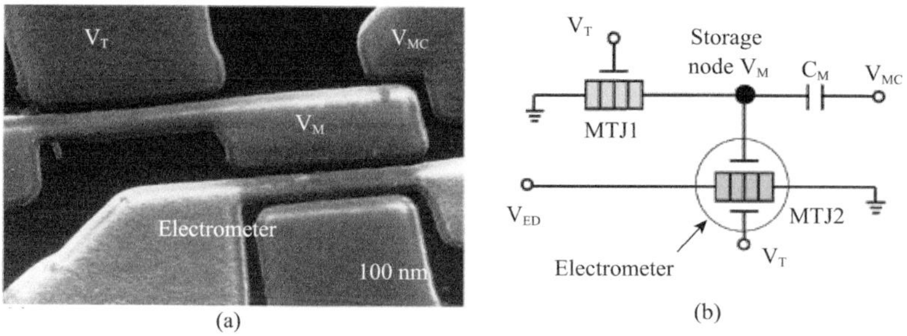

Figure 8.5 (a) Scanning electron micrograph of the compact silicon single-electron memory cell. (b) Equivalent circuit of the memory cell.

8.2 Si Set Used as Memory

Stone et al. [2] fabricated a compact single-electron memory in silicon, which is based on the CB effect observed in highly doped silicon nanowires. The circuit shows clear memory operation with a $>$ 100 mV gap between "0" and "1" levels when tested at a temperature of 4.2 K. The response of the circuit to write and erase pulse sequences is also presented. The silicon nanowires used as MTJs are 500 nm in length and have a width of 50 nm. The devices were fabricated in a silicon-on-insulator (SOI) material. Figure 8.5 shows a scanning electron micrograph of the compact cell after oxidation, and the equivalent circuit diagram is shown in Fig. 8.5b [2]. The memory cell consists of MTJ1 and a memory node that occupies an area of only 0.5 μm^2. The second device, MTJ2, forms an electrometer that has the function of detecting the state of the memory node. Each MTJ has a side gate to trim its operating point. The voltage V_{MC} controls the amount of charge on the memory node via capacitor C_M.

Memory operation is demonstrated by the two distinct levels in the memory node voltage as the voltage V_{MC} is scanned over a range as shown in Fig. 8.6a [2]. If V_{MC} is increased positively from zero the voltage on the memory node V_M increases until it reaches $+V_C$, the CB voltage. Further increase of V_{MC} results in a single electron flowing through MTJ1 onto the memory node, and the node voltage V_M then falls back below the Coulomb gap voltage $+V_C$ by

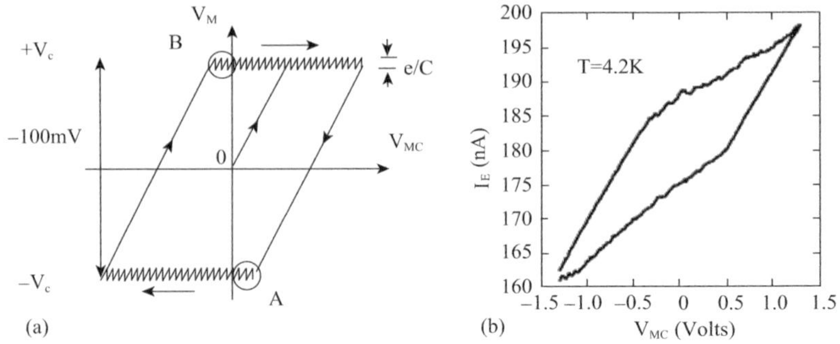

Figure 8.6 (a) Schematic representation of the hysteresis loop in a memory node voltage V_M as capacitor voltage V_{MC} is swept. (b) Hysteresis loop at 4.2 K. V_{MC} is swept in the range of ± 1.3 V.

an amount e/C (where C is the self-capacitance of the memory node). A subsequent increase of V_{MC} adds electrons to the memory node one by one. If V_{MC} is reduced toward zero the memory node voltage will fall finally to $-V_C$. The excess electrons at A in Fig. 8.6a leave the memory node via MTJ, and V_M is held at $-V_C$. The remaining electrons are trapped on the node. If a negative voltage V_{MC} is applied to the capacitor electrons are removed one by one from the node, which remains clamped at $-V_C$. The cyclic operation of V_{MC} results in the two levels as shown in Fig. 8.6a as parts of a hysteresis loop—the basis of memory operation. Figure 8.6b shows the measured current of the MTJ2 I_E as a function of V_{MC} at 4.2 K, showing the hysteresis loop.

The response of the memory to "write" and "erase" pulses is shown in Fig. 8.7 [2]. Positive and negative voltages were applied alternatively by V_{MC} separated by read periods of 0 V, while the memory current was monitored. During the read periods two distinct levels can be seen at B and D, signifying the "0" and "1" states. The change in current between B and D is determined by the CB gap, which is set by the trimming gate V_T of MTJ1. The self-capacitance of the island was estimated to be 100 aF. For a Coulomb gap of 0.05 V this would give an excess or shortfall of 30 electrons at two levels of V_M in Fig. 8.7b.

Takahashi et al. [3] fabricated a Si memory device composed of a small one-dimensional (1D) Si-wire MOSFET and a SET. The 1D

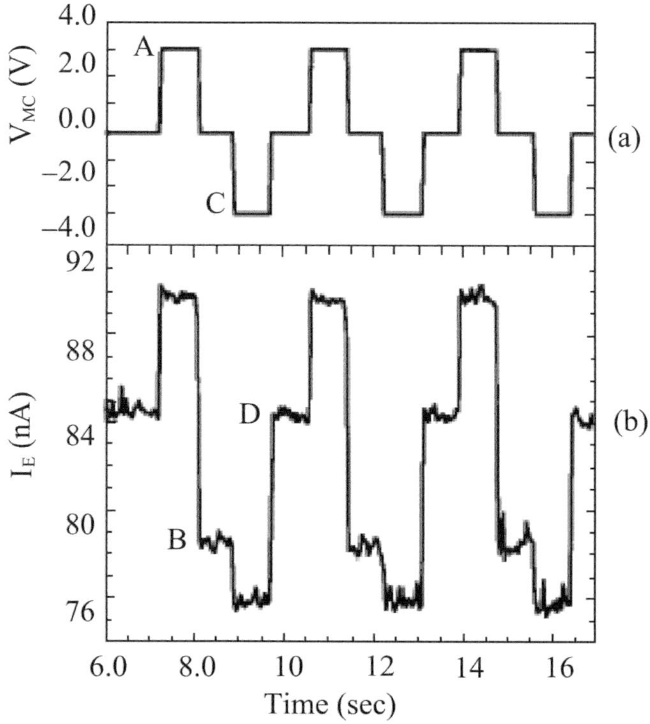

Figure 8.7 Memory read-write characteristics of single-electron memory. (a) Drive pulse wave form. (b) Electrometer output pulses monitoring the memory node.

MOSFET provides a very steep subthreshold slope that is very close to the physical limit at room temperature. This guarantees a low-voltage operation as well as a small size. The very small number of stored electrons is detected by a highly sensitive SET electrometer. The device can operate with an extremely small number of electrons, which ensures ultra-low-power and high-speed operation.

The SEM image of the Si memory device is shown in Fig. 8.8a, and its equivalent circuit is shown in Fig. 8.8b [3]. The memory node is connected via the 1D MOSFET to an electron reservoir (side electrode). By using the MOSFET to control electron transport between the memory node and the side electrode, high write/erase speed can be achieved without sacrificing retention time. This is

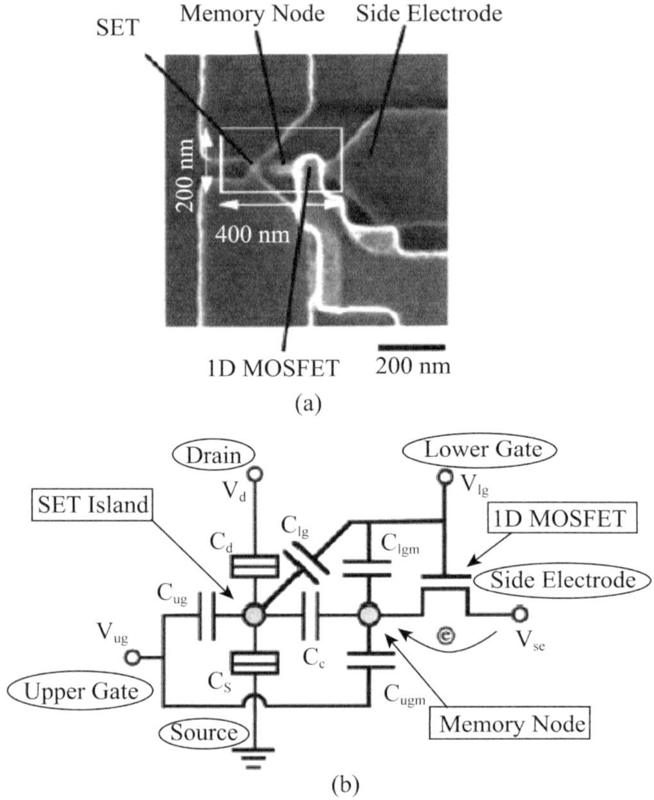

Figure 8.8 (a) SEM image of the Si memory device. (b) Its equivalent circuit.

in contrast to flash-memory-type single-electron memories, whose operating speed is limited by the low tunneling rate between the memory island and the channel to ensure sufficient storage time. The SET is used to sense a small number of electrons stored in the memory node. The $I_d - V_d$ characteristic of the device with the initially 30 nm wide 1D MOSFET with the 80 nm long gate shows that the conductance is remarkably high in spite of a channel width of only 20 nm. This guarantees a high driving capacity of the 1D MOSFET connected to the small memory node.

Figure 8.9 shows how the device state evolves as the lower-gate voltage (V_{lg}) is scanned in the positive direction and then in the reverse direction [3]. Initially, V_{lg} was set at a low voltage of −2.7 V,

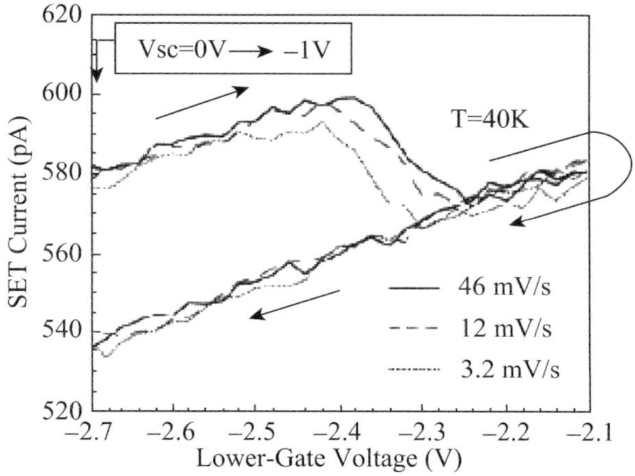

Figure 8.9 Hysteresis characteristics of the SET current representing the "write" and "erase" actions of the Si memory device.

at which the channel of the 1D MOSFET was closed. Then the side-electrode voltage V_{se} was changed from 0 V to −1 V before the start of the V_{lg} scan at various scan speeds. The rapid fall of the SET current at V_{lg} of about −2.3 V, where the 1D MOSFET is turned on, corresponds to the "write" action. The written information is held while V_{lg} is scanned back because the 1D MOSFET is turned off. The number of electrons in the memory node for a "write" voltage is estimated to be about 100. In Fig. 8.9 the threshold gate voltage at which the write operation occurs becomes lower as the scanning speed of V_{lg} decreases. This indicates that electron writing proceeds very slowly through the 1D MOSFET at gate voltages slightly lower than the threshold.

8.3 Floating Gate Memory Operating at Room Temperature

Guo et al. [4] fabricated a room-temperature silicon SET memory that consists of (i) a narrow channel MOSFET with a width (~10 nm) smaller than the Debye screening length of single electron and (ii) a nanoscale polysilicon dot (7 nm × 7 nm) as the floating gate

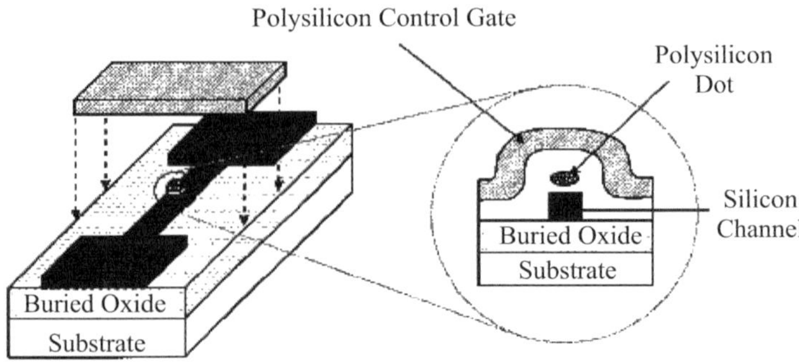

Figure 8.10 Schematic of a single-electron MOS memory that has a narrow silicon channel and a nanoscale polysilicon dot as the floating gate. The cross-section view illustrates the floating gate and the channel region.

embedded between the channel and the control gate. They observed that storing one electron on the floating gate can significantly screen the channel from the potential on the control gate, leading to a discrete shift in the threshold voltage, a staircase relationship between the charging voltage and the threshold shift.

Figure 8.10 shows the schematic of the single-electron MOS memory [4]. The polysilicon dot and the Si channel is formed by thermal oxidation, which would consume silicon, reducing the thickness of the polysilicon dot by about 9 nm and the lateral size of the dot and the silicon channel width by about 18 nm. It is noted that no tunnel oxide is intentionally added between the channel and the polysilicon floating gate. This will allow fast charging and minimize the potential difference between the channel and the floating dot during the charging process, so the CB effect can regulate the number of electrons stored on the floating dot gate for a given charging voltage. The potential barrier still exists between the channel and the floating gate because of the grain boundary and a thin native oxide.

First, a voltage pulse positive relative to the ground source was applied to the control gate and the drain voltage was maintained at 50 mV. This caused the electrons to tunnel from the channel to the floating dot. At various gate pulse voltages from 2 to 14 V the measured I–V characteristics are shown in Fig. 8.11 [4]. Despite a continuous charge voltage, the threshold voltage (defined as the

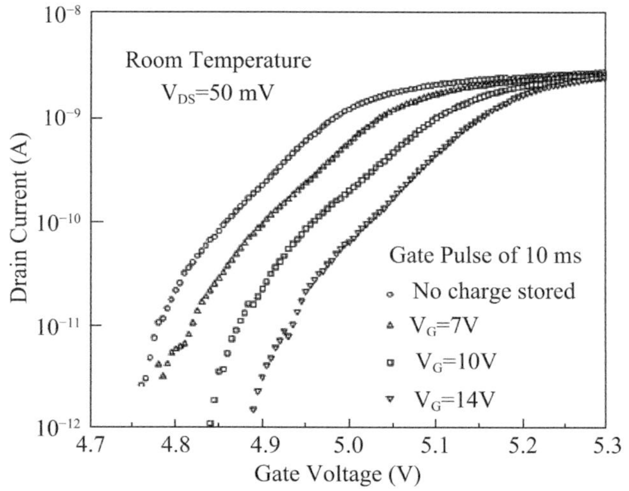

Figure 8.11 Room temperature I–V characteristics of the device before and after the charges are stored onto the floating dot.

gate voltage at which the drain current reaches 100 pA) always shifts a discrete increment of about 55 mV and each threshold shift corresponds to a charging voltage interval of ~4 V.

Therefore, there is a staircase relation between the threshold voltage shift and the charging voltage, as shown in Fig. 8.12a [4]. Moreover, for a given charging voltage, the threshold shift is self-

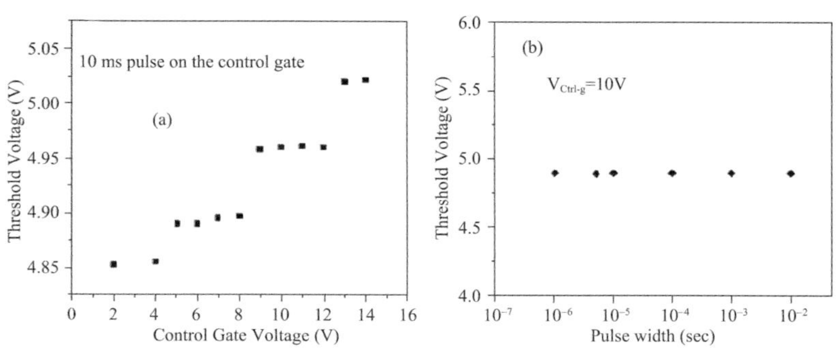

Figure 8.12 (a) Threshold voltage as a function of the charging voltage on the control gate, showing a staircase relation with ~4 V for each stair. (b) Threshold voltage as a function of the charging time, while the charging voltage is fixed at 10 V.

limited, the threshold voltage shift is independent of the charging time, as shown in Fig. 8.12b. Since there is no intentional tunnel oxide, the charge stored at the floating gate can be held for an average of 5 s after the control-gate potential is set back to the ground, and after 5 s the threshold voltage of the device returns to its origin value (the first I–V on the left in Fig. 8.11).

Nakajima et al. [5] fabricated seemingly at the same time a Si single-electron field-effect transistor (FET) memory device having a self-aligned floating dot gate. This device demonstrates single-electron memory operation at room temperature. Figure 8.13a shows the structure of this FET memory and 8.13b the cross-section view of the structure along the $a - a'$ line in Fig. 8.13a [5]. By use of the self-aligned manufacture a floating dot having a side length of about 30 nm is located on a narrow channel wire. The channel width is equal to or slightly larger than the floating dot. Finally, a second gate oxide 200 nm thick and a poly-Si control gate were deposited by low-pressure chemical vapor deposition (LPCVD).

Figure 8.14 shows the typical drain current I_d versus gate voltage V_g characteristics of the device [5]. For the device having a floating dot (Fig. 8.14a), an increasing V_g produced an I_d shift at a V_g of 3.9 V. However, with a decrease of V_g, the I_d shift did not occur, indicating a hysteresis of the $I_d - V_g$ curve. It means that the V_g reaches the CB boundary and an electron enters into the floating gate dot, resulting in the I_d shift. With a subsequent V_g increase (Fig. 8.14b), an additional I_d shift occurred at 5.1 V, the V_g, at which the I_d shift occurred, is called threshold voltage V_{th} (3.9, 5.1 V). For the device without a floating dot (Fig. 8.14c) such a quantized V_{th} shift and the $I_d - V_g$ curve hysteresis could not be observed.

From Figs. 8.14b and 8.14a the neighboring threshold voltages V_{th} separation ΔV_w is 1.2 V and the quantized threshold voltage ΔV_{th} is 0.1 V, which are related to the capacitances,

$$\Delta V_w = \frac{e}{C_{gt}},$$

and

$$\Delta V_{th} = \frac{e}{C_{gc} + C_{gt}}, \tag{8.3}$$

where C_{gt} is the capacitance between the control gate and the floating dot gate, which is estimated to be 0.16 aF from the

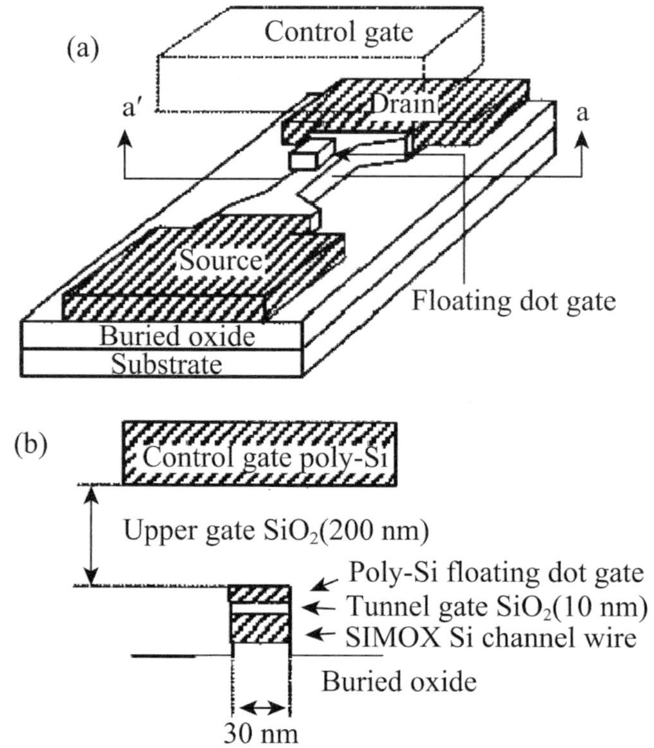

Figure 8.13 (a) Schematic diagram of Si single-electron FET memory with a self-aligned floating gate. (b) Cross-section view along the $a - a'$ line.

geometrical factors of the device. The ΔV_W is calculated to be 1.0 V, which is close to that obtained from measurement 1.2 V. C_{gc} is the capacitance between the control gate and the narrow channel wire, which is estimated to be 1.4 aF, giving ΔV_{th} a value of 0.1 V, in agreement with the measurement result.

8.4 Silicon Nanocrystal-Based Memory

The floating gates introduced in above sections are fabricated by lithography technology. The size of dots is limited by the manufacture condition. This section introduces a new kind of quantum dot (nanocrystals) formed by the chemical method. Tiwan

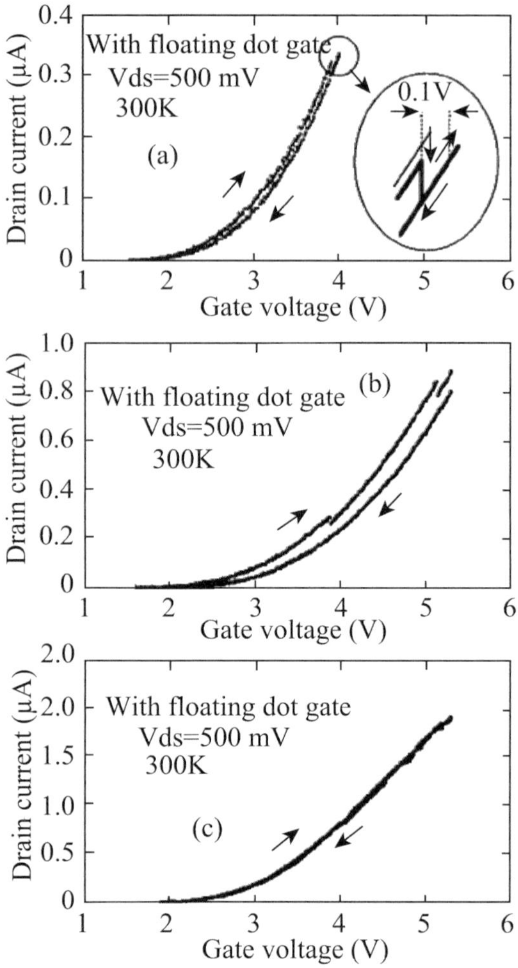

Figure 8.14 Drain current versus control gate voltage at room temperature for (a) and (b) FET with a floating dot gate and (c) FET with no floating dot gate.

et al. [6] fabricated a new memory structure using threshold shifting from charge stored in nanocrystals of silicon (~5 nm in size). A threshold voltage shift of 0.2–0.4 V with read and write times less than hundreds of a nanosecond at an operating voltage below 2.5 V has been obtained experimentally. The retention times are measured

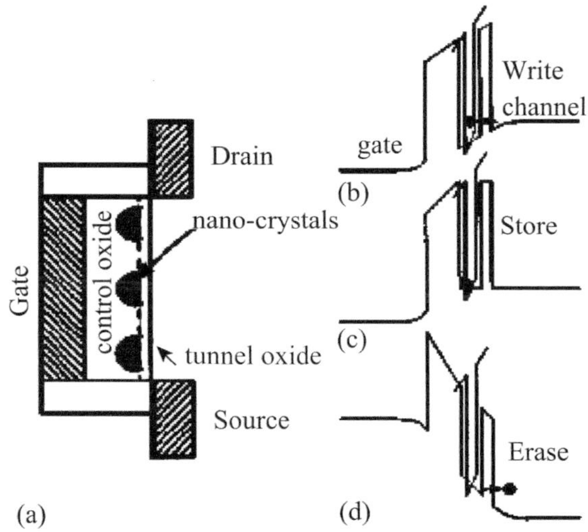

Figure 8.15 (a) Schematic cross section of the device. (b–d) Band diagram for (b) injection, (c) storage, and (d) removal of an electron from a nanocrystal.

in days and weeks, and the structure has been operated in excess of 10^9 cycles without degradation in performance.

Figure 8.15 shows a schematic cross section and band diagram during injection (write cycle), storage (store), and removal (erasure) of an electron in the device [6]. A thin tunneling oxide (1.1–1.8 nm thick) separates the inversion surface of an n-channel silicon FET from a distributed film of nanocrystals of silicon that covers the entire surface channel region. A thicker tunneling oxide (4.5 nm or thicker) separates the nanocrystals from the control gate of the FET. An injection of an electron occurs from the inversion layer via direct tunneling when the control gate is forward biased with respect to the source and drain. The resulting stored charge screens the gate charge and reduces the conduction in the inversion layer, that is, it effectively shifts the threshold voltage of the device to be more positive, whose magnitude for a single electron per nanocrystal is approximately given by

$$\Delta V_{\text{th}} = \frac{en_{\text{w}}}{\varepsilon_{\text{ox}}} \left(t_{\text{c}} + \frac{1}{2} \frac{\varepsilon_{\text{ox}}}{\varepsilon_{\text{Si}}} t_{\text{w}} \right), \tag{8.4}$$

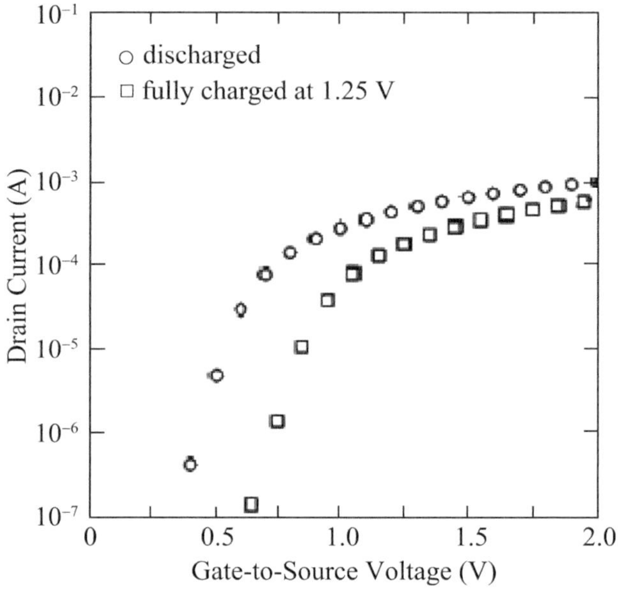

Figure 8.16 A demonstration of the bistability as a result of injection of electrons into nanocrystals. The characteristics are measured with a 20 ns gate voltage pulse.

where $\Delta V_{\rm th}$ is the threshold voltage shift, $t_{\rm c}$ is the thickness of the control oxide under the gate, $t_{\rm w}$ is the linear dimension of the nanocrystal well, the ε's are the permittivities, q is the magnitude of the electronic charge, and $n_{\rm w}$ is the density of nanocrystals. For $t_{\rm w} =$ 5 nm, 5 nm apart, $n_{\rm w} = 1 \times 10^{12}$ cm^{-2}, $t_{\rm c} = 7$ nm, and the threshold shift is nearly 0.36 V for one electron per nanocrystal. This is easily current-sensed.

Figure 8.16 shows the drain current-gate voltage characteristics in the presence and absence of electrons in nanocrystals [6]. The characteristics are obtained using 20 ns pulses, applied between gate and grounded source, with the drain maintained at 100 mV. The threshold shift of ~0.25 V under a 1.25 V static gate bias shows the effect of the storage of electrons, estimated to be one electron per nanocrystal. At a 3 V static gate bias the threshold shift is nearly 1.2 V, corresponding to a larger number of electrons per nanocrystal.

The refresh times for the structures are very large, that is, the stored charge does not leak out in accurately measurable times at room temperature. Leakage of the charge occurs either laterally by conduction between the nanocrystals or from the silicon surface with an eventual path to the n^+-doped source and drain regions. The band alignment is unfavorable for tunneling, and the charge movement within the bulk silicon requires diffusion of carriers. Thus, the nanomemory has a long retention time. The write time constants are about 100 ns, which are in the range of theoretical expectations, and should scale exponentially with a reduction in the tunneling oxide thickness. The corollary to the long refresh time is an increase in erase times for the charge. Complete removal of the charge requires milliseconds at voltages of 3 V and above.

Applying a pulse voltage of ± 2.5 V can complete a circle of writing (1 μs) and erasing (1 ms). The test shows that after 10^9 circles the threshold voltages persist basically unchanged. It is due to direct tunneling in the charging and discharging of the structure to prevent hot-carrier degradation. Besides, the test used a smaller nanocrystal density to limit power dissipation.

8.5 Retention Property of Nanocrystal Floating Memory

The key of the non-volatility memory of the nanocrystal floating memory is the long period retention property of the electrons in the memory. When some electrons are stored in the nanocrystals, under a selected voltage the stored electrons have a probability to tunnel out of the nanocrystal, leading to a gradual change of the channel current. This gradual change reflects directly the information of the potential height/width, internal electric field, defects, interface states, etc. Thus, research of the time variation of the memory window and the charging and discharging behavior of the Si nanocrystal will clearly explain the retention property of the Si nanocrystal memory devices.

Hinds et al. [7] studied the lifetime of the emission of a single electron stored in a nanocrystalline Si (nc-Si) dot in order to

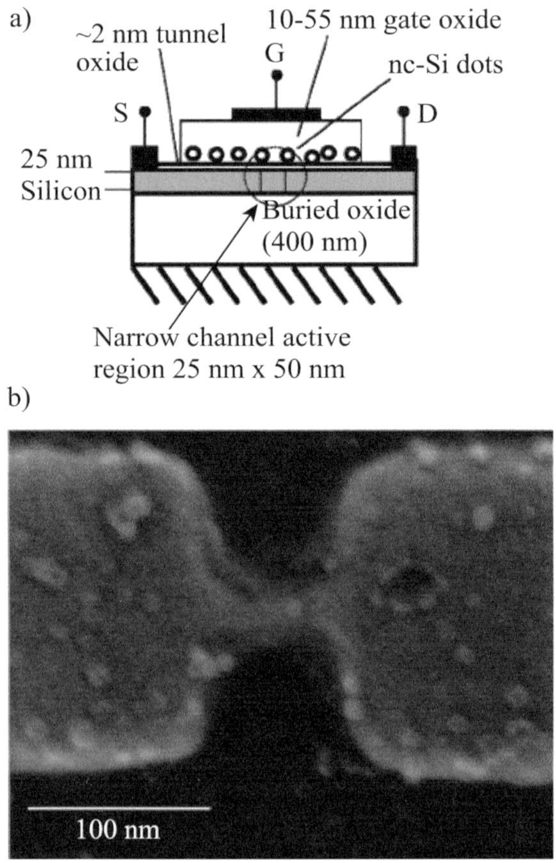

Figure 8.17 (a) Schematic of a single-electron memory device isolating a single nc-Si and a floating gate memory node. (b) Planar scanning electron microscopy view of a representative memory device showing a single nc-Si dot near the center of the narrow channel active region.

understand the physical processes for memory application. The schematic of the FET with nc-Si floating gate nodes is shown in Fig. 8.17a [7]. A small active area FET channel (50 $\times$ 25 nm) is defined by electron beam lithography on a thin (20 nm) SOI channel and allows for the electric isolation of a single nc-Si dot. Remote plasma-enhanced chemical vapor deposition is used to form 8 $\pm$

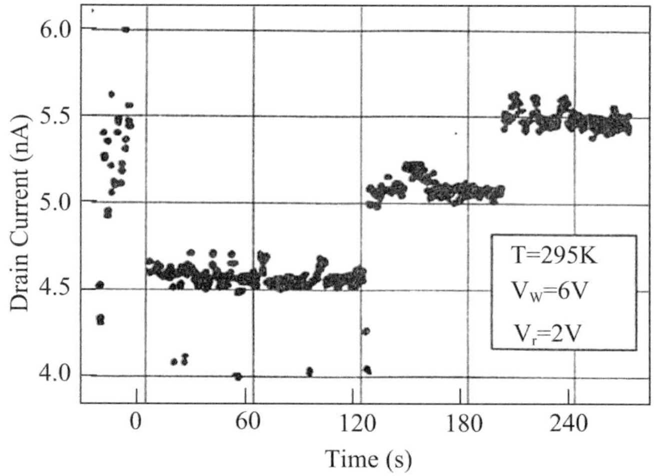

Figure 8.18 Time dependence of the channel current after the writing process.

1 nm diameter nc-Si dots. Electrons stored in a dot result in an observed discrete threshold shift of 90 mV.

For a 25 nm wide channel and a gate oxide thickness of 50 nm the observed shift in V_{th} is 90 mV. Notable is the discrete step in drain current at a constant gate bias after a writing pulse as shown in Fig. 8.18 [7]. A stepwise increase in drain current clearly shows the discrete emission of an electron from the nc-Si dot. The data shown in Fig. 8.18 is a representative example of memory lifetime, and there was a significant distribution of lifetimes for other pulses under the same conditions. For the erase voltage $V_e = -2.25$ to -0.25 V and $T = 295$ K the distribution of the measured memory lifetime is quite broad, with the mean value being nearly equal to the standard deviation. The median lifetime as a function of gate bias is shown in Fig. 8.19 [7]. The shown curves are calculated from a polarizable Schrodinger equation for a 1D confined potential well at the given nc-Si dot diameter. Most notable is the rather small dependence of lifetime on bias over a large voltage range and different slopes for positive and negative biases. The temperature dependence of lifetime is seen to be proportional to $1/T^2$, which is consistent with a direct tunneling process.

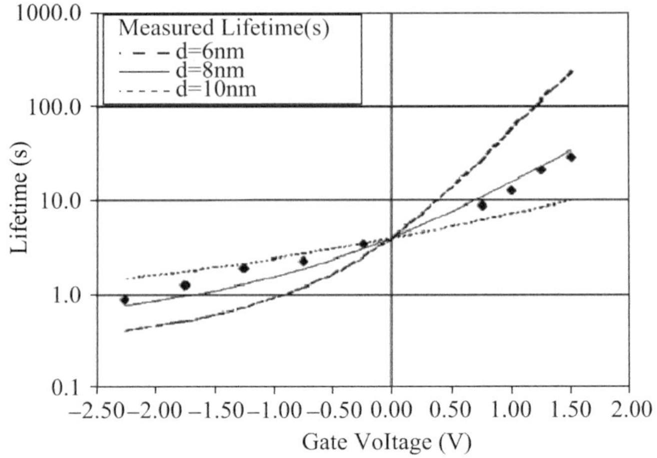

Figure 8.19 Observed median lifetime (solid points) of a nc-Si memory device as a function of applied gate bias.

Because the lifetime has a weak dependence on the gate bias, it is able to infer that the observed electron emission in the nc-Si memory is not from a defect site on the SOI channel/SiO_2 interface or from the nc-Si/SiO_2 interface. Thus, the stored electron behaves as if it were delocalized over the whole dot. To explain the lifetime as a function of applied gate bias, an important concept is that the electron will be attracted to the top of the dot with a positive bias, thus reducing the tunnel current from the bottom of the dot into the channel. And in the negative bias the electron will be attracted to the bottom of the dot, increasing the tunnel current. Figure 8.19 shows the calculated lifetimes as a function of gate bias for three different nc-Si dot diameters. The lifetime can be improved by increasing the tunnel oxide thickness, but this would also increase the write and erase times in a similar fashion. To improve the retention time while not increasing write and erase times it is possible to take advantage of the shallow interfacial traps on the nc-Si interface, for example, the Si_3N_4 trap.

Saitoh et al. [8] proposed an ultra-narrow silicon floating gate memory, in which the channel width is scaled to sub-10 nm. In this ultra-narrow-channel memory a larger threshold voltage shift has been observed than in the wide-channel memory. From the

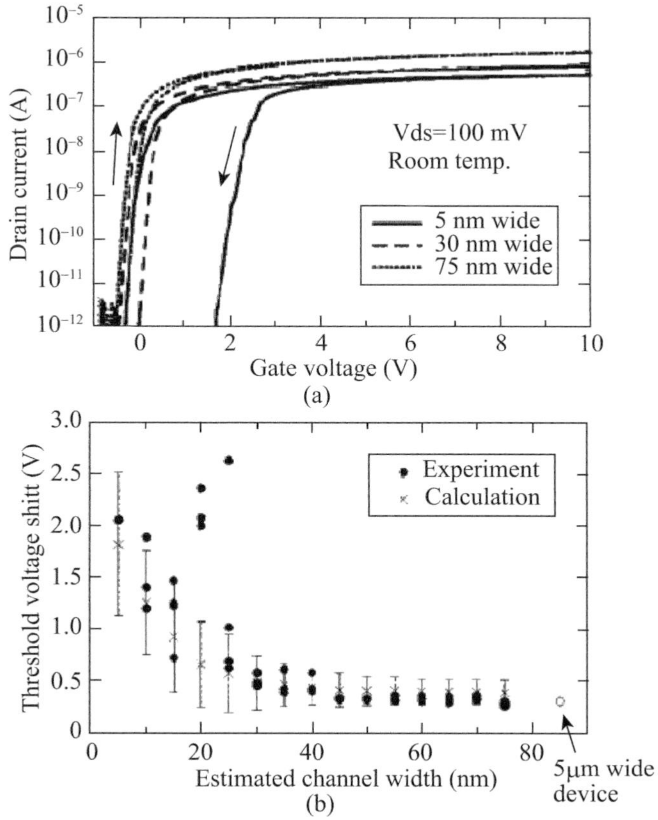

Figure 8.20 (a) Room-temperature hysteresis characteristics of the devices with various channel widths. (b) Measured and calculated results of the channel width dependence of ΔV_{th}.

numerical calculation, it turns out that this is caused by bottleneck regions that dominate the conductance of the whole channel in the ultra-narrow channel. Moreover, a longer charge-retention time has been also obtained in this device.

Figure 8.20a shows the drain current I_{d} as a function of the gate voltage V_{g} with a drain voltage $V_{\text{ds}} = 100$ mV for different channel widths at room temperature [8]. V_{g} scan proceeds from -1 to 10 V and back to -1 V. Hysteresis loops are observed in all the devices, and ΔV_{th} increases as the channel width decreases. Figure 8.20b shows measured ΔV_{th} as a function of estimated channel width.

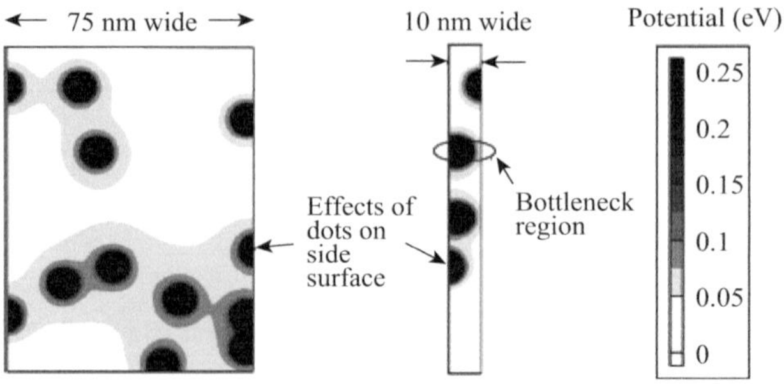

Figure 8.21 Calculated results of the potential distribution in a 75 nm wide channel and a 10 nm wide channel.

When the channel is narrower than 25 nm, ΔV_{th} rapidly increases. The characteristics of 5 μm wide devices (also shown in Fig. 8.20b) are almost the same as 75 nm wide channel devices. From ΔV_{th} in the wide-channel devices, the average number of electrons per dot is estimated to be approximately 0.25.

To understand the mechanism of the large ΔV_{th} in the ultra-narrow channel, the potential distributions are calculated in a 75 nm wide channel and a 10 nm wide channel when one-fourth of the dots are occupied by one electron, and the results are shown in Fig. 8.21 [8]. In the wide channel, the current can flow through the wide and low-potential region. In the ultra-narrow channel, the current path is completely blocked at the bottleneck region where one dot covers almost the entire channel (bottleneck effect). Moreover, the average potential in the ultra-narrow channel is higher than in the wide channel because of the effects of dots on the side surfaces. These effects are the origins of large ΔV_{th} in the ultra-narrow devices.

Figure 8.22a shows the measured retention characteristics [8]. First change the gate voltage from 0 to 10 V (write) or -10 V (erase), and then suddenly change the gate voltage to 0 at a time $t = 0$. V_{th} was measured several times afterward. It is apparent that the ultra-narrow channel (5 nm) device has a wider memory window than the wide channel (5 μm) device at 10^4 s after programming. Even in comparison to normalized ΔV_{th}, the ultra-narrow channel device has longer retention time, as shown in Fig. 8.22b. In the time scale, the retention time in the 5 nm wide channel device is five times longer.

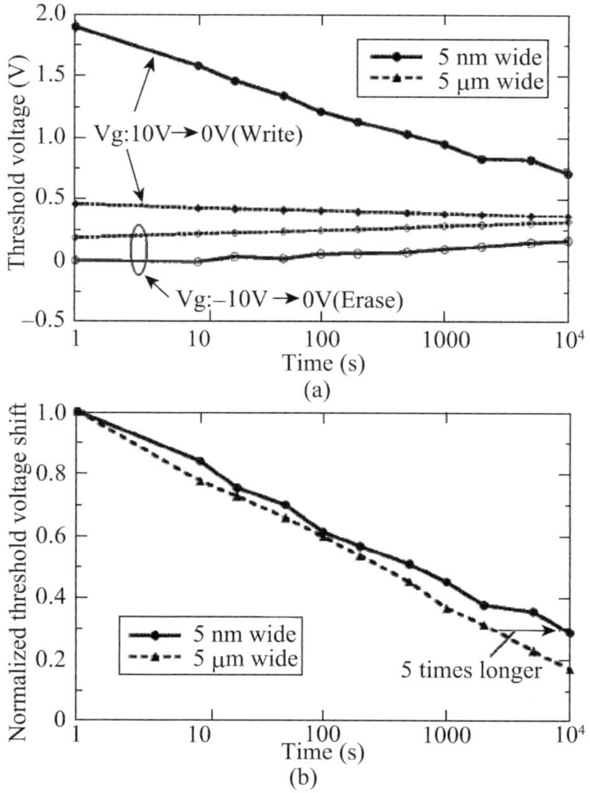

Figure 8.22 (a) Measured retention characteristics in a 5 nm wide device and a 5 μm wide device. (b) Retention characteristics when ΔV_{th} is normalized to the value at 1 s.

Longer retention time in the ultra-narrow-channel devices can be explained by the following two mechanisms. One is a classical bottleneck effect. The linear relationship between ΔV_{th} and the total number of electrons in dots is valid for the wide channel device, while a strong nonlinear relationship exists for the ultra-narrow-channel device. This is because a small number of electrons in several bottleneck regions dominate the conductance of the whole channel and the influence of one electron is much larger in the ultra-narrow-channel devices. It is found that when electrons are emitted from dots and the total number of electrons is decreased by 90%,

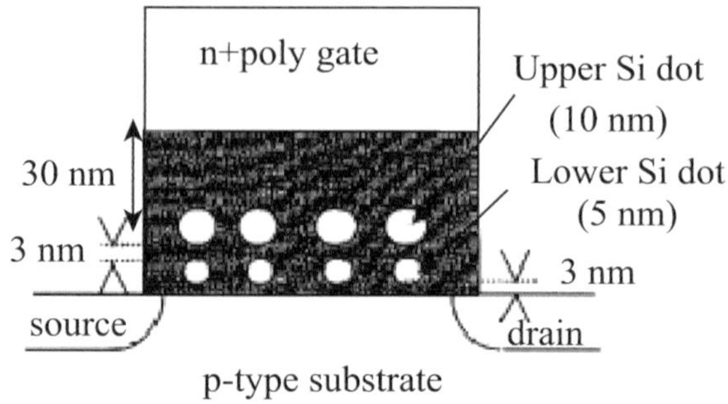

Figure 8.23 Schematic diagram of Si self-aligned doubly stacked dot memory.

ΔV_{th} decreases by 90% in the 75 nm wide channel device and by 61% in the 5 nm wide channel device. The other mechanism is the quantum confinement effect. The tunneling rate from a dot to the ultra-narrow channel is lower than that to the wide channel because the ground energy level in the ultra-narrow channel is lifted up due to the increased quantum confinement.

The manufacture of an ultra-narrow channel is difficult. Ohba et al. [9] proposed a Si dot memory whose floating gate consists of self-aligned doubly stacked Si dots, as shown in Fig. 8.23. A smaller Si dot exists immediately below an upper dot and lies over the thin tunnel oxide. It is experimentally shown that charge retention is improved compared to the usual single layer Si dot memory. Due to the quantum confinement effect and the CB effect of the lower layer dots, the energy of electronic states increases and the tunneling rate of electrons in the upper layer dots through the lower layer dots decreases, leading to increase of the retention time.

The upper diagram in Fig. 8.24 shows the energy band diagrams in a low-gate-voltage (V_g) region [9], and V_{eff} represents the voltage difference between the channel and the charged Si dot. In the usual single-dot memory the electrons only pass through a thin tunnel oxide. However, in the doubly stacked dot memory, the tunnel probability P_t will decrease by the energy barrier height ΔE, which is due to CB and quantum confinement in the lower Si dot. Since the lower dot is smaller than the upper dot, ΔE will be determined by the lower dot size effects. The ratio of tunnel probability $\alpha_{ret} = P_{tsingle}/P_{tdouble}$ in a low-voltage region is a measure of

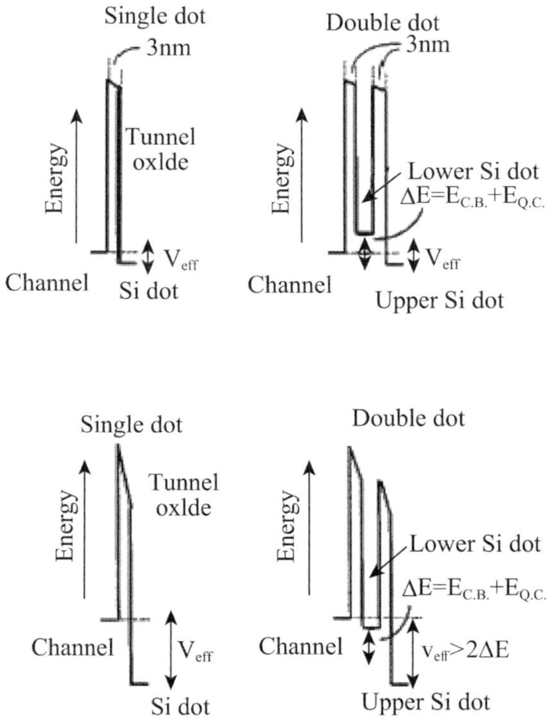

Figure 8.24 Upper diagram: In a low V_g region. Lower diagram: In write operation the energy band diagrams for single and double-dot memory.

retention improvement of the double-dot memory compared to the single-dot one. The retention improvement factor is obtained as follows:

$$\alpha_{\text{ret}} = \alpha\left(V_{\text{eff}} \approx 0\right) = \left(\frac{2k_{\text{B}}T}{\Delta E}\right)\sinh\left(\frac{\Delta E}{k_{\text{B}}T}\right). \tag{8.5}$$

The exponential dependence on ΔE is very important.

Figure 8.25 shows the retention factor α_{ret} as a function of ΔE at 300 K. With a decrease of the lower dot diameter the α_{ret} increases in orders of magnitude. Though the w/e speed reduction factor $\alpha_{\text{w/e}}$ is small and less than 10, the retention is improved exponentially.

The lower diagram of Fig. 8.24 shows the band diagram in write operations. When the write/erase process is in the $|V_{\text{eff}}| > 2\Delta E$, a high speed will be possible in the double-dot memory because there is no energy barrier in the tunneling process. As in the retention

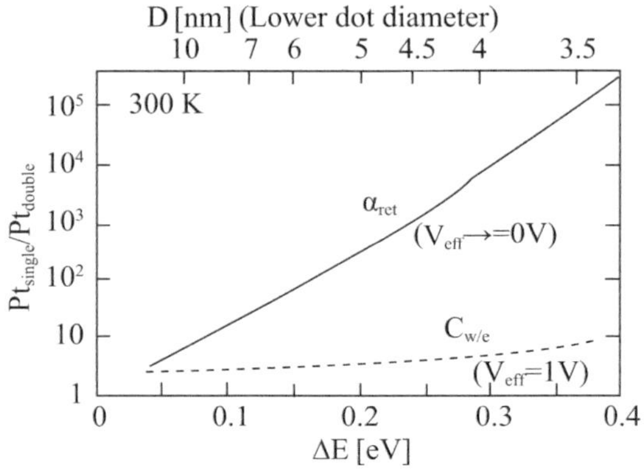

Figure 8.25 Retention improvement factor α_{ret} as a function of ΔE. The corresponding lower Si dot diameter D is plotted in the upper abscissa. $\alpha_{\text{w/e}}$ is the velocity reduction factor.

case, the current ratio $\alpha_{\text{w/e}}$ is a measure of the speed reduction. The speed reduction factor is approximately expressed as

$$\alpha_{\text{w/e}} = \alpha\left(V_{\text{eff}} > \frac{2\Delta E}{e}\right) = \frac{2}{\left[1 - \left(\frac{2\Delta E}{V_{\text{eff}}}\right)\right]}. \quad (8.6)$$

It is important that $\alpha_{\text{w/e}}$ have no exponential dependence on ΔE.

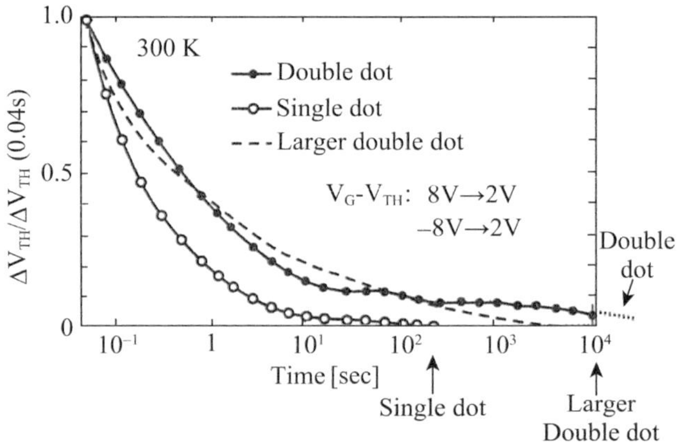

Figure 8.26 Time dependence of the memory window at the same bias condition for the double-dot memory, the double-dot memory with larger dots, and the single-dot memory.

Figure 8.26 shows the time dependence of the memory window at the same bias condition for the double-dot memory, the double-dot memory with larger dots, and the single-dot memory [9]. The value for a 5 nm lower dot memory (solid dots) is larger than that for the single-dot memory (empty circle) by more than 10^2 times, consistent with the retention factor $\alpha_{\text{ret}} \approx 300$ in Fig. 8.25. For the larger double-dot memory (dashed line) ΔV_{th} disappears at 10^4 s, corresponding to $\alpha_{\text{ret}} \approx 30$ in Fig. 8.25.

References

1. Takahashi, Y., Ono, Y., Fujiwara, A., and Inokawa, H. (2002). Silicon single-electron devices. *J. Phys. C*, **14**, R995–R1033.
2. Store, N. J., and Ahmed, H. (1998). Silicon single electron memory cell. *Appl. Phys. Lett.*, **73**, 2134–2136.
3. Takahashi, Y., Fujiwara, A., Yamazaki, K., et al. (1999). A Si memory device composed of a one-dimensional metal-oxide-semiconductor field effect-transistor switch and a single-electron-transistor detector. *Jpn. J. Appl. Phys.*, **38**, 2457–2461.
4. Guo, L., Leobandung, E., and Chou, S. Y. (1997). A room-temperature silicon single-electron metal-oxide-semiconductor memory with nanoscale floating-gate and ultranarrow channel. *Appl. Phys. Lett.*, **70**, 850–852.
5. Nakajima, A., Futasugi, T., Kosemura, K., Fukano, T., and Yokoyama, N. (1997). Room temperature operation of Si single-electron memory with self-aligned floating dot gate. *Appl. Phys. Lett.*, **70**, 1742–1744.
6. Tiwari, S., Rana, F., Hanafi, H., Harstein, A., Crabbe, E. F., and Chan, K. (1996). A silicon nanocrystals based memory. *Appl. Phys. Lett.*, **68**, 1377–1379.
7. Hinds, B. J., Yamanaka, T., and Oda, S. (2001). Emission lifetime of polarizable charge stored in nano-crystalline Si based single-electron memory. *J. Appl. Phys.*, **90**, 6402–6408.
8. Saitoh, M., nagata, E., and Hiramoto, T. (2003). Large memory window and long charge-retention time in ultranarrow-channel silicon floating-dot memory. *Appl. Phys. Lett.*, **82**, 1787–1789.
9. Ohba, R., Sugiyama, N., and Uchida, K. (2002). Nonvolatile Si quantum memory with self-aligned doubly-stacked dots. *IEEE Trans. Electron Devices*, **49**, 1392–1398.

Part II

Quantum Waveguide Theory in Mesoscopic Systems

Chapter 9

Properties of Quantum Transport

9.1 Characteristic Length [1]

In traditional transistors, the electron's movement is classical. But when the scale of the device is so small that its scale is about equal to or smaller than some characteristic lengths, there will be distinct quantum effects of current.

Electrons in two-dimensional electron gas (2DEG) degenerate at a low temperature, so their Fermi wave vector can be written as

$$k_{\mathrm{F}} = \sqrt{2\pi n_{\mathrm{s}}}, \tag{9.1}$$

where n_{s} is the area density of 2DEG. The electrons' Fermi wavelength is

$$\lambda_{\mathrm{F}} = \frac{2\pi}{k_{\mathrm{F}}} = \sqrt{\frac{2\pi}{n_{\mathrm{s}}}}. \tag{9.2}$$

For electrons with an area density $n_{\mathrm{s}} = 5 \times 10^{11}\ \mathrm{cm}^{-2}$, the Fermi wavelength is about 35 nm, so it is obviously comparable to the scale of mesoscopic devices. At a low temperature, the electric current is contributed mostly by electrons near the Fermi energy; other electrons with energy less than the Fermi energy do not contribute to the conductance. Therefore the Fermi wavelength is associated with the quantum effect of current.

Quantum Waveguide in Microcircuits (Second Edition)
Jian-Bai Xia, Duan-Yang Liu, and Wei-Dong Sheng

ISBN 978-981-4968-30-0 (Hardcover), 978-1-003-65146-8 (eBook)
www.jennystanford.com

Electrons that move in a semiconductor will be scattered by impurities, defects, or phonons. Their momentum relaxation time τ_m and scattering time τ_s have the relationship

$$\frac{1}{\tau_m} = \frac{1}{\tau_s}\alpha_m, \tag{9.3}$$

where α_m is a constant between 0 and 1, and it denotes effectiveness for momentum relaxation of different scatterings.

The mean free path L_m is defined as the average distance traveled by an electron before it loses its momentum. So it can be written as

$$L_m = k_F \tau_m, \tag{9.4}$$

where k_F is the Fermi velocity, and for electrons with an area density $n_s = 5 \times 10^{11}$ cm^{-2}, on the basis of Eq. 9.1, we obtain:

$$k_F = \frac{\hbar k_F}{m^*} = \frac{\hbar}{m^*}\sqrt{2\pi n_s} = 3 \times 10^7 \text{ cm/s}, \tag{9.5}$$

where m^* is the electron's effective mass and m_0 is the rest mass of an electron. If $m^* = 0.067\ m_0$ and $\tau_m \cong 100$ ps, then from Eq. 9.4 $L_m = 30$ μm can be obtained.

The phase relaxation length is defined as the average distance traveled by an electron before the electron wave loses its phase. Similar to the momentum relaxation time τ_m, the phase relaxation time τ_φ has the following relationship with the scattering time τ_s:

$$\frac{1}{\tau_\varphi} = \frac{1}{\tau_c}\alpha_\varphi, \tag{9.6}$$

where α_φ is a constant between 0 and 1, and it denotes the effectiveness for phase destruction of different scattering.

To understand the concept of phase destruction, we imagine an experiment of an Aharonov–Bohm (AB) ring. The incident electron beam splits into two at one end of the ring, moves up and down the two arms, then meets at the other end of the ring. If the lengths of both arms are the same and the phases of the two waves at the meeting point are equal, then the amplitude rises. Assume that there are impurities or defects (named scatterers) randomly distributed in each arm. They will scatter the passing electron wave. Because these scatterers are static and elastic, they do not change the electron's energy, just change its phase. When the two electron waves meet, their phases are unequal and so interference reduces the amplitude.

Because the scatterers are static, the phase relationship between the two paths is definite.

If we add a magnetic field perpendicular to the ring, the electron wave moving up or down an arm get a extra positive or negative phase shift, which is in proportion to the magnetic flux Φ through the ring. Therefore the amplitude of the output electron wave oscillates with Φ, and it is the AB effect. In the case that there are scatterings in the arms, although the amplitude reduces when $\Phi = 0$ (the magnetic field is zero), when the strength of the field increases, the amplitude's oscillation with Φ doesn't change. When Φ reaches a certain value, the amplitude reaches its maximum. We can say that the loss of the phase at the scatterings was compensated by the magnetic field. In this condition, we can consider

$$\alpha_\varphi = 0, \ \tau_\varphi \to \infty, \tag{9.7}$$

that is, the elastic scattering of static scatterings doesn't affect the phase relaxation time. This fact has been proved by the experiment. Experiments found that the length of two arms of an AB ring is much larger than the mean free path L_{m} in general, that is, an electron has experienced repeatedly the momentum elastic scattering but experiments still observe the AB oscillation.

A major factor that affects phase relaxation is the inelastic scattering of electron-phonon. A phonon is one quantum mode of the lattice vibrations, which is not a static point as an impurity or a defect, so its scattering on electrons has the property of being random. As it is inelastic scattering, in every scattering the electron's energy will increase or decrease the energy of a phonon. When two electron waves scattered by phonons meet, the relationship of their phases is irregular, so the average amplitude of the electron wave reduces. In summary, rigid scatterers do not contribute to phase relaxation, only fluctuating scatterers do.

Assume that after time τ_φ, the mean square energy that the electron gains in phonon scattering is equal to the square of energy that the electron gains each time multiplied by the number of scattering:

$$(\Delta\varepsilon)^2 = (\hbar\omega)^2 \left(\tau_\varphi / \tau_{\mathrm{s}}\right), \tag{9.8}$$

where $\hbar\omega$ is the energy of a phonon. The phase relaxation time is defined as that after time τ_φ the mean square value of an electron's phase change is roughly equal to 1:

$$\Delta\varphi \sim (\Delta\varepsilon)\,\tau_\varphi/\hbar \sim 1. \tag{9.9}$$

It can be obtained from Eq. 9.8 and Eq. 9.9:

$$\tau_\varphi = \left(\frac{\tau_S}{\omega^2}\right)^{1/3}. \tag{9.10}$$

Therefore for the low-frequency phonon (acoustic phonon), its influence on the phase relaxation is small. A major factor of phase relaxation is the optical phonon.

At a low temperature, a major factor of the phase relaxation is electron–electron scattering. The frequency of electron–electron scattering is determined by the difference of electron energy E and Fermi energy E_F: $\Delta = E - E_F$. Because Δ is smaller, the states that can be scattered are very few because of the Pauli exclusion principle. So the probability of scattering goes to zero. In 2DEG, it has been proved that

$$\frac{\hbar}{\tau_\varphi} \sim \frac{\Delta^2}{E_F}\left[\ln\left(\frac{E_F}{\Delta}\right) + \text{const.}\right]. \tag{9.11}$$

Because the hot electron's average energy $\Delta \approx k_B T$, the relationship between τ_φ and T is the formula when Δ is replaced by $k_B T$ in Eq. 9.11.

9.2 Phase-Coherent Effect [2]

When electrons pass through a device, if the size of the device is smaller than electrons' coherent length (inelastic mean free path), different waves will interfere. The coherent effect will result in additional scattering, then reduce the conductivity. In addition, the coherent effect will result in the AB effect, universal conductance fluctuation, etc.; all these are named mesoscopic effects.

One of important problems is the conductance fluctuation caused by the quantum effects. Assume the action area of the gate is 0.1 $\times$ 0.05 μm^2 and the carrier concentration in the inversion layer is 2×10^{12} cm^{-2}. Then there are only 100 electrons under the gate. The conductance's change caused by the phase coherence is on the order of e^2/h, about 40 μS. If the normal conductivity is 1000 mS/mm, then for a device with a gate width 0.1 μm, the total conductance is 100 μS. Therefore the conductance fluctuation caused by the phase coherence is 40% of total conductance, so it will limit the performance of the device greatly.

In addition, if the orbit of a ballistic electron surrounds an obstacle, such as an impurity atom, then interference between two orbits will also result in the phase coherent effect. Especially when there is a perpendicular magnetic field in the area surrounded by two orbits, the AB effect will occur. Here the wave function in the closed loop can be expressed by the integral along the loop:

$$\Psi \sim \exp\left[\frac{e}{\hbar}\left(\int_1 \mathbf{A}\cdot d\mathbf{l} - \int_2 \mathbf{A}\cdot d\mathbf{l}\right)\right] = \exp\left[\frac{e}{\hbar}\oint \mathbf{B}\cdot d\mathbf{S}\right] = \exp\left(\frac{\Phi}{\Phi_0}\right), \qquad (9.12)$$

where $\Phi_0 = h/e$ is the magnetic flux quantum. The electric current passing through the closed loop will oscillate periodically with the magnetic flux. When a current of 10 mA pass through the loop, if there is a magnetic field of 0.02 T, then the AB effect can result in distinct fluctuation.

9.3 Coulomb Blockade Effects

Experiments found that for a small quantum dot, only when the electric voltage is larger than a certain value, the electron can pass through the dot. The reason is that the Coulomb interaction between the electron in a quantum dot and the electron in a circuit will reject the entry of the second electron to the quantum dot. Only when the energy of the electron exceeds the Coulomb interaction energy, the second electron can enter the quantum dot and the current can occur. This appearance is named Coulomb blockade (CB) effects, and the relationship curve between voltage and current has the shape of a step, which is named the Coulomb staircase. The CB effects have been observed in 2DEG systems in Si metal-oxide semiconductor field-effect transistor (Si-MOSFET) and GaAs heterojunction.

From the point of view of a large-scale integrated circuit, all effects above are unfavorable, and we should avoid these in design.

On the other hand, these mesoscopic effects may be utilized in the new generation of electronic devices. For example, in a device that has a size smaller than the inelastic scattering mean free path, electrons follow the motion law of quantum mechanics, so we can utilize properties of quantum interference and quantum waveguide to design the device and to control the movement of electrons. And we can design a single-electron transistor by using CB effects, which can greatly reduce the number of electrons stored in one bit and can reduce power dissipation of memory devices. Because spin of an electron has a much longer scattering time and distance than that of electric charge, we can utilize spin as the information carrier to design spintronics devices and so on.

9.4 Landauer–Büttiker Formula

Datta first proposed the concept of a quantum device in 1989 [3], which can only be understood by considering an electron's wave property in simple terms. Concepts that can describe an electron's movement are no longer scattering probability, relaxation time, mobility. and so on. For a two-terminal device, assume that a^+ and a^- are amplitudes of the electron wave in the income circuit and b^+ and b^- are amplitudes of the electron wave in the outcome circuit. Then for a definite energy E, they can be related by a scattering matrix

$$\begin{pmatrix} a^- \\ b^- \end{pmatrix} = \begin{pmatrix} r(E) & t'(E) \\ t(E) & r'(E) \end{pmatrix} \begin{pmatrix} a^+ \\ b^+ \end{pmatrix}, \tag{9.13}$$

where $r(E)$ and $r'(E)$ denote reflection coefficients and $t(E)$ and $t'(E)$ denote transmission coefficients, which are determined by the internal potential and structure.

The circuit has a certain width. Then there are transverse modes with different energies. In common, the transport in the longitudinal direction is multimode, so the amplitudes a^+ and a^- are column matrices of size $(M_a \times 1)$ and b^+ and b^- are column matrices of size $(M_b \times 1)$. Similarly, r is a matrix with size $(M_a \times M_b)$, t is a matrix with size $(M_b \times M_a)$, and so on. M_a and M_b are the number of transverse modes in the income circuit and outcome circuit, respectively.

In Landauer–Büttiker formula, conductance is expressed by the scattering performance, so this theory is suitable for solving conductance of a system that is made of different materials or conductors of different shapes in contact with each other. This theory is the basic theory in the research on mesoscopic transport.

Suppose an electron moves in an infinite ideal conductor along the z axis and ignore the positive ions in lattice points and the scattering of positive electric charge on the electron. In the absence of a magnetic field (so take no account of the spin), the Hamiltonian of the electron is

$$H = \frac{p^2}{2m} + V(x, y), \tag{9.14}$$

where $V(x, y)$ is the lateral confinement potential in the wire. On the basis of the Hamiltonian (Eq. 9.14), we can obtain the system's eigenstate

$$\Psi_{\alpha k}(x, y, z) = \frac{1}{\sqrt{2\pi}} e^{ikz} \phi_\alpha(x, y), \tag{9.15}$$

where k is the wave vector along the z direction of the wire, α denotes the α-th transverse eigenstate, and the eigenenergy corresponding to the α-th eigenstate is $E_{\alpha k} = E_\alpha + \hbar^2\ \mathrm{k}^2/2\ \mathrm{m}$. The first term and the second term are lateral and longitudinal energy, respectively, and the lateral energy E_α is determined by the lateral confinement potential $V(x, y)$. We can use the square potential well to express the lateral confinement potential of an ideal wire. Suppose the well in one direction is very narrow and the width in another direction is W. Then the lateral energy $E_\alpha = (\mathrm{n}\hbar\pi)^2/2\ mW^2$ has discrete energy levels. According to this, the energy spectrum of the system is made up of a series of parabolic and discrete energy bands and the minimum value of the corresponding curve is E_α. As shown in Fig. 9.1, every parabola denotes a lateral energy band, which is named the allow mode.

When $T = 0$ K, the current in the α-th channel of the wire can be expressed as

$$I_\alpha = \frac{2}{2\pi}\int_0^{k_{\max}} ev_{\alpha k} dk = \frac{2}{2\pi}\int_0^{k_{\max}} ev_{\alpha k}\frac{dk}{dE_{\alpha k}} dE_{\alpha k} = \frac{2e}{h}(\mu - E_\alpha), \tag{9.16}$$

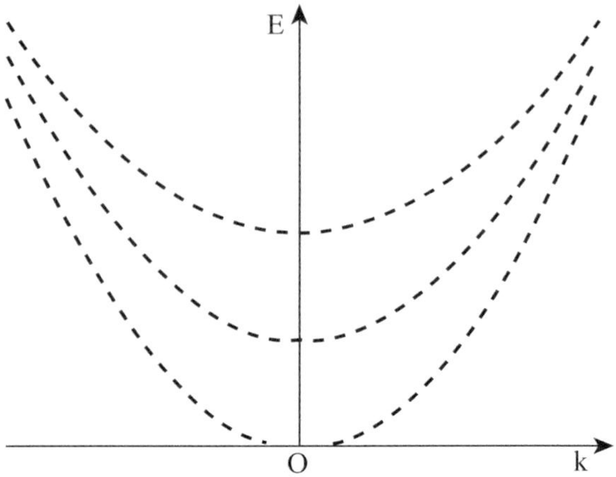

Figure 9.1 Dispersion relation $E(k)$ as the function of k.

where $v_{\alpha k} = 1/\hbar \times dE_{\alpha k}/dk$ is electron's group velocity and μ is the chemical potential (Fermi energy level).

If the width of the ideal conductor is small enough, then the energy gap between two energy bands will be large enough for the electron to just occupy the lowest energy band, and then this energy band will become the one and only allow channel. At present we can calculate the net current in the wires with the chemical potentials μ_1 and μ_2 connected to each other:

$$I = I_1 - I_2 = \left(\frac{2e}{h}\right)\mu_1 - \left(\frac{2e}{h}\right)\mu_2 = \left(\frac{2e^2}{h}\right)(V_1 - V_2). \quad (9.17)$$

Therefore the conductance of a two-terminal, single-channel device is

$$G = \frac{I}{V_1 - V_2} = \frac{2e^2}{h}. \quad (9.18)$$

This is the Landauer formula of a two-terminal, single-channel system; $2e^2/h$ is called the quantum conductance.

If the wire has a finite width, the electron may occupy several energy bands. Assume that N channels have been filled in and $f(E)$ denotes the probability that the channel has been occupied by the electron. At present the current can be written as

$$I = \frac{2e}{h}\int_0^{\infty} dE\left[f_1(E)\sum_i T_i(E) - f_2(E)\sum_i T_i(E)\right], \quad (9.19)$$

where $T_i = \sum_j T_{ij} = \sum_j |t_{ij}|^2$ denotes the total transmission probability of an electron transport from all channels of wire 1 to the i-th channel of wire 2. $f_1(E)$ and $f_2(E)$ are the Fermi–Dirac distribution function in wire 1 and wire 2, respectively. Therefore the current of the system is

$$I = \frac{2e}{h}\int dE\,[f(E) - f(E + V_{\mathrm{D}})]T(E), \tag{9.20}$$

where $f(E)$ is the Fermi–Dirac distribution function, $V_D = V_1 - V_2$ is the bias voltage of the system, and $T(E) = \Sigma_i T_i(E)$ is the total transmission probability. For a small bias, Eq. 9.20 can be simplified as

$$G = \frac{I}{V_D} = \frac{2e^2}{h}\int dE\left(-\frac{\partial f}{\partial E}\right)T(E). \tag{9.21}$$

At a low temperature, $-\partial f/\partial E = \delta(E - E_{\mathrm{F}})$. So in the end, we obtain

$$G = \frac{2e^2}{h}T(E_{\mathrm{F}}) = \frac{2e^2}{h}\sum_{i=1}^{M_{\mathrm{a}}}\sum_{j=1}^{M_{\mathrm{b}}}\left|t_{ij}(E_{\mathrm{F}})\right|^2, \tag{9.22}$$

where $t_{ij}(E_{\mathrm{F}})$ is the transmission coefficient shown in Eq. 9.13, and Eq. 9.22 is the Landauer–Büttiker formula for the two-terminal, multichannel device. It follows that the key point of utilizing the Landauer–Büttiker formula is solving the transmission coefficient matrix t. Most research on mesoscopic transport focuses on this problem, so next we will introduce three methods that are widely used to calculate the transmission coefficient matrix.

The two-terminal Landauer formula can be extended to the multi-terminal device. Considering a three-terminal device, we can write its scattering matrix functions similar to Eq. 9.13:

$$\begin{pmatrix} a^- \\ b^- \\ c^- \end{pmatrix} = \begin{pmatrix} r_{aa} & t_{ab} & t_{ac} \\ t_{ba} & r_{bb} & t_{bc} \\ t_{ca} & t_{cb} & r_{cc} \end{pmatrix}\begin{pmatrix} a^+ \\ b^+ \\ c^+ \end{pmatrix}. \tag{9.23}$$

Büttiker has proved that the current I_i $(i = a, b, c)$ has the following relationship with the chemical potential of every circuit μ_i:

$$I_i = \frac{2e^2}{h}\sum_{j=a,b,c}(T_{ij}\mu_j - T_{ji}\mu_i), \tag{9.24}$$

where

$$T_{ij} = \sum_{m,n=1}^{M} \left|\left(t_{ij}\right)_{mn}\right|^2. \tag{9.25}$$

When there is no external field, $T_{ij} = T_{ji}$ and Eq. 9.24 can be written as

$$I_i = \frac{2e^2}{h} \sum_{j=a,b,c} T_{ij}\left(\mu_j - \mu_i\right), \tag{9.26}$$

where $\mu_j - \mu_i$ corresponds to the bias voltage between two circuits. The conductance between each pair of circuits can be defined as

$$G_{ij} = \frac{2e^2}{h} T_{ij}, \quad i, j = a, b, c\,(i \neq j). \tag{9.27}$$

Equations 9.22 and 9.27 are the fundamental formulas of quantum conductance. They form the theoretical basis of future quantum devices. For a certain device, we should obtain scattering matrices $r_{aa}(E)$, $t_{ab}(E)$, ... It is shown in Eq. 9.24 that the quantum device is non-local, and change of any terminal's chemical potential will influence currents in all circuits.

9.5 Spintronics

Electron spin, or intrinsic electronic angular momentum, is another intrinsic property of an electron besides electric charge. The electron spin has two discrete values $\pm\hbar/2$, which are commonly named the spin-up state and the spin-down state. This is very similar to the high and low levels or switch off and switch on used to realize logic '1' and '0' states. Moreover, a 1/2-spin system is an ideal two-level system, and spin-up and spin-down states are an ideal choice to realize the quantum bit (qubit) in quantum calculation. Recently in the traditional electronic industry, devices are getting more and more integrated and the size of devices is getting smaller and smaller, which results in distinct quantum effects and too much energy consumption. Therefore, naturally, people think of utilizing spin-free degree, which always was neglected formerly, to replace or to combine the charge-free degree. Therefore, the research on how to control spin-free degree effectively has attracted wide attention of scientists and was named "spintronics" by Wolf et al. formally in 2001 [4]. Spintronics researches a particle's spin in a solid and applies it in devices, and it contains spin polarization's production, control, transport, and exploration. It is an interdiscipline field that involves electronics, magnetism, optics, and micro- and nanoprocessing technology. According to the materials researched, it can be grouped into two areas: metal spintronics and semiconductor spintronics.

Ideal spintronics devices realize functions by coupling of spin orientation and spin, so they may not be restricted by the number of electric charges and energy dissipation, and they have a lower power dissipation and a faster running speed. Moreover, electronic storage devices based on spin have the advantage of nonvolatility. This shows clearly that spintronics will have wide prospects of development and application in areas of quantum calculation, communication, molecules, chemistry, and so on. Therefore people expect that spintronics can construct the physical foundation of a new generation of nanoelectronics and the spintronics device will replace the current microelectronics devices.

The symbol of spintronics' birth is the discovery of the giant magnetoresistance (GMR) of the metallic multilayer structure in 1988 [5, 6]. The proposition of spin-vales and room-temperature magnetic sensor utilizing anisotropic magnetoresistance makes for great success of GMR in commerce. Nowadays a new generation of the magnetic heads of high-density disks is using the GMR principle. Now thanks to the high sensitivity to the magnetic field of GMR, each GMR magnetic head has to be reduced to 30–50 nm and more than 4 T volume in desktops and more than 1 T volume in laptops have been realized.

Now spintronics devices are mostly made of metal magnetic materials, which belong to the first category of spintronics, that is, magnetoelectronics. In 1995 scientists found a new phenomenon, tunneling magnetoresistance (TMR) [7], which has larger magnetoresistance and smaller coercive force. The magnetic tunnel junction (MTJ) utilizing the TMR technology has been created, and new magnetoresistive random access memory (MRAM) is expected to replace nonvolatile memory based on the complementary metal-oxide semiconductor (CMOS). This new generation of memory has advantages such as no data loss on power failure, a rapid reading speed comparable to static random access memory (SRAM), and large memory space comparable to dynamic random access memory (DRAM), so it has wide application prospects.

The second category of spintronics mainly studies semiconductors, and it hopes to create an imbalance of the spin number to realize spin transistors and spin valves and then replace traditional electronic devices. These new spintronics devices have characteristics such as low energy consumption and rapid switching speed. Because ready-made mature semiconductor technology and equipment can be used to reduce the developing cost, these devices have attracted wide attention. To utilize spin-free degree in semiconductors, we must produce, keep, control, and detect the spin polarization of carriers.

There are several methods that can produce spin current: ohmic injection, tunnel injection, ballistic electron injection, utilizing the giant Zeeman splitting of a dilute magnetic semiconductor (DMS) in a magnetic field, utilizing a ferromagnetic semiconductor (FMS) as a spin calibrator, the optical method, and so on. Ohmic injection means using a ferromagnetic metal (FM) as an electrode to inject spin-polarized electrons. If the ohmic contact is created in the junction between the FM and the semiconductor, then it can be expected that the spin-polarized current in the FM can be injected into the semiconductor. However, the ohmic contact by heavy doping produces scattering of the spin flip, which reduces spin polarization. At $T < 10$ K, from FM-InAs ohmic contact, spin-polarized electrons of 4.5% have been obtained [8]. At room temperature, from the Fe-GaAs contact, spin-polarized electrons of 2% have been obtained [9].

The scanning tunneling microscopy (STM) experiment has proved that a process of vacuum tunnel can inject spin-polarized electrons into a semiconductor effectively [10]. The FM-insulator-FM tunnel junction with a high magnetoresistance experiment has determined that the tunneling barrier can remain spin polarization of electrons in the tunneling process. This means tunneling is a more effective method of spin injection than diffusive transport. The two methods above both rely on the ferromagnetic film's epitaxial growth on a semiconductor and demand a sharp interface and a high-quality Schottky potential barrier between the metal and the semiconductor.

The third method is using ballistic transport. From the energy difference of the conduction bands of two kinds of spin in a ferromagnetic material and the conduction band of a semiconductor, we can determine the ballistic electron's spin transmissivity and reflectivity at the junction. If we create a point contact between a FM and a nonferromagnetic metal, experiments have proved that current with spin polarization up to 40% is realized in a nonferromagnetic metal by using a ballistic injection [11]. Using the DMS $Be_{0.07}Mn_{0.03}Zn_{0.9}Se$ as the spin aligner, at a low temperature ($\sim$10 K) and an external magnetic field (>1 T), spin polarization injection efficiency up to 90% is realized when electrons pass through the $Be_{0.07}Mn_{0.03}Zn_{0.9}Se$ with a width of 300 nm [12]. This is because in a magnetic field, a low concentration of Mn ions creates giant Zeeman splitting in the conduction and valence bands by the sp-d interaction and the electron's effective Lande g-factor may be up to 100. At a low temperature, electrons only occupy the lowest $S = -1/2$ band, so a spin polarization current is produced. Similarly, a FMS can be used as a spin aligner.

The optical method is an important method to produce a spin current, and it is also an important method to detect spin polarization of carriers. Assume that there is a sample of a GaAs quantum well, and its direction of growth is along the z axis. In the GaAs quantum well, the energies of the heavy hole and the light hole are split and the heavy hole state is the ground state, while the light hole state is the excited state. Using a beam of circularly polarized light (σ^+) with energy equal to a light exciton to radiate on the quantum well sample along its growth direction, due to the transition selection rule of heavy holes, light holes, and electrons, this beam of light can just excite electrons in light hole states of $L_z = -1/2$ to electron states of $S_z = 1/2$ ($\Delta L_z = +1$), so in the conduction band electrons with $S_z = 1/2$ are much more than electrons with $S_z = -1/2$. After the laser stops, electrons in the conduction band will recombine with the heavy hole and produce light with σ^+ polarization and σ^- polarization, which correspond to electron–heavy hole transition ($S_z = -1/2 \rightarrow L_z = -3/2$) and ($S_z =$

$+1/2 \rightarrow L_z = 3/2$), respectively. Therefore the intensity of light with σ^- polarization I_- is much larger than that of light with σ^+ polarization I_+, and I_+/I_- is equal to the ratio of numbers of electrons with $S_z = 1/2$ and electrons with $S_z = -1/2$.

About how to control spin polarization of carriers in a semiconductor, various semiconductor spintronics devices have been proposed. The original one is a spin field-effect transistor (SFET) proposed by Datta and Das in 1989 [13], which is also the most famous one. We will introduce it in detail in the next part. Other important ones are the graphene spin field-effect transistor, the metal-oxide-silicon spin field-effect transistor, the Johnson spin switch, the unipolar spin transistor, and so on. Although making great efforts, people have not succeeded even in the original Datta–Das transistor yet, which shows that we have a long way to go and there is much work to do.

The third category of spintronics is very different from the first two. It attempts to control spin the quantum state of a single electron or several electrons and then realize quantum calculation and quantum communication. In other words, it attempts to produce a qubit by using the two-quantum-state system of spin. The study of this kind of spintronics devices is in the very early stages. Due to the advantage that a qubit can be laid in solid materials, it has a better prospect than some other schemes of quantum calculation, such as iron trap and cavity quantum electrodynamics.

9.6 Rashba Spin-Orbit Interaction

Datta and Das first proposed the concept of the spin transistor, which is based on controlling the spin of an electron [13]. The paper's title is "Electronic Analog of the Electro-optic Modulator," and its principle is as shown in Fig. 9.2. Figure 9.2a is an electro-optic modulator, and Fig. 9.2b is a spin transistor similar to an electro-optic modulator.

The principle of the electro-optic modulator should be introduced first. A linearly polarized light at 45° to the y axis is radiated from the left of the electro-optic modulator along the x direction, and it can be represented as a combination of the z- and y-polarized

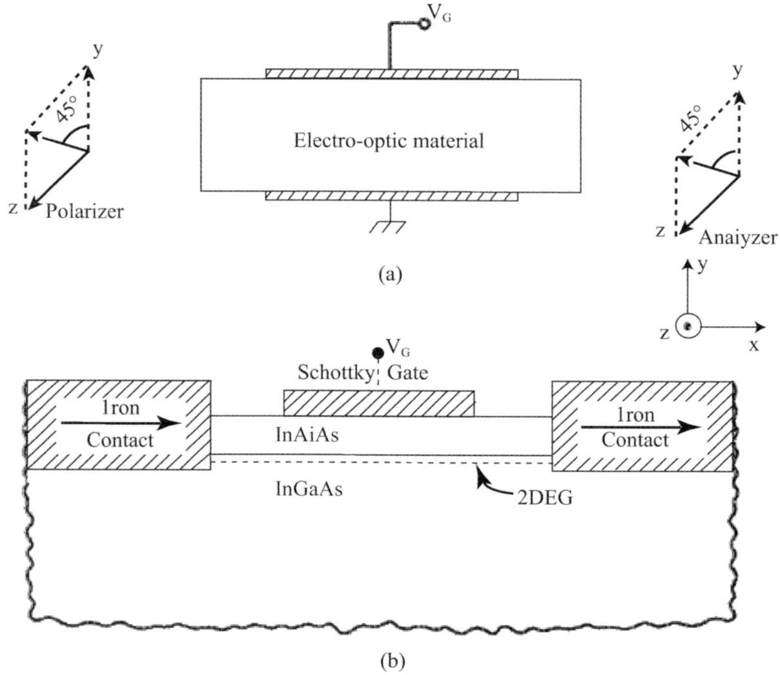

Figure 9.2 (a) Electro-optic modulator; (b) a spin transistor similar to an electro-optic modulator.

light

$$\begin{pmatrix} 1 \\ 1 \end{pmatrix}_{(45^\circ \text{ pol})} = \begin{pmatrix} 1 \\ 0 \end{pmatrix}_{(z \text{ pol})} + \begin{pmatrix} 0 \\ 1 \end{pmatrix}_{(y \text{ pol})} . \tag{9.28}$$

Because the electro-optic effect makes the dielectric constant ε_{zz} slightly different from ε_{yy}, as this light passes through the electro-optic material, the two polarizations suffer different phase shifts $k_1 L$ and $k_2 L$, where L denotes the length of the crystal. The light emerging from the electro-optic material is represented as $((e^{ik_1^L})/(e^{ik_2^L}))$. The analyzer at the output lets the light with polarization (1/1) pass through, so the output power P_o is given by

$$P_o \propto \left| \begin{pmatrix} 1 & 1 \end{pmatrix} \begin{pmatrix} e^{ik_1 L} \\ e^{ik_2 L} \end{pmatrix} \right|^2 = 4\cos^2 \frac{(k_1 - k_2)\, L}{2}. \tag{9.29}$$

The light output is modulated by a gate voltage that controls the difference of the phase shifts $\Delta\theta = (k_1 - k_2)L$.

The theorem of the spin transistor is based on the interaction of the electron orbit and spin, that is, the Rashba interaction [14],

$$H_{\mathrm{R}} = \frac{\alpha}{\hbar}(\sigma \times p)_y = i\alpha\left(\sigma_x\frac{\partial}{\partial z} - \sigma_z\frac{\partial}{\partial x}\right), \tag{9.30}$$

where α is named as the Rashba coefficient and generally is $(1-10) \times 10^{-10}$ eVcm and σ_x and σ_z are the Pauli matrices.

In the spin transistor as shown in Fig. 9.2b, both ends are FM iron contacts, which are magnetized in the x direction and correspond to the polarizer and analyzer in the electro-optic modulator of Fig. 9.2a. Electrons injected through the source electrode are spin-polarized along the x direction, which can be represented as a linear combination of positive z-polarized and negative z-polarized electrons

$$\begin{pmatrix}1\\1\end{pmatrix}_{(+z\ \mathrm{pol})} = \begin{pmatrix}1\\0\end{pmatrix}_{(+z\ \mathrm{pol})} + \begin{pmatrix}0\\1\end{pmatrix}_{(-z\ \mathrm{pol})}. \tag{9.31}$$

Due to the Rashba term in the effective mass Hamiltonian (Eq. 9.30) $H_{\mathrm{R}} = \alpha(\sigma_z k_x - \sigma_x k_z)$, in the absence of a magnetic field, the Rashba term causes the spin-up and spin-down states to split. If an electron is traveling in the x direction with $k_x \neq 0$ and $k_z = 0$, then considering the Rashba term, the energies of the electron can be written as

$$E\ (+z\ \mathrm{pol}) = \frac{\hbar^2 k_{x1}^2}{2m^*} + \alpha k_{x1},$$

and

$$E\ (-z\ \mathrm{pol}) = \frac{\hbar^2 k_{x2}^2}{2m^*} - \alpha k_{x2}. \tag{9.32}$$

Electrons with two kinds of spin polarizations have the same energy, so

$$k_{x2} - k_{x1} = \frac{2m^*\alpha}{\hbar^2}. \tag{9.33}$$

Therefore there is a differential phase shift for spin-up and spin-down electrons after they pass through the junction area, which is shown as

$$\Delta\theta = (k_{x2} - k_{x1})\,L = \frac{2m^*\alpha L}{\hbar^2}. \tag{9.34}$$

For InGaAs/InAlAs heterostructures, from the experimentally observed zero-field spin splitting, α was estimated to be $\sim 3.9^{10}$ eV·cm.

To make a phase difference of π, L is about 0.67 μm, which is less than the mean free path (≥1 μm) in high-mobility semiconductors at low temperatures.

All the above discussion is in the condition $k_x \neq 0$ and $k_z = 0$. For a general case, both k_x and k_z should be taken into account, so the Hamiltonian is more complex. And If $k_z \neq 0$ and $k_x = 0$, the eigenstates are $(1/1)$ and $(1/(-1))$ so the incoming beam is not split at all. It goes entirely into one of the eigenstates, and consequently no current modulation is expected. As the angle θ of propagation of the electrons with the x axis is increased, it can be shown that the effect is reduced gradually to zero at $\theta = 90°$. To prevent this effect, Datta and Das suggested a confining potential $V(z)$, which makes electrons form a series of discrete subbands in the z direction. If the confining potential is narrow enough so the different subbands are sufficiently far apart in energy, then subband mixing can be neglect. Then we return to the case $k_z = 0$, when the current modulation is the strongest.

Since Datta's paper was reported, many kinds of semiconductor spintronics devices have been proposed, almost all of which are based on the spin-orbit interaction (SOI), especially the Rashba spin-orbit interaction (RSOI) [14]. The SOI researched by spintronics can be classified into two classes: the RSOI and the Dresselhaus spin-orbit interaction (DSOI) [15]. The RSOI causes a spin split of electron states in the absence of an external field, that is, 1 order term of momentum arises in the Hamiltonian of the system, which makes the dispersion relation of the electron energy split from the single parabolic curve to two ones.

Both SOIs can be obtained from the Dirac relativistic wave equation and are related to the spin-orbit coupling term in the nonrelativistic limit, that is, the Thomas term. For a semiconductor heterojunction, if $V(z)$ is not inversion symmetric at the junction of 2DEGs, the Thomas term can be written as

$$H_{\mathrm{R}} = i\alpha \left(\sigma_y \frac{\partial}{\partial x} - \sigma_x \frac{\partial}{\partial y} \right),$$

which is just Eq. 9.30, and the Rashba coefficient α is related to the symmetry breaking of $V(z)$. On the other hand, if the system studied has the bulk crystal inversion asymmetry, such as diamond or zincblende semiconductor structures, the spin-orbit coupling term can be written as

$$H_{\mathrm{D}} = \beta \left[\sigma_x k_x \left(k_y^2 - k_z^2\right) + \sigma_y k_y \left(k_z^2 - k_x^2\right) + \sigma_z k_z \left(k_x^2 - k_y^2\right)\right], \tag{9.35}$$

where β is the strength coefficient of spin-orbit coupling, which is related to a specific material. This expression of spin-orbit coupling was first proposed by Dresselhaus in 1955 [15], so it is known as Dresselhaus spin-orbit interaction.

9.7 Quantum Waveguide Theory

It follows that the key point of utilizing the Landauer–Büttiker formula is solving the transmission coefficient matrix t. Actually most research on mesoscopic transport focuses on this problem. To calculate the scattering matrix, authors proposed and developed quantum waveguide theory, including one-dimensional quantum waveguide theory [16], two-dimensional quantum waveguide theory [17], and theory of one-dimensional quantum waveguide taking account of the Rashba interaction [18, 19]. In other chapters we will introduce these theories, related calculation methods, and some results.

References

1. Datta, S. (1995). *Electron Transport in Mesoscopic Systems* (Cambridge, London), p. 16.
2. Ferry, D. K., Takagaki, Y., and Zhou, J. R. (1994). *Jpn. J. Appl. Phys.*, **33**, 873.
3. Datta, S. (1989). *Superlatt. Microstruct.*, **6**, 83.
4. Wolf, S. A., Awschalom, D. D., Buhrman, R. A., et al. (2001). *Science*, **294**, 1488.
5. Baibich, M. N., Broto, J. M., Fert, A., et al. (1988). *Phys. Rev. Lett.*, **61**, 2472.
6. Binasch, G., Grunberg, P., Saurenbach, F., et al. (1989). *Phys. Rev. B*, **39**, 4828.

7. Moodera, J. S., Kinder, L. R., Wong, T. M., et al. (1995). *Phys. Rev. Lett.*, **74**, 3273.
8. Hu, C. M., et al. (2001). *Phys. Rev. Lett.*, **63**, 125333.
9. Zhu, H. J., et al. (2001). *Phys. Rev. Lett.*, **87**, 016601.
10. Alvorado, S. F., and Renand, Ph. (1992). *Phys. Rev. Lett.*, **68**, 1387.
11. Upadhyay, S. K., et al. (1998). *Phys. Rev. Lett.*, **81**, 3247.
12. Fiederling, R., Keim, M., Reuscher, G., et al. (1999). *Nature*, **402**, 787.
13. Datta, S., and Das, B. (1990). *Appl. Phys. Lett.*, **58**, 665.
14. Rashba, E. I. (1960). *Sov. Phys. Solid State*, **2**, 1109.
15. Dresselhaus, G. (1955). *Phys. Rev.*, **100**, 580.
16. Xia, J. B. (1992). *Phys. Rev. B*, **45**, 3593.
17. Sheng, W. D., and Xia, J. B. (1996). *J. Phys. C*, **8**, 3635.
18. Liu, D. Y., Xia, J. B., and Chang, Y. C. (2009). *J. Appl. Phys.*, **106**, 093705.
19. Liu, D. Y., and Xia, J. B. (2010). *J. Appl. Phys.*, **108**, 053717.

Chapter 10

One-Dimensional Quantum Waveguide Theory

In the 1980s, most of the initial work on electron transport in a small system dealt with metal samples, in which many transverse subbands were involved and the transport was diffusive. At the same time, advances in semiconductor microtechnology made it possible to fabricate extremely high mobility quantum wires with narrow widths, in which only a few of the lowest subbands are occupied and the transport is ballistic. The allowed modes in the channel are then the "waveguide" modes.

It follows from Chapter 9 that the key point of utilizing the Landauer–Büttiker formula is solving the transmission coefficient matrix t. Here we introduce the one-dimensional quantum waveguide theory [1, 2], which can be applied to one-dimensional quantum waveguide circuits of any shape and any structure. The so-called one-dimensional circuit means that the width of the circuit is narrow enough for the energy spacing between subbands produced by the transverse confinement to be much larger than the electron's longitudinal kinetic energy and there is only one mode of electron moving in the circuit. The single-mode movement of an electron is

Quantum Waveguide in Microcircuits (Second Edition)
Jian-Bai Xia, Duan-Yang Liu, and Wei-Dong Sheng

ISBN 978-981-4968-30-0 (Hardcover), 978-1-003-65146-8 (eBook)
www.jennystanford.com

described by the plane wave function with the wave vector k along the circuit's direction, which follows the quantum mechanics law.

10.1 Two Basic Equations

Let ψ_i be the wave function in the i-th circuit,

$$\psi_i = c_{1i}e^{ikx} + c_{2i}e^{-ikx}, \tag{10.1}$$

where the coordinate x axis is along the longitudinal direction of the circuit.

The core of this theory is giving boundary conditions at each intersection crossed by more than two circuits, which is described by two basic equations. The first one is the continuity of the wave functions at the intersection, which can be written as

$$\psi_1 = \psi_2 = \psi_3 = \cdots = \psi_n, \tag{10.2}$$

where ψ_i denotes the value of wave function in the i-th circuit at the intersection. The second equation is the conservation of the current density,

$$\sum_i \frac{\partial \psi_i}{\partial x_i} = 0, \tag{10.3}$$

where all the coordinates x point to or back to the intersection.

In the general case, especially when there is spin-orbit interaction, we introduce a current density operator L_i, which has a more complex form than $\partial/\partial x_i$, as in Eq. 10.3. Generally we can obtain it from the Hamiltonian

$$L_i = -\frac{i}{\hbar}\left[l, H\right]. \tag{10.4}$$

The second equation can be written as

$$\sum_i L_i \psi_i = 0. \tag{10.5}$$

The completeness of Eqs. 10.2 and 10.3: From Eq. 10.1 there are $2n$ unknown coefficients for the n circuits crossing the intersection, among which the n coefficients can be determined by the n equations Eqs. 10.2 and 10.3. The other half of the coefficients will be determined by the boundary conditions at the other intersections or the conditions at the input or output terminals. Hence the set of equations Eqs. 10.2 and 10.3 at all intersections is complete for determining the wave function of the whole system.

10.2 Ring with Two Arms

To illustrate the application of the above theory, we consider the structure of a ring with two arms as shown in Fig. 10.1a in the absence of a magnetic field. The two arms of the ring have different lengths L_1 and L_2. We introduce the local coordinate system for each circuit such that the direction is along the electron-current direction and the origin is taken at the intersection. The choice of the coordinate origin is noncritical; it only affects a phase factor on the wave function.

We assume that an electron with a wave vector k enters through circuit 1 and departs from circuit 4. In the local coordinate system, the wave functions in circuits 1–4 shown in Fig. 10.1a can be written as

$$\begin{aligned}\psi_1 &= e^{ikx} + ae^{-ikx},\\ \psi_2 &= c_1e^{ikx} + c_2e^{-ikx},\\ \psi_3 &= d_1e^{ikx} + d_2e^{-ikx},\end{aligned} \tag{10.6}$$

and

$$\psi_4 = ge^{ikx},$$

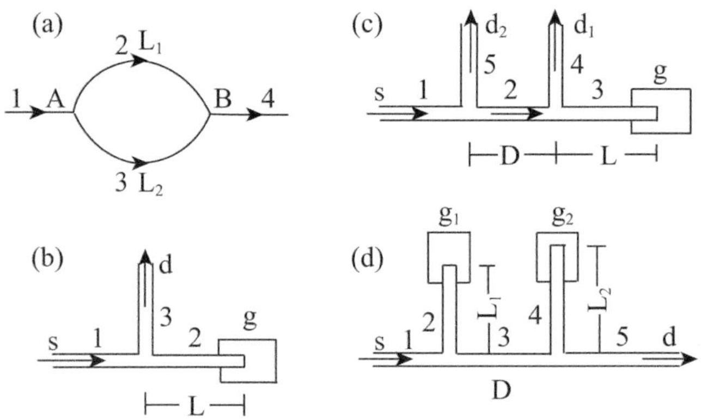

Figure 10.1 Various configurations of mesoscopic structures: (a) a ring with two arms, (b) a quantum interference transistor, (c) a quantum interference device with two drains, and (d) a quantum interference device with two gates. s, g, and d represent the source, gate, and drain, respectively.

where in ψ_1 the coefficient of the injected wave is 1, denoting that one electron enters, and in circuit 4, ψ_4 has only one term, denoting the output wave without the input wave. Thus the coefficients a and g are the reflection and transmission amplitudes, respectively.

The boundary condition equations Eq. 10.2 and Eq. 10.3 for the wave function (Eq. 10.6) can be written at the A and B points,

$$\begin{aligned} 1+a &= c_1+c_2, \\ 1+a &= d_1+d_2, \\ 1-a &= c_1-c_2+d_1-d_2, \\ c_1e^{ikL_1}+c_2e^{-ikL_1} &= g, \\ d_1e^{ikL_2}+d_2e^{-ikL_2} &= g, \end{aligned} \tag{10.7}$$

and

$$c_1e^{ikL_1}-c_2e^{-ikL_1}+d_1e^{ikL_2}-d_2e^{-ikL_2}=g.$$

Solving the set of linear algebraic equations (10.7), we obtain the transmission function [1]

$$\begin{aligned} T &= |g|^2 = \frac{64}{\Delta_L^2}\left[1-\cos(kL)\right]\left[1+\cos(k\Delta L)\right], \\ \Delta_L^2 &= 4\left\{\left[4-5\cos(kL)+\cos(k\Delta L)\right]^2+\left[4\sin(kL)\right]^2\right\}, \end{aligned} \tag{10.8}$$

and

$$L=L_1+L_2, \qquad \Delta L=L_2-L_1.$$

From Eq. 10.8 we see that the amplitude of the output wave changes periodically not only with L but also with ΔL. It is caused by the phase-coherent effect. The T as a function of kL for different $k\Delta L$ and T as a function of $k\Delta L$ for different kL are shown in Figs. 10.2a and 10.2b, respectively.

10.3 Aharonov–Bohm Ring

The structure of the Aharonov–Bohm (AB) ring is the same as that in Fig. 10.1a. When a magnetic field is applied perpendicularly to the ring, in the magnetic field the Schrödinger equation is

$$\left[\frac{1}{2m^*}\left(\mathbf{p}+\frac{e}{c}\mathbf{A}\right)^2+V(\mathbf{r})\right]\psi(\mathbf{r})=E\psi(\mathbf{r}), \tag{10.9}$$

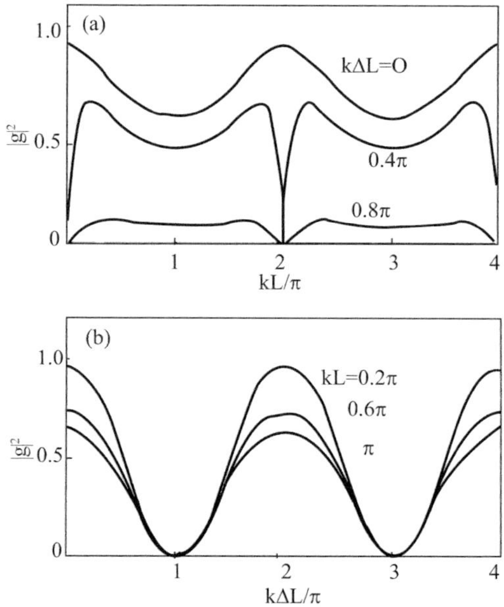

Figure 10.2 (a) T as a function of kL for different $k\Delta L$; (b) T as a function of $k\Delta L$ for different kL.

where **A** is the vector potential of the magnetic field **B**,

$$\mathbf{A} = \nabla \times \mathbf{B}. \tag{10.10}$$

As the magnetic field **B** is perpendicular to the ring plane, according to the Gauss theorem the vector potential **A** is along the ring direction and its magnitude

$$\mathbf{A} = \frac{\Phi}{L}, \tag{10.11}$$

where $\Phi = B \cdot S$ is the magnetic flux through the ring section area S and L is the ring round length.

Inserting Eq. 10.11 into Eq. (10.9), we obtain the one-dimensional (1D) Schrödinger equation,

$$\left[\frac{1}{2m^*}\left(\frac{\hbar}{i}\frac{d}{dx} - \frac{e}{c}\frac{\Phi}{L}\right)^2 + V(x)\right]\psi(x) = E\psi(x). \tag{10.12}$$

The wave function $\psi(x)$ is still a plane wave with wave vector k_1, its eigenenergy

$$E = \frac{\hbar^2}{2m^*}\left(k_1 - \frac{e\Phi}{\hbar c L}\right)^2, \tag{10.13}$$

which should be equal to the energy of the injected electron, $\hbar^2 k^2/2m^*$. Thus we have

$$k_1 = k + \frac{e\Phi}{\hbar c L}. \tag{10.14}$$

For the electron moving in the opposite direction in the ring, we obtain the wave vector of the electron,

$$k_2 = k - \frac{e\Phi}{\hbar c L}. \tag{10.15}$$

The wave functions in circuits 1-4 shown in Fig. 10.1a are written as

$$\begin{aligned} \psi_1 &= e^{ikx} + ae^{-ikx}, \\ \psi_2 &= c_1 e^{ik_1 x} + c_2 e^{-ik_2 x}, \\ \psi_3 &= d_1 e^{ik_2 x} + d_2 e^{-ik_1 x}, \end{aligned} \tag{10.16}$$

and

$$\psi_4 = g e^{ikx},$$

where k_1 and k_2 are given in Eqs. 10.14 and 10.15, respectively. Similarly, we can write down the boundary condition equations at the A and B points and obtain a set of linear algebraic equations for the coefficients a, c_1, c_2, d_1, d_2, and g. For the case $L_1 = L_2 = L_0$, we obtain [1]

$$g = \frac{8i}{\Delta_k}(\sin k_1 L_0 + \sin k_2 L_0)$$

and

$$\Delta_k = 2 + 8\cos\phi + 9e^{-ikL} - e^{ikL}, \tag{10.17}$$

where

$$\phi = (k_2 - k_1)\frac{L}{2} = \frac{e\Phi}{\hbar c}. \tag{10.18}$$

From Eq. 10.17 we obtain the transmission function

$$T = |g|^2 = \frac{64}{\Delta_k^2}(1 - \cos kL)(1 + \cos\phi),$$

where

$$\Delta_k^2 = 4\left[(1 + 4\cos\phi - 5\cos kL)^2 + (4\sin kL)^2\right]. \tag{10.19}$$

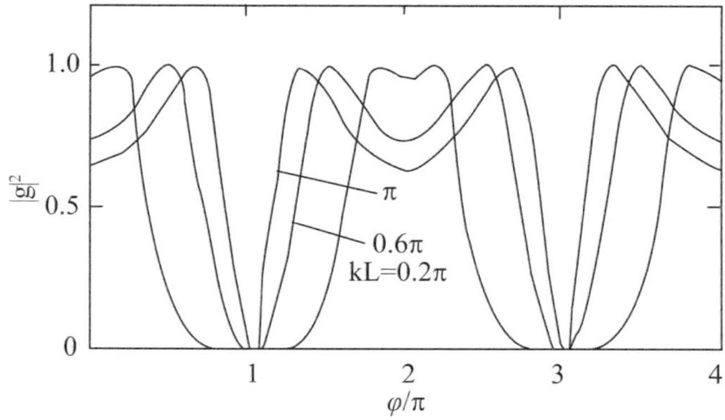

Figure 10.3 T as a function of ϕ/π for different kL values in the AB ring.

From Eq. 10.19 we see that the T varies periodically as the magnetic flux Φ changes with the period

$$\Phi = \frac{hc}{e}. \tag{10.20}$$

This is the basic result of the AB effect, which is in agreement with the results of Datta and Bandyopadhyay [3]. The T as a function of ϕ/π for several kL values is shown in Fig. 10.3. From the figure we see that when kL is close to 0 (or integer times of 2π) the wave's shape is close to the harmonic wave and when kL is close to π the wave's shape deviates from the harmonic wave, indicating there are components of higher harmonics.

10.4 Quantum Interference Devices

The quantum interference transistor [4, 5] as shown in Fig. 10.1b differs from the ordinary field-effect transistor (FET) in that the gate lies outside the classical path of electrons. Conductance oscillations as a function of the gate potential have been observed in such a structure. The wave functions in circuits 1–3 (Fig. 10.1b) can be written as

$$\psi_1 = e^{ikx} + ae^{-ikx},$$
$$\psi_2 = c\sin\left[k(x-L)\right], \tag{10.21}$$

and

$$\psi_3 = ge^{ikx},$$

where it is assumed that the wave function at the gate is zero, that is, it is a standing wave in circuit 2, and L is the distance between the intersection and the gate.

Applying the boundary conditions (Eqs. 10.2 and 10.3), we obtain

$$1 + a = -c \sin kL,$$
$$1 + a = g,$$

and

$$1 - a + ic \cos kL = g. \tag{10.22}$$

From Eq. 10.22 it is easy to obtain

$$a = -\frac{i \cos kL}{2 \sin kL + i \cos kL}$$

and

$$g = \frac{2 \sin kL}{2 \sin kL + i \cos kL}. \tag{10.23}$$

The $T = |g|^2$ as a function of kL is shown in Fig. 10.4 (dashed line). It can be seen that the conductance curve is in good agreement with the experimental single-mode results [4]. This conductance oscillation is also caused by the phase-coherent effect. When the injected electron wave arrives at the intersection, one part goes to the gate and is reflected from the gate. On this path the phase change is $2kL$. This part of the wave meets the wave at the intersection, and there is an interference effect. Consequently, the amplitude of the output wave varies periodically with $2kL$, as shown in Fig. 10.4 (dashed line).

As a development of the interference device, we consider the structure with two drains controlled by one gate as shown in Fig. 10.1c, and the two drains are separated by a distance D. The wave functions in circuits 1–5 can be written as

$$\psi_1 = e^{ikx} + ae^{-ikx},$$
$$\psi_2 = c_1 e^{ikx} + c_2 e^{-ikx},$$
$$\psi_3 = d \sin[k(x - L)],$$
$$\psi_4 = g_1 e^{ikx}, \tag{10.24}$$

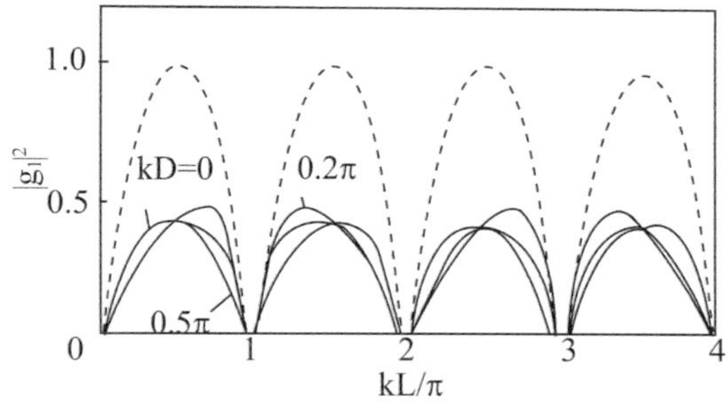

Figure 10.4 T as function of kL; dashed line for the single-drain structure in Fig. 10.1b; solid lines for different kD in the first drain of the structure in Fig. 10.1c.

and

$$\psi_5 = g_2 e^{ikx}.$$

Similarly applying the boundary conditions Eq. 10.2 and Eq. 10.3 we obtain the T_1 and T_2 as functions of kL for three kD values shown in Figs. 10.4 and 10.5, respectively. From Fig. 10.4 we see that in the drain near the gate, the conductance oscillations are nearly the same, independent of kD, and their magnitudes are about half of that in the single-drain structure (dashed line). In the drain far away from the gate, the conductance oscillations critically depend on the kD. When $kD = 0.5\,\pi$, the $T_2 = 1$ at $kL = \pi$. The results of the two-gate structure Fig. 10.1d are not given here; the reader can consult Ref. [1].

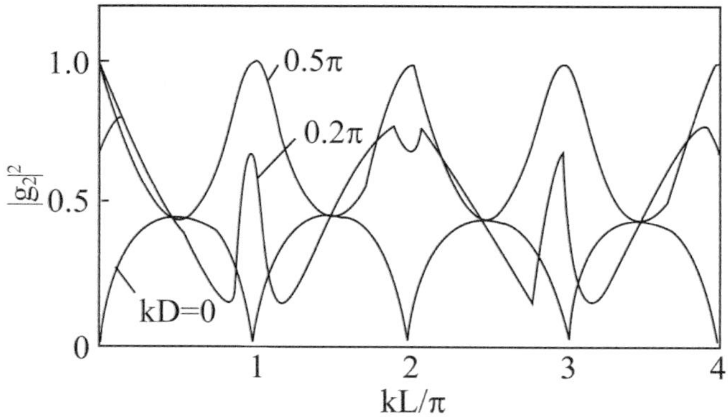

Figure 10.5 T_2 as a function of kL for different kD values in the second drain of the structure in Fig. 10.1c.

10.5 Stub Model

For comparison with the theoretical results of the ideal two-dimensional (2D) electron waveguide model [6], we calculate the transmission function T for the single- and double-gate (stub) structures (as shown in Fig. 10.1d) with the same parameters as in Ref. [6]. Of course, the width effect is neglected in our model. We take the electron effective mass $m^* = 0.05\ m_0$, the electron energy $E = 0.08$ eV, and the separation between two stubs $D = 95$ Å. In Fig. 10.6 the T for the structures of a single stub, two identical stubs, and two stubs with length difference are given, respectively. From the figure we see that there is qualitative agreement with the 2D theoretical results for the structure with equal length [6]. In the case of a single stub, the transmission valley is narrow, while in the case of two identical stubs the valley becomes broader. In the case of two stubs of different lengths, there appears an additional peak at the transmission valley. It is found that the height of the additional peak is sensitive to the kD, and the T as a function of kL^* (length of the shorter one of the two stubs) is shown in Fig. 10.6d. From the figure

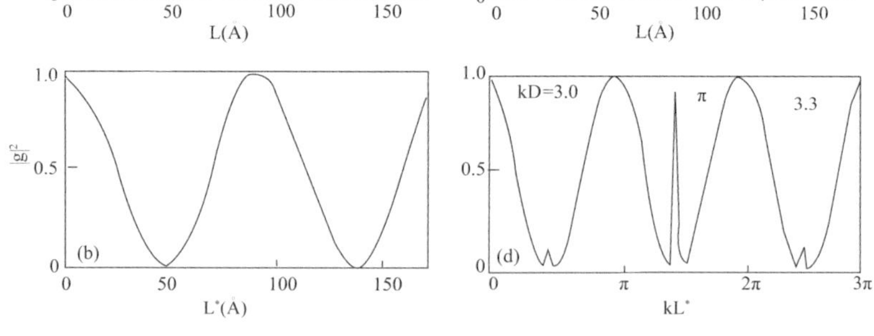

Figure 10.6 T as a function of L for (a) a single stub, (b) two identical stubs, and (c) two stubs with a length difference $\Delta L = 10$ Å. (d) T as a function of kL^* for $kD = 3.0, \pi$, and 3.3.

we see that if $kD = \pi$, there is a strong peak at the transmission valley; if kD deviates from π, the resonant peak decreases greatly.

10.6 One-Dimensional Waveguide Theory of Holes

On the basis of the 1D waveguide theory of electrons we developed the 1D waveguide theory of holes [2]. The hole has an orbital angle momentum $l = 1$ (p state), which is coupled with spin $s = 1/2$ to get the total angle momentum $J = 3/2$ and $1/2$. Generally, the $J = 3/2$ state is the basic state, that is, at the top of the valence band. Thus the hole wave function has four components, corresponding to $J = 3/2, 1/2, -1/2$, and $-3/2$ states. The motion of the hole can be described by the following 4×4 Luttinger Hamiltonian [7]

$$H = \frac{1}{2m_0} \begin{pmatrix} P_1 & Q & R & 0 \\ Q^* & P_2 & 0 & R \\ R^* & 0 & P_2 & -Q \\ 0 & R^* & -Q^* & P_1 \end{pmatrix}, \tag{10.25}$$

where

$$\begin{aligned} P_1 &= (\gamma_1 + \gamma_2)\left(p_x^2 + p_y^2\right) = (\gamma_1 - \gamma_2)\, p_z^2, \\ P_2 &= (\gamma_1 - \gamma_2)\left(p_x^2 + p_y^2\right) = (\gamma_1 + 2\gamma_2)\, p_z^2, \\ Q &= -i2\sqrt{3}\gamma_3\left(p_x - ip_y\right) p_z, \end{aligned} \tag{10.26}$$

and

$$R = \sqrt{3}\gamma_2\left(p_x^2 - p_y^2\right) - 2i\gamma_3 p_x p_y,$$

where γ_1, γ_2, and γ_3 are Luttinger effective mass parameters.

We assume that the hole is confined in the x–y plane; then the momentum p_z can be set to 0. Therefore, the Q term in Eq. 10.26 vanishes, and the Hamiltonian (Eq. 10.24) splits into two 2×2 equivalent blocks,

$$H = \frac{1}{2m_0} \begin{pmatrix} P_1 & R \\ R^* & P_2 \end{pmatrix}. \tag{10.27}$$

In the axial approximation $\gamma_2 = \gamma_3$ and $R = \sqrt{3}\gamma_2\left(\mathbf{p}_x - i\mathbf{p}_y\right)^2$.

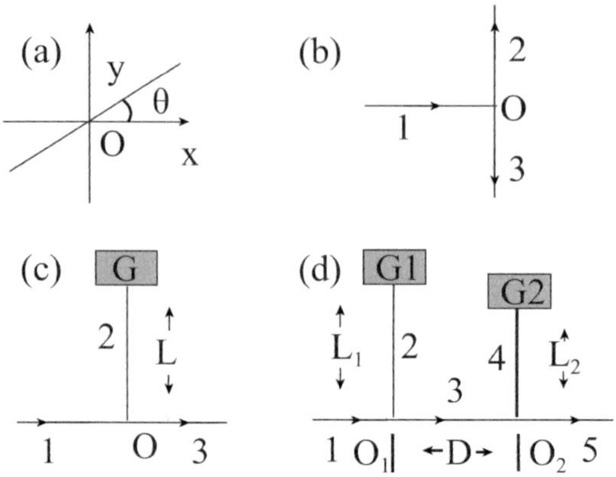

Figure 10.7 (a) Circuit 1 with a polar angle θ; (b) branch structure; (c) quantum-interference device with one gate; (d) quantum interference device with two gates.

The wave function of a hole in the 1D circuit is different from that of the electron. The electron wave function is a plane wave independent of the direction of the circuit, while the hole wave function is two components and dependent on the direction of the circuit. We study the hole wave function in the circuit shown in Fig. 10.7a with a polar angle of θ. The Hamiltonian can be written as

$$H = -\frac{\hbar^2}{2m_0}\begin{pmatrix} (\gamma_1+\gamma_2) & \sqrt{3}\gamma_2 e^{-2i\theta} \\ \sqrt{3}\gamma_2 e^{2i\theta} & (\gamma_1-\gamma_2) \end{pmatrix}\frac{d^2}{dl^2}, \tag{10.28}$$

where l is the coordinate along the circuit. We take the wave function as a plane wave form,

$$\phi = \begin{pmatrix} c_1 \\ c_2 \end{pmatrix} e^{ikl}, \tag{10.29}$$

where k is the wave vector and c_1 and c_2 are the coefficients to be determined. Substituting Eq. 10.29 into the effective mass equation $H\phi = E\phi$ will give rise to the following secular equation:

$$\begin{vmatrix} (\gamma_1+\gamma_2)-\varepsilon & \sqrt{3}\gamma_2 e^{-2i\theta} \\ \sqrt{3}\gamma_2 e^{2i\theta} & (\gamma_1-\gamma_2)-\varepsilon \end{vmatrix} = 0, \quad \varepsilon = E\Big/\left(\frac{\hbar^2k^2}{2m_0}\right). \tag{10.30}$$

From Eq. 10.30 we obtain

$$\varepsilon = \gamma_1 \pm 2\gamma_2. \tag{10.31}$$

From the definition of ε in Eq. 10.30, m_0/ε corresponds to the effective mass of a hole. Then the values of ε in Eq. 10.31 correspond to the light hole state and the heavy hole state, respectively. The wave functions of the heavy hole state and light hole state are

$$\phi_{\rm h}(\theta, k) = \phi_{\rm h}(\theta)\, e^{ikl} = \begin{pmatrix} 1/2 \\ -\sqrt{3}e^{2i\theta}/2 \end{pmatrix} e^{ikl}$$

and

$$\phi_{\rm l}(\theta, k) = \phi_{\rm l}(\theta)\, e^{ikl} = \begin{pmatrix} \sqrt{3}e^{-2i\theta}/2 \\ 1/2 \end{pmatrix} e^{ikl}. \tag{10.32}$$

For the hole state with a given energy E, the general wave function in the circuit Fig. 10.7a can be written as

$$\Phi = c_1\phi_{\rm h}(\theta)\, e^{ik_{\rm h}l} + c_2\phi_{\rm l}(\theta)\, e^{ik_{\rm l}l} + c_3\phi_{\rm h}(\theta)\, e^{-ik_{\rm h}l} + c_4\phi_{\rm l}(\theta)\, e^{-ik_{\rm l}l}, \tag{10.33}$$

where

$$k_{\rm h} = \sqrt{2m_{\rm h}E}, \qquad k_{\rm l} = \sqrt{2m_{\rm l}E}$$

$$m_{\rm h} = \frac{m_0}{\gamma_1 - 2\gamma_2}, \quad \text{and} \quad m_{\rm l} = \frac{m_0}{\gamma_1 + 2\gamma_2} \tag{10.34}$$

are the wave vectors of the heavy hole and light hole, respectively; and c_1, c_2, c_3, and c_4 are the coefficients to be determined by the boundary conditions.

The first boundary condition for the hole wave functions are the same as that for the electron wave function (Eq. 10.2). The second boundary condition is given by Eqs. 10.4 and 10.5, namely the current density operator

$$L_{\rm l} = \frac{1}{m_0}\begin{pmatrix} (\gamma_1 + \gamma_2) & \sqrt{3}\gamma_2 e^{-2i\theta} \\ \sqrt{3}\gamma_2 e^{2i\theta} & (\gamma_1 - \gamma_2) \end{pmatrix} p_{\rm l}. \tag{10.35}$$

The corresponding boundary conditions at the intersection are

$$\Phi_1 = \Phi_2 = \cdots = \Phi_n$$

and

$$\sum_{i=1}^{n} L_i \Phi_i = 0. \tag{10.36}$$

10.7 Quantum Interference Device of a Hole

Obviously, the hole case is more complicated than the electron case, there are four unknown coefficients in each circuit, and generally we cannot obtain the analytical results from the set of linear equations of coefficients. In the following we consider the hole transport in the quantum interference transistors [4, 5] with one gate, as shown in Fig. 10.7c. We assume that an incident heavy hole wave with a wave vector k enters into source circuit 1 and departs from the drain circuits 3. Stub 2 is the gate, whose length L is controlled by the gate voltage. The wave function in stub 2 is a standing wave with its zero pint at the gate. If we choose the node O as the origin of all three circuits, then the wave functions can be written as

$$\Phi_1 = \phi_{\mathrm{h}}(0)\, e^{ikl_1} + a_1\phi_{\mathrm{l}}(0)\, e^{ik'l_1} + a_2\phi_{\mathrm{l}}(0)\, e^{-ik'l_1},$$

$$\Phi_2 = c_1\phi_{\mathrm{h}}(\pi/2)\sin\left[k\left(l_2 - L\right)\right] + c_2\phi_{\mathrm{l}}(\pi/2)\sin\left[k'\left(l_2 - L\right)\right],$$

and

$$\Phi_3 = d_1\phi_{\mathrm{h}}(0)\, e^{ikl_3} + d_2\phi_{\mathrm{l}}(0)\, e^{ik'l_3}, \tag{10.37}$$

where k and k' are the wave vectors of a heavy hole and a light hole, respectively (see Eq. 10.34).

Applying the boundary conditions (Eq. 10.36) we obtain a set of linear algebraic equations of six coefficients in Eq. 10.37. The final results are given in Ref. [2]. It is noticed that the current densities of the heavy hole and light hole equal $\hbar k/m_{\mathrm{h}}$ and $\hbar k'/m_{\mathrm{l}}$ multiplied by their corresponding coefficients, respectively. The transmission and reflection functions can be written as

$$T_{\mathrm{hh}} = |d_1|^2, \quad T_{\mathrm{hl}} = |d_2|^2\frac{k}{k'},$$

$$R_{\mathrm{hh}} = |a_1|^2, \quad \text{and} \quad R_{\mathrm{hl}} = |a_2|^2\frac{k}{k'}, \tag{10.38}$$

where T_{hh} represents the transmission probability of one heavy hole to one heavy hole, T_{hl} the transmission probability of one heavy hole to one light hole, R_{hh} the reflection probability of one heavy hole to one heavy hole, and R_{hl} the reflection probability of one heavy hole to one light hole.

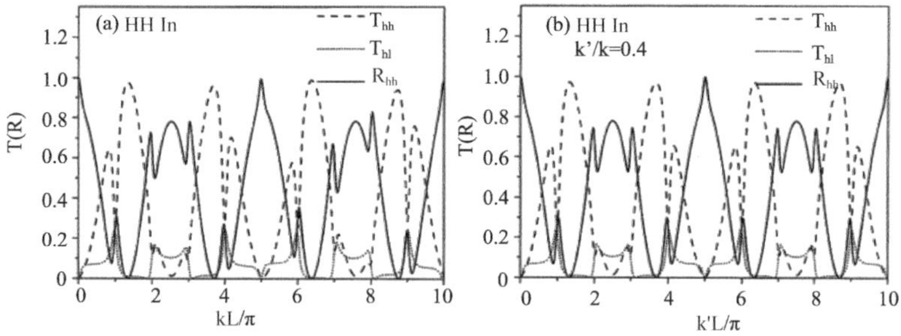

Figure 10.8 Transmission and reflection functions as functions of kL for the one gate structure shown in Fig. 10.7c. (a) $k'/k = 0.395$; (b) $k'/k = 0.4$.

The numerical calculation results for T_{hh}, T_{hl}, and R_{hh} as functions of kL are shown in Fig. 10.8a. In calculation we take the GaAs effective mass parameters $\gamma_1 = 6.85$ and $\gamma_2 = 10.5$. From the figure we see that the T_{hh}, T_{hl}, and R_{hh} all oscillate with kL in a period of about 5 π. With these parameters the ratio between the light hole and heavy hole wave vector k'/k is approximately 0.395. If we substitute it with a close value 0.4, then all the trigonometric functions in the results have a common period 5 π and we obtain a more regular oscillation shape, as shown in Fig. 10.8b.

The oscillation period for the electrons is π in the same single-stub structure [1]. The T_{hh} curve in Fig. 10.8b shows two main peaks and two main valleys in each period. Besides, at the positions of T_{hl} peaks, the T_{hh} curve shows a sharp dip around π and 3π. It means that one part of the heavy hole is converted to a light hole. This arises from the interference between the heavy hole and light hole waves. We also find that the peak value of T_{hl} is about 0.2, which means a relatively small fraction of the heavy hole wave can be converted into a light hole wave.

If an incident light hole wave with wave vector k' enters into source 1, the transmission and reflection functions can be calculated in a similar way and its numerical results are shown in Fig. 10.9. The ratio k'/k is chosen as 0.4 so that these functions have a period of 2π with respect to $k'L$. The peaks of R_{ll} are much sharper than that

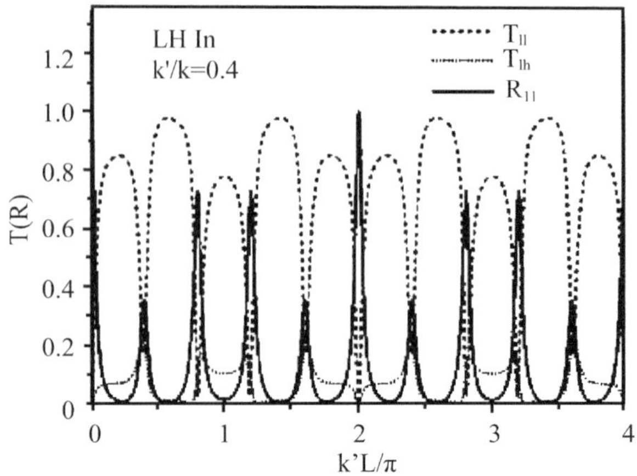

Figure 10.9 Transmission and reflection functions as functions of $k'L$ for the one-gate structure shown in Fig. 10.7c. k'/k is set to be 0.4.

of R_{hh} shown in Fig. 10.8b, and T_{ll} shows resonant plateaus rather than peaks.

References

1. Xia, J. B. (1992), *Phys. Rev. B*, **45**, 3593.
2. Wu, H. B., Xia, J. B., and Chang, K. (2003). *Solid State Commun.*, **128**, 125.
3. Datta, S., and Bandyopadhyay, S. (1987). *Phys. Rev. Lett.*, **58**, 717.
4. Datta, S. (1989). *Superlatt. Microstruct.*, **6**, 83.
5. Capasso, F., and Datta, S. (1990). *Phys. Today*, LB **(2)**, 74.
6. Sols, F., Macucci, M., Ravaioli, U., and Hess, K. (1989). *J. Appl. Phys.*, **66**, 3892.
7. Luttinger, J. M. (1956). *Phys. Rev.*, **102**, 1030.

Chapter 11

Two-Dimensional Quantum Waveguide Theory

Though many fundamental transport properties of quantum waveguides have been found in the study of idealized structures, in real waveguides the boundaries are usually defined via electrostatic confinement from metal gates and thus the geometries can be complicated. In this case, realistic lateral guiding potentials shall be taken into account, especially when the electrons are of high energies and would transport beyond the single-mode regime [1, 2]. When the width of the circuit is so large that the energy-level spacing between the transverse modes in the circuit is comparable to the electron kinetic energy, we should consider the transport of multiple transverse modes, that is, the two-dimensional waveguide theory. In this chapter, we will describe various theoretical approaches, such as the basic mode-matching technique, the transfer matrix, and the scattering matrix methods, to study electron transport through two-dimensional quantum waveguides. In the end, we will apply the theory to two-electron waveguide devices.

Figure 11.1 illustrates a typical two-dimensional quantum waveguide. The device can have several terminals, some of which may be used for input and the others for output. In general, one

Quantum Waveguide in Microcircuits (Second Edition)
Jian-Bai Xia, Duan-Yang Liu, and Wei-Dong Sheng

ISBN 978-981-4968-30-0 (Hardcover), 978-1-003-65146-8 (eBook)
www.jennystanford.com

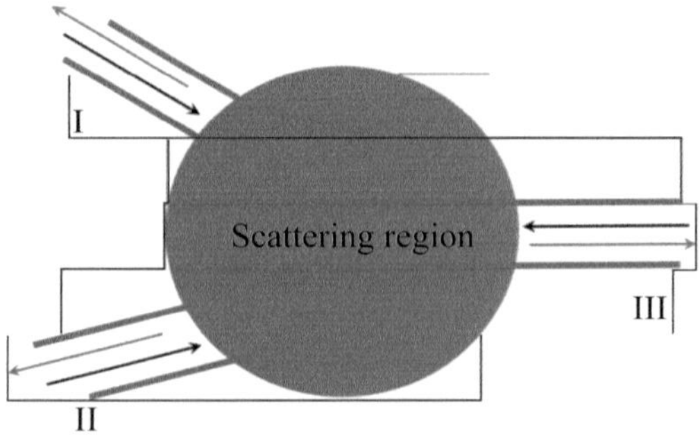

Figure 11.1 Schematic view of a two-dimensional multi-terminal quantum waveguide.

would like to know the output of the waveguide device if all the input information is given. Let $A_{\mathrm{I}}^{\mathrm{in}}$ and $A_{\mathrm{I}}^{\mathrm{out}}$ denote the incoming and outgoing electron waves to and from terminal I, respectively. All the outgoing waves can be related to the incoming waves by a matrix **S** in the linear response regime as follows:

$$\begin{bmatrix} A_{\mathrm{I}}^{\mathrm{out}} \\ A_{\mathrm{II}}^{\mathrm{out}} \\ A_{\mathrm{III}}^{\mathrm{out}} \end{bmatrix} = \begin{bmatrix} S_{11} & S_{12} & S_{13} \\ S_{21} & S_{22} & S_{23} \\ S_{31} & S_{32} & S_{33} \end{bmatrix} \begin{bmatrix} A_{\mathrm{I}}^{\mathrm{in}} \\ A_{\mathrm{II}}^{\mathrm{in}} \\ A_{\mathrm{III}}^{\mathrm{in}} \end{bmatrix}. \tag{11.1}$$

Here **S** is called the scattering matrix. For a scatterer, all the transport properties, including the transmission matrix, can be readily determined if the scattering matrix is obtained. In some simple cases, like for a two-terminal waveguide, it is common to use a transfer matrix method to obtain the transmission coefficient.

11.1 Transfer Matrix Method [3]

Let us start with a simple case, a two-terminal waveguide as illustrated in Fig. 11.2, as an example. The structure has two terminals L and R and a step-like interface between them. In terminals L and R, the electron wave functions can be written as

$$\varphi_{\mathrm{L}}(x, y) = \sum_{n=1}^{N_{\mathrm{L}}} \left(a_n^{\mathrm{L}} e^{ik_n^{\mathrm{L}} x} + b_n^{\mathrm{L}} e^{-ik_n^{\mathrm{L}} x} \right) \sqrt{\frac{2}{W_{\mathrm{L}}}} \sin\left(\frac{n\pi}{W_{\mathrm{L}}} y \right),$$

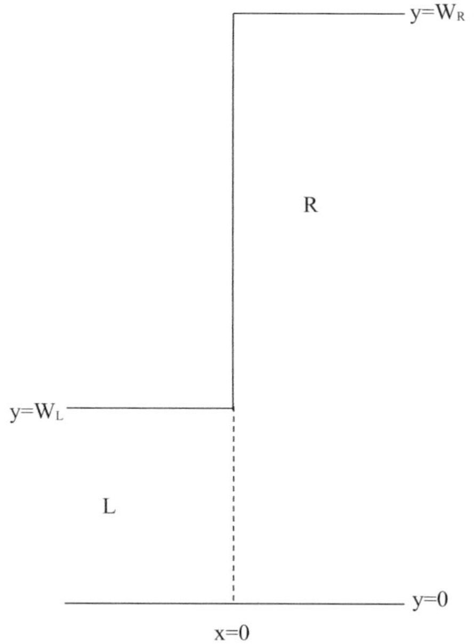

Figure 11.2 Schematic view of a two-terminal quantum waveguide.

and

$$\varphi_{\mathrm{R}}(x,y)=\sum_{n=1}^{N_{\mathrm{R}}}\left(a_n^{\mathrm{R}}e^{ik_n^{\mathrm{R}}x}+b_n^{\mathrm{R}}e^{-ik_n^{\mathrm{R}}x}\right)\sqrt{\frac{2}{W_{\mathrm{R}}}}\sin\left(\frac{n\pi}{W_{\mathrm{R}}}y\right), \qquad (11.2)$$

where the longitudinal wave numbers k_n^{L} and k_n^{R} satisfy

$$\frac{\left(\hbar k_n^{\mathrm{L}}\right)^2}{2m^*}+\frac{\hbar^2}{2m^*}\left(\frac{n\pi}{W_{\mathrm{L}}}\right)^2=E_{\mathrm{F}} \text{ and } \frac{\left(\hbar k_n^{\mathrm{R}}\right)^2}{2m^*}+\frac{\hbar^2}{2m^*}\left(\frac{n\pi}{W_{\mathrm{R}}}\right)^2=E_{\mathrm{F}}. \qquad (11.3)$$

Here E_{F} is the electron Fermi energy and N_{L} (N_{R}) is the number of transverse modes in terminal L(R). If one sets $x=0$ at the interface, the continuity of the wave function requires

$$\sum_{n=1}^{N_{\mathrm{L}}}\left(a_n^{\mathrm{L}}+b_n^{\mathrm{L}}\right)\sqrt{\frac{2}{W_{\mathrm{L}}}}\sin\left(\frac{n\pi}{W_{\mathrm{L}}}y\right)=\sum_{n=1}^{N_{\mathrm{R}}}\left(a_n^{\mathrm{R}}+b_n^{\mathrm{R}}\right)\sqrt{\frac{2}{W_{\mathrm{R}}}}\sin\left(\frac{n\pi}{W_{\mathrm{R}}}y\right). \qquad (11.4)$$

Similarly, matching the first derivatives of the wave functions at the interface gives

$$\sum_{n=1}^{N_\mathrm{L}} k_n^\mathrm{L}\left(a_n^\mathrm{L}-b_n^\mathrm{L}\right)\sqrt{\frac{2}{W_\mathrm{L}}}\sin\left(\frac{n\pi}{W_\mathrm{L}}y\right)=\sum_{n=1}^{N_\mathrm{R}} k_n^\mathrm{R}\left(a_n^\mathrm{R}-b_n^\mathrm{R}\right)$$
$$\times\sqrt{\frac{2}{W_\mathrm{R}}}\sin\left(\frac{n\pi}{W_\mathrm{R}}y\right). \quad (11.5)$$

Supposing $W_\mathrm{L} < W_\mathrm{R}$. We multiply both sides of Eqs. 11.4 and 11.5 by $\sqrt{2/W_\mathrm{R}}\sin(m\pi y/W_\mathrm{R})$ and $\sqrt{2/W_\mathrm{L}}\sin(m\pi y/W_\mathrm{L})$, respectively, integrate from 0 to W_L, and obtain

$$\sum_{n=1}^{N_\mathrm{L}} C_{nm}\left(a_n^\mathrm{L}+b_n^\mathrm{L}\right)=a_m^\mathrm{R}+b_m^\mathrm{R} \quad (11.6)$$

and

$$k_m^\mathrm{L}\left(a_m^\mathrm{L}-b_m^\mathrm{L}\right)=\sum_{n=1}^{N_\mathrm{R}} C_{mn}k_n^\mathrm{R}\left(a_n^\mathrm{R}-b_n^\mathrm{R}\right), \quad (11.7)$$

where

$$C_{nm}=\frac{2}{\sqrt{W_\mathrm{L}W_\mathrm{R}}}\int_0^{W_\mathrm{L}}\sin\left(\frac{n\pi}{W_\mathrm{L}}y\right)\sin\left(\frac{m\pi}{W_\mathrm{R}}y\right)dy. \quad (11.8)$$

If the widths of the two terminals are not much different, that is, $W_\mathrm{L} \approx W_\mathrm{R}$, one may set $N_\mathrm{L} = N_\mathrm{R}$ and the above equations can be rewritten as

$$\begin{bmatrix}\mathbf{L}_a\\ \mathbf{L}_b\end{bmatrix}=\mathbf{T}\begin{bmatrix}\mathbf{R}_a\\ \mathbf{R}_b\end{bmatrix}=\frac{1}{2}\begin{bmatrix}\mathbf{M}_1+\mathbf{M}_2 & \mathbf{M}_1-\mathbf{M}_2\\ \mathbf{M}_1-\mathbf{M}_2 & \mathbf{M}_1+\mathbf{M}_2\end{bmatrix}\begin{bmatrix}\mathbf{R}_a\\ \mathbf{R}_b\end{bmatrix}. \quad (11.9)$$

Here $\mathbf{L(R)}_a=\left[a_1^{\mathrm{L(R)}}, a_2^{\mathrm{L(R)}}, \ldots, a_{N_\mathrm{L(R)}}^{\mathrm{L(R)}}\right]^T$ and $\mathbf{L(R)}_b=\left[b_1^{\mathrm{L(R)}}, b_2^{\mathrm{L(R)}}, \ldots, b_{N_\mathrm{L(R)}}^{\mathrm{L(R)}}\right]^T$ are the vectors for terminal L and R, respectively; $\mathbf{M}_1 = \left(\mathbf{C}^T\right)^{-1}$; and $\mathbf{M}_2 = \left(\mathbf{K}^\mathrm{L}\right)^{-1}\mathbf{C}\mathbf{K}^\mathrm{R}$. $\mathbf{C}$ is the matrix with elements C_{nm} as defined in Eq. 11.8, while $\mathbf{K}^\mathrm{L}$ and $\mathbf{K}^\mathrm{R}$are diagonal matrices with elements $(\mathbf{K}^\mathrm{L})_{nn} = k_n^\mathrm{L}$ and $(\mathbf{K}^\mathrm{R})_{nn} = k_n^\mathrm{R}$

T is commonly known as the transfer matrix that relates all the wave components in the output port to those in the input terminal. Once it is obtained, the transmission and reflection amplitudes can be calculated in a straightforward manner. For an example, supposing electrons propagate in the ground transverse mode from the left terminal, we have the incident wave $\mathbf{L}_a = \left[1/k_1^{\mathrm{L}},\ 0,\ \ldots,\ 0\right]^T$ and $\mathbf{R}_b = 0$. The transmitting wave can then be obtained by $\mathbf{R}_a = \left[t_{11}/k_1^{\mathrm{R}},\ t_{21}/k_2^{\mathrm{R}},\ \ldots,\ t_{n1}/k_n^{\mathrm{R}}\right]^T = 2\left(\mathbf{M}_1 + \mathbf{M}_2\right)^{-1}\mathbf{L}_a$, followed by the reflecting wave $\mathbf{L}_b = \left[r_{11}/k_1^{\mathrm{L}}, r_{21}/k_2^{\mathrm{L}},\ \ldots, r_{n1}/k_n^{\mathrm{L}}\right]^T = \frac{1}{2}\left(\mathbf{M}_1 - \mathbf{M}_2\right)\mathbf{R}_a$.

Beyond the single-mode regime, the total transmission and reflection amplitudes T and R are given by

$$T = \sum_{\text{real } k_i^{\mathrm{R}}, k_j^{\mathrm{L}}} \left|t_{ij}\right|^2 \quad \text{and} \quad R = \sum_{\text{real } k_i^{\mathrm{L}}, k_j^{\mathrm{L}}} \left|r_{ij}\right|^2. \tag{11.10}$$

It is noted that one needs to include only the propagating modes with real wave numbers as all the evanescent states would eventually decay in the terminals [4]. The unitary condition is now given by $T + R = N_{\mathrm{i}}$, where N_{i} is the number of propagating modes in the incident port.

For a general waveguide with two aligning terminals, it is possible to divide the structure into a number of uniform sections (see Fig. 11.3). In this way, the total transfer matrix can be constructed as a product of $\mathbf{T}_m$, which describes the scattering by the interface between the $(m-1)$-th and m-th section and $\mathbf{T}_m^{\mathrm{f}}$, which

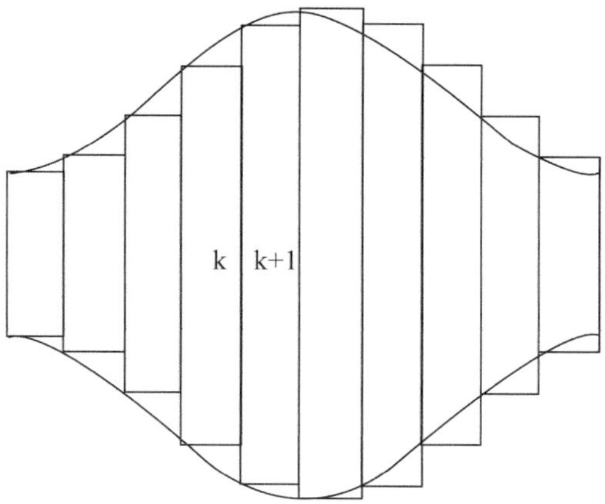

Figure 11.3 Schematic view of a two-terminal quantum waveguide that is divided into a number of transverse slices along its longitudinal direction.

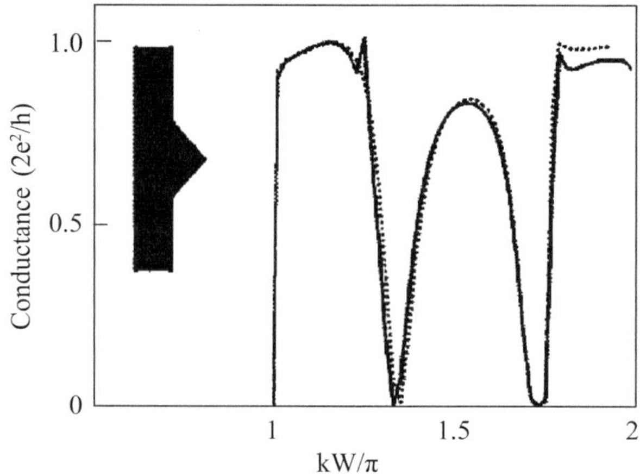

Figure 11.4 The conductance G in units of $2e^2/h$ versus kW/π for a triangular cavity structure as shown in the inset, where the solid line was calculated by the transfer method and the dotted line wa obtained by the recursive Green's function method.

describes free propagation from across the m-th section, that is,

$$\mathbf{T} = \prod_m \mathbf{T}_m \mathbf{T}_m^{\mathrm{f}} = \prod_m \mathbf{T}_m \begin{bmatrix} \mathbf{P}_m^- & 0 \\ 0 & \mathbf{P}_m^+ \end{bmatrix}, \tag{11.11}$$

where $\mathbf{P}_m^{\pm}$ are diagonal matrices with elements $\left(\mathbf{P}_m^{\pm}\right)_{nn} = e^{\pm i k_n^m L_m}$. L_m is the dimension of the m-th section along the propagating direction.

Figure 11.4 is the conductance G in units of $2e^2/h$ versus kW/π for a triangular cavity structure as shown in the inset, where the solid line was calculated by the above transfer matrix method and the dotted line was obtained by the recursive Green's function method [3]. From the figure we see that the agreement between the two results is good. In the numerical calculation of the transfer method, we divide the cavity into 20 segments along the longitudinal direction, and the transverse modes N is chosen to be 5. From Fig. 11.3 we see that when $kW/\pi = 1.3$ or 1.73, the conductance $G \approx 0$, that is, the two-dimensional (2D) circuit is blocked for electrons.

Figures 11.5a and 11.5b are the three-dimensional (3D) plots of the electron probability density ($|\Psi|^2$) in the triangular cavity

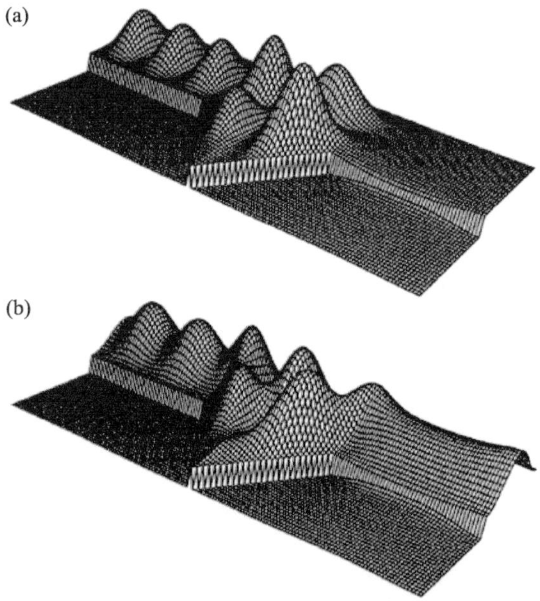

Figure 11.5 A three-dimensional plots of the electron probability density ($|\Psi|^2$) in the triangular cavity structure (a) for $kW/\pi = 1.73$ and $G = 0.04$ and (b) for $kW/\pi = 1.57$ and $G = 0.972$.

structure for which the transmission profile has been shown in Fig. 11.4 [3]. In the two figures, the electron is incident from the upper left. Figure 11.5a is for $kW/\pi = 1.73$ and $G = 0.04$, that is, the blocked case, and the electron density is concentrated mainly in the cavity, forming the standing wave (confined state), while on the right side of the cavity, there is nearly zero electron density. Figure 11.5b is for $kW/\pi = 1.57$ and $G = 0.972$, that is, the transmitted case, and the electron density at the input end (left) and the output end (right) are nearly same; the electron does not form the confined state in the cavity. This gives a physical picture of conductance in the 2D waveguide: if the electron forms a confined state in the cavity, the circuit is blocked and vice versa.

The transfer matrix method can be applied to many waveguides with different geometries, which most of the time gives reasonable results. However, the free propagation part $\mathbf{T}_m^{\mathrm{f}}$ contains both wave components $e^{ik_n^m L_m}$ and $e^{-ik_n^m L_m}$. For evanescent mode m, the wave number k_n^m becomes imaginary and thus $e^{ik_n^m L_m} = e^{-|k_n^m| L_m}$ can be very small while $e^{-ik_n^m L_m} = e^{|k_n^m| L_m}$ can be very big. Numerical errors would therefore inevitably accumulate during the calculation, which eventually destroys the unitary condition and renders the result useless [4]. In the next section, we will introduce a more stable method to calculate the scattering matrix directly.

11.2 Scattering Matrix Method [5]

For general waveguide devices, especially those with multiple terminals, each and every port can be used for input or output purpose. In this sense, it is the scattering matrix instead of the transfer matrix that represents its most fundamental property. Let us take an example of the two-terminal device as illustrated in Fig. 11.2. $\mathbf{L}_a(\mathbf{L}_b)$ and $\mathbf{R}_b(\mathbf{R}_a)$ denote the incoming (outgoing) wave components to ports L and R, respectively. Hence we may define $\mathbf{L}_{\text{in}} = \mathbf{L}_a$, $\mathbf{L}_{\text{out}} = \mathbf{L}_b$, $\mathbf{R}_{\text{in}} = \mathbf{R}_b$, and $\mathbf{R}_{\text{out}} = \mathbf{R}_a$ and look for the scattering matrix to relate every outgoing wave to all the incoming waves, that is,

$$\begin{bmatrix} \mathbf{L}_{\text{out}} \\ \mathbf{R}_{\text{out}} \end{bmatrix} = \mathbf{S} \begin{bmatrix} \mathbf{L}_{\text{in}} \\ \mathbf{R}_{\text{in}} \end{bmatrix} = \begin{bmatrix} \mathbf{S}_{11} & \mathbf{S}_{12} \\ \mathbf{S}_{21} & \mathbf{S}_{22} \end{bmatrix} \begin{bmatrix} \mathbf{L}_{\text{in}} \\ \mathbf{R}_{\text{in}} \end{bmatrix}. \tag{11.12}$$

For the interface between the two channels as illustrated in Fig. 11.2, the scattering matrices be derived from Eq. 11.9 in a straightforward manner as follows:

$$\mathbf{S}_{11} = -\left(\mathbf{M}_1^{-1} + \mathbf{M}_2^{-1}\right)^{-1}\left(\mathbf{M}_1^{-1} - \mathbf{M}_2^{-1}\right), \quad \mathbf{S}_{12} = 2\left(\mathbf{M}_1^{-1} + \mathbf{M}_2^{-1}\right)^{-1},$$
$$\mathbf{S}_{21} = 2\left(\mathbf{M}_1 + \mathbf{M}_2\right)^{-1}, \quad \text{and} \quad \mathbf{S}_{22} = -\left(\mathbf{M}_1 + \mathbf{M}_2\right)^{-1}\left(\mathbf{M}_1 - \mathbf{M}_2\right). \tag{11.13}$$

Free propagation across a uniform channel of length L is described by the scattering matrix

$$\mathbf{S}_{\text{f}} = \begin{bmatrix} 0 & \mathbf{P} \\ \mathbf{P} & 0 \end{bmatrix}, \tag{11.14}$$

where $\mathbf{P}$ is a diagonal matrix with elements $P_{nn} = e^{ik_n L}$.

The total scattering matrix for a general waveguide structure can be calculated in a way similar to what is depicted in Fig. 11.3. When combining two consecutive sections A and B into a single section A+B, the overall transfer matrix is just a simple matrix product of the two individual matrices, that is, $\mathbf{T}_{\text{A+B}} = \mathbf{T}_{\text{A}}\mathbf{T}_{\text{B}}$. However, the overall scattering matrix $\mathbf{S}_{\text{A+B}}$ can't be written as a simple function of $\mathbf{S}_{\text{A}}$ and $\mathbf{S}_{\text{B}}$. Let us start with the definition of $\mathbf{S}_{\text{A}}$ and $\mathbf{S}_{\text{B}}$,

$$\begin{bmatrix} \mathbf{L}_{\text{A}}^{o} \\ \mathbf{R}_{\text{A}}^{o} \end{bmatrix} = \mathbf{S}_{\text{A}} \begin{bmatrix} \mathbf{L}_{\text{A}}^{i} \\ \mathbf{R}_{\text{A}}^{i} \end{bmatrix} = \begin{bmatrix} \mathbf{S}_{\text{A}}^{11} & \mathbf{S}_{\text{A}}^{12} \\ \mathbf{S}_{\text{A}}^{21} & \mathbf{S}_{\text{A}}^{22} \end{bmatrix} \begin{bmatrix} \mathbf{L}_{\text{A}}^{i} \\ \mathbf{R}_{\text{A}}^{i} \end{bmatrix} \tag{11.15}$$

and

$$\begin{bmatrix} \mathbf{L}_{\text{B}}^{o} \\ \mathbf{R}_{\text{B}}^{o} \end{bmatrix} = \mathbf{S}_{\text{B}} \begin{bmatrix} \mathbf{L}_{\text{B}}^{i} \\ \mathbf{R}_{\text{B}}^{i} \end{bmatrix} = \begin{bmatrix} \mathbf{S}_{\text{B}}^{11} & \mathbf{S}_{\text{B}}^{12} \\ \mathbf{S}_{\text{B}}^{21} & \mathbf{S}_{\text{B}}^{22} \end{bmatrix} \begin{bmatrix} \mathbf{L}_{\text{B}}^{i} \\ \mathbf{R}_{\text{B}}^{i} \end{bmatrix}$$

and remind that the right port of section A aligns with the left port of section B, that is, $\mathbf{R}_{\text{A}}^{i} = \mathbf{L}_{\text{B}}^{o}$ and $\mathbf{R}_{\text{A}}^{o} = \mathbf{L}_{\text{B}}^{i}$. We may eliminate the connecting port and obtain the overall scattering matrix as follows:

$$\begin{bmatrix} \mathbf{L}_{\text{A}}^{o} \\ \mathbf{R}_{\text{B}}^{o} \end{bmatrix} = \mathbf{S}_{\text{A+B}} \begin{bmatrix} \mathbf{L}_{\text{A}}^{i} \\ \mathbf{R}_{\text{B}}^{i} \end{bmatrix} = \begin{bmatrix} \mathbf{S}_{\text{A+B}}^{11} & \mathbf{S}_{\text{A+B}}^{12} \\ \mathbf{S}_{\text{A+B}}^{21} & \mathbf{S}_{\text{A+B}}^{22} \end{bmatrix} \begin{bmatrix} \mathbf{L}_{\text{A}}^{i} \\ \mathbf{R}_{\text{B}}^{i} \end{bmatrix}, \tag{11.16}$$

where

$$\begin{aligned} \mathbf{S}_{\text{A+B}}^{11} &= \mathbf{S}_{\text{A}}^{11} + \mathbf{S}_{\text{A}}^{12}\mathbf{S}_{\text{B}}^{11}\left(1 - \mathbf{S}_{\text{A}}^{22}\mathbf{S}_{\text{B}}^{11}\right)^{-1}\mathbf{S}_{\text{A}}^{21}, \\ \mathbf{S}_{\text{A+B}}^{12} &= \mathbf{S}_{\text{A}}^{12}\left(1 - \mathbf{S}_{\text{B}}^{11}\mathbf{S}_{\text{A}}^{22}\right)^{-1}\mathbf{S}_{\text{B}}^{12}, \\ \mathbf{S}_{\text{A+B}}^{21} &= \mathbf{S}_{\text{B}}^{21}\left(1 - \mathbf{S}_{\text{A}}^{22}\mathbf{S}_{\text{B}}^{11}\right)^{-1}\mathbf{S}_{\text{A}}^{21}, \end{aligned} \tag{11.17}$$

and

$$\mathbf{S}_{\text{A+B}}^{22} = \mathbf{S}_{\text{B}}^{22} + \mathbf{S}_{\text{B}}^{21}\mathbf{S}_{\text{A}}^{22}\left(1 - \mathbf{S}_{\text{B}}^{11}\mathbf{S}_{\text{A}}^{22}\right)^{-1}\mathbf{S}_{\text{B}}^{12}.$$

By defining a composition operator $\mathbf{S}_{\text{A+B}} = \mathbf{S}_{\text{A}} \otimes \mathbf{S}_{\text{B}}$, we can express the overall scattering matrix for a waveguide as depicted in Fig. 11.3 as [5, 6]

$$\mathbf{S} = \mathbf{S}_{\text{f}}^{1} \otimes \mathbf{S}_{12} \otimes \mathbf{S}_{\text{f}}^{2} \ldots \otimes \mathbf{S}_{m,m+1} \otimes \mathbf{S}_{\text{f}}^{m+1} \otimes \ldots \otimes \mathbf{S}_{\text{f}}^{N}, \tag{11.18}$$

where $\mathbf{S}_{m,m+1}$ describes the scattering across the interface between the m-th and $(m+1)$-th sections, $\mathbf{S}_{\text{f}}^{m}$ is for free propagation across section m, and N is the total number of sections. It is noted that the composition operator, like the matrix multiplication, satisfies the associative law but not the commutative law, that is,

$$(\mathbf{S}_{\mathrm{A}} \otimes \mathbf{S}_{\mathrm{B}}) \otimes \mathbf{S}_{\mathrm{C}} = \mathbf{S}_{\mathrm{A}} \otimes (\mathbf{S}_{\mathrm{B}} \otimes \mathbf{S}_{\mathrm{C}}), \ \mathbf{S}_{\mathrm{A}} \otimes \mathbf{S}_{\mathrm{B}} \neq \mathbf{S}_{\mathrm{B}} \otimes \mathbf{S}_{\mathrm{A}}.$$

The overall scattering matrix for a two-terminal waveguide device is commonly written in an intuitive way as $\mathbf{S} = \begin{bmatrix} \mathbf{r} & \mathbf{t}' \\ \mathbf{t} & \mathbf{r}' \end{bmatrix}$, where **t** and **r** are the transmission and reflection matrices, respectively.

Although the transfer matrix method is very efficient for structures whose dimensions are not very large, it is numerically singular for structures with dimensions much larger than the electron de Broglie wavelength, while the same problem does not exist in the scattering matrix method, and we found that numerical errors would not accumulated in the composition of scattering matrices. Figure 11.6 shows the conductance G versus kW/π for a multiple-stub structure shown in the inset [5]. The length of the channel linking the two adjacent stubs on one side of the main wire is set to be $2W$, where W is the width of the terminal as well as the width and height of the stub. Similar structures are found to exhibit perfect quantum-modulated transistor action.

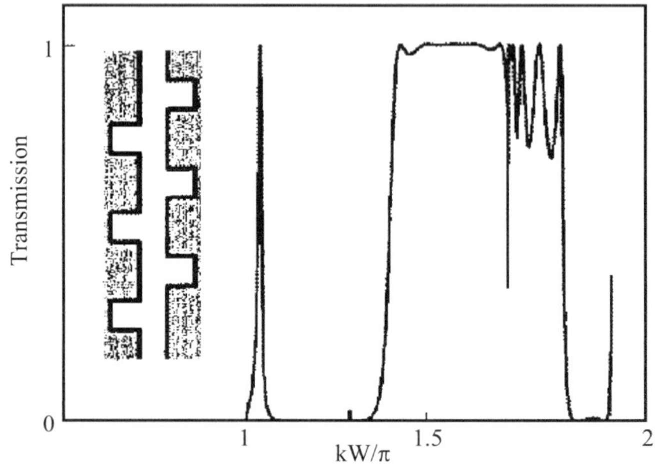

Figure 11.6 Conductance G versus kW/π for a multiple-stub structure shown in the inset.

11.3 Waveguide with Multiple Terminals

In many applications, such as quantum directional coupler [7], the waveguide device comes with multiple terminals. In this section, we will illustrate how to construct scattering matrices for two typical four-terminal structures.

The first device is depicted in Fig. 11.7, which can be regarded as two coupled parallel waveguides. The structure as shown is divided into three sections, the left two terminals L_1 and L_2, the connecting section W, and the right two terminals R_1 and R_2. Correspondingly, the overall scattering matrix can be expressed as

$$\mathbf{S} = \mathbf{S}_{L_1,L_2;W} \otimes \mathbf{S}^f_{\mathrm{W}} \otimes \mathbf{S}_{W;R_1,R_2}. \tag{11.19}$$

The scattering by the interface between L_1, L_2 and W is described by $\mathbf{S}_{\mathrm{L}_1,\mathrm{L}_2;\mathrm{W}}$, which is defined by $\left[\mathbf{A}^o_{\mathrm{L}_1};\mathbf{A}^o_{\mathrm{L}_2};\mathbf{A}^o_{\mathrm{W}}\right] = \mathbf{S}_{\mathrm{L}_1,\mathrm{L}_2;\mathrm{W}} \left[\mathbf{A}^i_{\mathrm{L}_1};\mathbf{A}^i_{\mathrm{L}_2};\mathbf{A}^i_{\mathrm{W}}\right]$. By using the mode-matching technique, we have $\mathbf{M}_1\left[\mathbf{A}^i_{\mathrm{L}_1};\mathbf{A}^o_{\mathrm{L}_1};\mathbf{A}^i_{\mathrm{L}_2};\mathbf{A}^o_{\mathrm{L}_2}\right] = \mathbf{M}_2\left[\mathbf{A}^i_{\mathrm{W}};\ \mathbf{A}^o_{\mathrm{W}}\right]$, where

$$\mathbf{M}_1 = \begin{bmatrix} \mathbf{C}_1^T & \mathbf{C}_1^T & \mathbf{C}_2^T & \mathbf{C}_2^T \\ \mathbf{K}_{\mathrm{L}} & -\mathbf{K}_{\mathrm{L}} & 0 & 0 \\ 0 & 0 & \mathbf{K}_{\mathrm{L}} & -\mathbf{K}_{\mathrm{L}} \end{bmatrix}, \ \mathbf{M}_2 = \begin{bmatrix} 1 & 1 \\ \mathbf{C}_1\mathbf{K}_{\mathrm{W}} & -\mathbf{C}_1\mathbf{K}_{\mathrm{W}} \\ \mathbf{C}_2\mathbf{K}_{\mathrm{W}} & -\mathbf{C}_2\mathbf{K}_{\mathrm{W}} \end{bmatrix}. \tag{11.20}$$

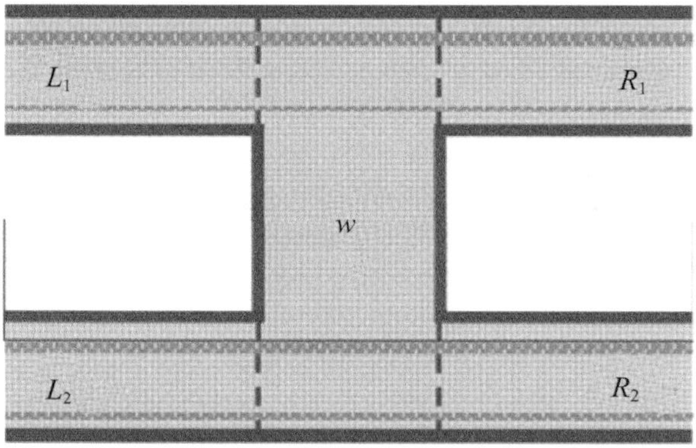

Figure 11.7 Schematic view of a four-terminal quantum waveguide.

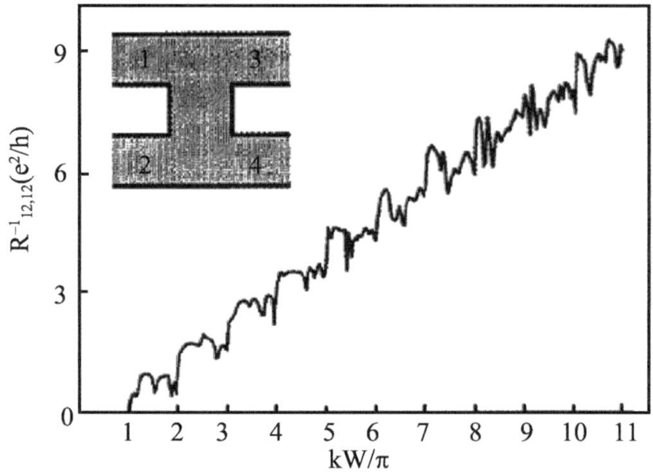

Figure 11.8 The inverse resistance $R^{-1}_{12,12}$ versus kW/π for the four-terminal structure shown in the inset.

Rearranging the above matrix equation, we can obtain $\mathbf{S}_{\mathrm{L}_1,\mathrm{L}_2;\mathrm{W}}$ as follows:

$$\mathbf{S}_{\mathrm{L}_1,\mathrm{L}_2;\mathrm{W}} = \begin{bmatrix} C_1^{\mathrm{T}} & C_2^{\mathrm{T}} & -1 \\ -K_{\mathrm{L}} & 0 & -C_1 K_{\mathrm{W}} \\ 0 & -K_{\mathrm{L}} & -C_2 K_{\mathrm{W}} \end{bmatrix}^{-1} \begin{bmatrix} -C_1^{\mathrm{T}} & -C_2^{\mathrm{T}} & 1 \\ -K_{\mathrm{L}} & 0 & -C_1 K_{\mathrm{W}} \\ 0 & K_{\mathrm{L}} & -C_2 K_{\mathrm{W}} \end{bmatrix}. \quad (11.21)$$

$\mathbf{S}_{\mathrm{W};\mathrm{R}_1,\mathrm{R}_2}$ can be obtained in a similar way.

Figure 11.8 shows the calculated Büttiker resistance $R_{12,12}$ for the four-terminal structure as shown in the inset [5]. The parameters are taken to be the same as those adopted in Fig. 11.8 of Ref. [6]. A comparison of our results with those in Ref. [6] shows that the two results are consistent each other as $kW/\pi < 7$. When $kW/\pi > 7$, the profile of our results becomes more complicated while that in Ref. [6] becomes simpler. This is because when $kW/\pi > 7$, the mode-mixing effect is enhanced and the interference between various modes becomes more complicated for higher electron energies, which have not been considered in Ref. [6].

The second device, as illustrated in Fig. 11.9, is a cross structure with four ports. As the terminals don't align with each other like that shown in Fig. 11.7, it is not possible to construct the overall

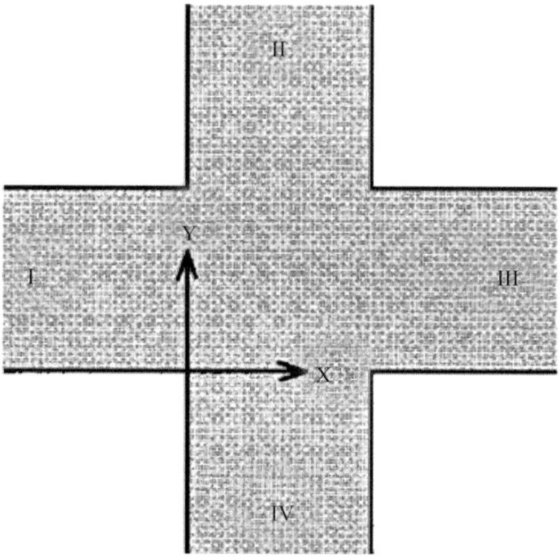

Figure 11.9 Schematic view of a quantum waveguide with a cross junction.

scattering matrix from those for the individual sections. Let us first expand the wave functions in I and III as well as II and IV onto the transverse modes $\phi_n(y) = \sqrt{\frac{2}{W}}\sin\left(\frac{n\pi}{W}y\right)$ and $\phi_n(x) = \sqrt{\frac{2}{W}}\sin\left(\frac{n\pi}{W}x\right)$, respectively, as in Eq. 11.2. Here W is the width of each of the four terminals. Next we introduce four auxiliary functions $\zeta_n^\alpha(xy)$ ($\alpha =$ I, II, III and IV) in the internal cross junction so that each of them matches the transverse mode in the corresponding terminal, that is, $\zeta_n^\alpha|_{\sigma_\alpha} = \phi_n^\alpha$. Thanks to the simple device geometry, ζ_n^α's are found to be [8]

$$\zeta_n^1(x, y) = \left(\cos k_n x - \frac{\sin k_n x}{\tan k_n W}\right)\varphi_n(y),$$

$$\zeta_n^2(x, y) = \frac{\sin k_n y}{\sin k_n W}\phi_n(x),$$

$$\zeta_n^3(x, y) = \frac{\sin k_n x}{\sin k_n W}\phi_n(y), \text{ and}$$

$$\zeta_n^4(x, y) = \left(\cos k_n y - \frac{\sin k_n y}{\tan k_n W}\right)\varphi_n(x). \qquad (11.22)$$

Hence the wave function in the cross junction is given by

$$\psi(x, y) = \sum_{\alpha, n} \left(a_n^\alpha + b_n^\alpha\right) \zeta_n^\alpha (x, y) \qquad (11.23)$$

and continuity of the first-order derivatives gives

$$\sum_{n} i k_n \left(a_n^\alpha - b_n^\alpha\right) \nabla \phi_n^\alpha \big|_{\sigma_\alpha} = \sum_{\alpha', n} \left(a_n^{\alpha'} + b_n^{\alpha'}\right) \nabla \zeta_n^\alpha \big|_{\sigma_{\alpha'}} . \qquad (11.24)$$

By proper integration of the above equations, we may eliminate ζ_n^α and ϕ_n^α and obtain the overall scattering matrix for the cross junction, $\mathbf{S}_+ = (i\mathbf{K}_4 + \mathbf{D})^{-1} (i\mathbf{K}_4 - \mathbf{D})$ where

$$\mathbf{K}_4 = \begin{bmatrix} \mathbf{K}' & 0 & 0 & 0 \\ 0 & \mathbf{K}' & 0 & 0 \\ 0 & 0 & \mathbf{K}' & 0 \\ 0 & 0 & 0 & \mathbf{K}' \end{bmatrix}, \ D = \begin{bmatrix} -\mathbf{C}_1 & \mathbf{FA} & \mathbf{C}_2 & \mathbf{A} \\ \mathbf{AF} & -\mathbf{C}_1 & \mathbf{FAF} & \mathbf{C}_2 \\ \mathbf{C}_2 & \mathbf{FAF} & -\mathbf{C}_1 & \mathbf{AF} \\ \mathbf{A} & \mathbf{C}_2 & \mathbf{FA} & -\mathbf{C}_1 \end{bmatrix}. \qquad (11.25)$$

with the matrix elements given by $\mathbf{K}'_{mn} = k'_n \delta_{mn}$ $(k'_n = k_n W/\pi)$, $\mathbf{F}_{mn} = (-1)^{n+1} \delta_{mn}$, $\mathbf{A}_{mn} = 2mn/\left[\pi m^2 - \pi \left(k'_n\right)^2\right]$, $(\mathbf{C}_1)_{mn} = k'_n tg k'_n \pi \delta_{mn}$, and $(\mathbf{C}_2)_{mn} = \frac{k'_n}{\sin k'_n \pi} \delta_{mn}$. Similarly, we may obtain the scattering matrix for the L-shaped junction, which consists of ports I and II, as follows:

$$\mathbf{S}_L = \begin{bmatrix} i\mathbf{K}' - \mathbf{C}_1 & \mathbf{FA} \\ \mathbf{AF} & i\mathbf{K}' - \mathbf{C}_1 \end{bmatrix}^{-1} \begin{bmatrix} i\mathbf{K}' + \mathbf{C}_1 & -\mathbf{FA} \\ -\mathbf{AF} & i\mathbf{K}' + \mathbf{C}_1 \end{bmatrix}. \qquad (11.26)$$

Further, the scattering matrix for the T-shaped junction, which consists of ports I, II, and III, can be given by

$$\mathbf{S}_T = \begin{bmatrix} i\mathbf{K}' - \mathbf{C}_1 & \mathbf{FA} & \mathbf{C}_2 \\ \mathbf{AF} & i\mathbf{K}' - \mathbf{C}_1 & \mathbf{FAF} \\ \mathbf{C}_2 & \mathbf{FAF} & i\mathbf{K}' - \mathbf{C}_1 \end{bmatrix}^{-1} \begin{bmatrix} i\mathbf{K}' + \mathbf{C}_1 & -\mathbf{FA} & -\mathbf{C}_2 \\ -\mathbf{AF} & i\mathbf{K}' + \mathbf{C}_1 & -\mathbf{FAF} \\ -\mathbf{C}_2 & -\mathbf{FAF} & i\mathbf{K}' + \mathbf{C}_1 \end{bmatrix}. \qquad (11.27)$$

Figure 11.10 shows the transmission coefficients versus kW/π for the structure with rounded corners shown in the inset [8]. The solid line is for forward transmission, and the dotted line is for side transmission. From the figure we see that the forward transmission profile has a sharp dip (anti-resonance). It is attributed to the quantum bound state. The bound energy of the quantum bound state becomes larger with enlargement of the internal junction area.

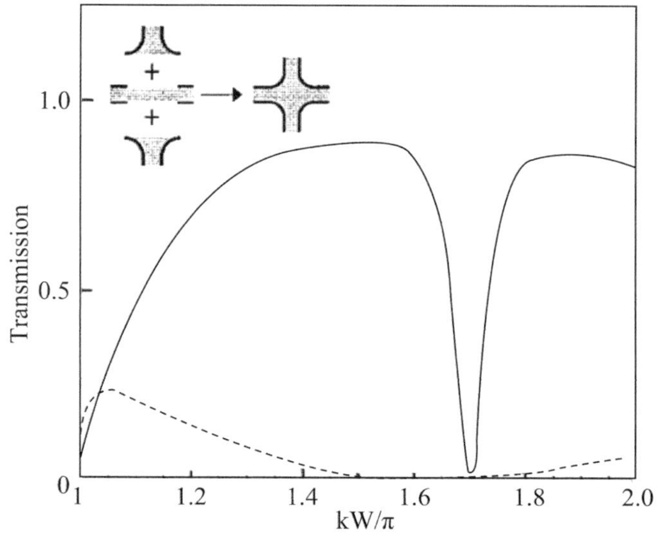

Figure 11.10 Transmission coefficients versus kW/π for the structure with rounded corners shown in the inset. The solid line is for forward transmission, and dotted line for side transmission.

Therefore the scattering of the propagating states with the bound state is enhanced. We also find the side transmission is much smaller due to the enhanced reflection.

References

1. Sheng, W.-D., and Xia, J. B. (1996). *Phys. Letts. A*, **220**, 268.
2. Xia, J. B., and Sheng, W.-D. (1996). *J. Appl. Phys.*, **79**, 7780.
3. Sheng, W.-D., and Xia, J. B. (1996). *J. Phys.: Condens. Matter*, **8**, 3635.
4. Sheng, W.-D., and Xia, J. B. (1997). *J. Appl. Phys.*, **81**, 734.
5. Sheng, W.-D. (1997). *J. Phys.: Condens. Matter*, **9**, 8369.
6. Wang, J., Wang, Y. J., and Guo, H. (1992). *Phys. Rev. B*, **46**, 2420.
7. Sheng, W.-D., and Xia, J. B. (1997). *Appl. Phys. A*, **64**, 167.
8. Sheng, W.-D. (1997). *J. Appl. Phys.*, **81**, 6210.

Chapter 12

One-Dimensional Quantum Waveguide Theory of a Rashba Electron [1]

Some of the first observations of quantum interference were made using thin metal films and silicon inversion layers [2, 3]. Since then most of the quantum interferences and ballistic transports are observed in two-dimensional systems. The physics of quantum interference includes the Aharonov–Bohm effect, quantum interference transistors, universal conductance fluctuations, ballistic electron transport and Landauer–Büttiker formula, quantized conductance in point contacts, multi-terminal devices, quantum dot resonant tunneling devices, etc. In these structures the electron movement is dominated by quantum mechanics, not classical mechanics. In the past 20 years, mesoscopic physics and spintronics have become two most active areas in condensed matter physics.

The nature of the spin of carriers in semiconductors has received considerable attention. With the rapid progress in semiconductor fabrication techniques, it is now possible to dope magnetic impurities inside semiconductors to produce diluted magnetic semiconductors (DMSs). The doped Mn ions interact with the carriers through sp-d exchange interaction, which results in a giant Zeeman splitting of the spin sublevels of the electron and hole.

Quantum Waveguide in Microcircuits (Second Edition)
Jian-Bai Xia, Duan-Yang Liu, and Wei-Dong Sheng

ISBN 978-981-4968-30-0 (Hardcover), 978-1-003-65146-8 (eBook)
www.jennystanford.com

A spin-polarized current can be produced by DMS-semiconductor heterostructures [4] or resonant tunneling with a DMS as a potential well [5] due to the giant Zeeman splitting. Now the spin coherence relaxation time has reached the nanosecond level at room temperature and the coherence distance exceeds 100 μm. The ability to preserve coherence spin states in semiconductors may enable quantum computing in the solid state. On the other hand, it was found that the Curie temperature T_C in a DMS in excess of 100 K has been realized in (GaMn)As systems [6]. The high T_C DMSs can be used as an efficient spin injector from a ferromagnetic DMS into a semiconductor without applying a magnetic field. Datta [7] has proposed a spin transistor device.

For a conduction-band electron in zinc-blende semiconductors the dominant spin-orbit interaction (SOI) in low-dimensional semiconductor structures comes from two physical factors: one is the Rashba SOI [8], which arises from the structure inversion asymmetry of the confined potential, for example, in a triangle well formed in modulated-doped heterojunctions or the external electric field. The other is the Dresselhaus SOI [9] induced by the lack of bulk inversion symmetry, for example, in III-V compounds. Sheng et al. [10] studied theoretically electron spin states in one-dimensional ring in the presence of both the Rashba SOI and the Dresselhaus SOI in a perpendicular magnetic field. They found that persistent charge current, persistent spin current, etc., are very sensitive to the strength of the Rashba and Dresselhaus SOI.

In this chapter, we apply the one-dimensional quantum theory [11] to the case of an electron with a spin in the presence of a Rashba SOI [8].

12.1 Rashba State Wave Function

In 1960 Rashba proposed Hamiltonian about an electron's spin in a semiconductor [8]

$$H_{\mathrm{R}} = \alpha\,[\sigma \times \mathbf{k}] \cdot \mathbf{n}, \tag{12.1}$$

where σ is the Pauli matrix of an electron's spin, $\mathbf{k}$ is the wave vector of the electron, $\mathbf{n}$ is the unit vector perpendicular to the interface, and α is the Rashba coefficient. Andrada et al. [12] proved that when the potential $V(z)$ is asymmetric along the **n** direction, that is, $dV(z)/dz \neq 0$, α is not equal to zero, that is, there is a Rashba effect. In most two-dimensional electron gases (2DEGs) formed in the modulation doping heterojunction structure $V(z)$ is asymmetric, so in many modulation doping field-effect transistors (MODFET) there is a Rashba effect. Moreover, in the ordinary quantum well, if there is a, external electric field perpendicular to the interface, the Rashba effect will arise and the strength of the Rashba effect is proportional to $dV(z)/dz$. To fabricate spin-controllable devices by utilizing the Rashba effect, we have studied quantum waveguide theory of Rashba electrons [1].

According to Eq. 12.1, the Hamiltonian of an electron in the two-dimensional (2D) system with the Rashba spin-orbit interaction (RSOI) is

$$H = \begin{pmatrix} -\frac{\hbar^2}{2m^*}\nabla^2 + V(x, y) & \frac{\alpha}{\hbar}(ip_x + p_y) \\ \frac{\alpha}{\hbar}(-ip_x + p_y) & -\frac{\hbar^2}{2m^*}\nabla^2 + V(x, y) \end{pmatrix}, \tag{12.2}$$

where m^* is the electron effective mass and α is the Rashba coefficient. If the electron moves in a one-dimensional (1D) straight circuit with a polar angle θ(see Fig. 12.1a) and the potential is zero in the circuit, then the Hamiltonian can be written as

$$H = \begin{pmatrix} -\frac{\hbar^2}{2m^*}\frac{\partial^2}{\partial l^2} & \alpha e^{-i\theta}\frac{\partial}{\partial l} \\ -\alpha e^{i\theta}\frac{\partial}{\partial l} & -\frac{\hbar^2}{2m^*}\frac{\partial^2}{\partial l^2} \end{pmatrix}, \tag{12.3}$$

where l is the coordinate along the circuit. The electron wave function has the plane wave form,

$$\Phi = \begin{pmatrix} c_1 \\ c_2 \end{pmatrix} e^{ikl}. \tag{12.4}$$

Then the eigenenergies are determined as

$$E = \frac{\hbar^2}{2m^*}k^2 \pm \alpha k, \tag{12.5}$$

and the wave functions are calculated to be

$$\phi_1(\theta) = \frac{1}{\sqrt{2}}\begin{pmatrix} 1 \\ ie^{i\theta} \end{pmatrix} \text{ and } \phi_2(\theta) = \frac{1}{\sqrt{2}}\begin{pmatrix} 1 \\ -ie^{i\theta} \end{pmatrix}, \tag{12.6}$$

corresponding to the $-\alpha k$ and $+\alpha k$ terms in Eq. 12.5, respectively.

The spin orientations of the states ϕ_1 and ϕ_2 can be determined by $\phi_i^+ s_x \phi_i$ and $\phi_i^+ s_y \phi_i$, respectively, where s_x and s_y are the spin component operators along the x and y directions, respectively.

$$\begin{aligned} \bar{S}_{x1} &= -\sin\theta, \quad \bar{S}_{x2} = \sin\theta, \\ \bar{S}_{y1} &= \cos\theta, \quad \bar{S}_{y2} = -\cos\theta, \\ \bar{S}_{z1} &= 0 \quad \text{and} \quad \bar{S}_{z2} = 0. \end{aligned} \tag{12.7}$$

From Eq. 12.7 we found that the spin orientation of the ϕ_1 and ϕ_2 states are perpendicular to the circuit and the angles between the spin and the circuit are $+\pi/2$ and $-\pi/2$, respectively. Afterward, we call the spin with the angle $+\pi/2$ to the circuit as spin-up and that with the angle $-\pi/2$ as spin-down.

If the electron states ϕ_1 and ϕ_2 have the same energy E, from Eq. 12.5 we obtain their wave vectors,

$$\begin{aligned} &k_1 = k_0 + k_\delta, \quad k_2 = k_0 - k_\delta, \\ &k_0 = \frac{m^*}{\hbar^2}\sqrt{\alpha^2 + \frac{2\hbar^2}{m^*}E}, \quad \text{and} \quad k_\delta = \frac{m^*}{\hbar^2}\alpha. \end{aligned} \tag{12.8}$$

Therefore, due to the RSOI the electrons with the same energy will have different wave vectors and spin orientations. Note that there is another pair of solutions associated with $-k_0$.

Up to now, we have only considered that the electron can pass through the circuit freely. Now we consider the case with a ferromagnetic contact [7] (see Fig. 12.1b) or a gate [11] (see Fig. 12.1c) at the end of the circuit, and either the spin-up (or spin-down) electrons or all electrons cannot pass through the circuit. If the spin-up electrons are reflected, then the wave function in the circuit will not have the plane wave form as in Eq. 12.4. It has the form of a standing wave, as

$$\begin{aligned} \Phi &= \phi_1(\theta)\, e^{ik_1 l} - \phi_2(\theta + \pi)\, e^{-ik_2 l} \\ &= \phi_1(\theta)\, e^{ik_\delta l} \sin(k_0 l), \end{aligned} \tag{12.9}$$

where the origin of the coordinate l is at the contact or gate and we have used the relation $\phi_2(\theta + \pi) = \phi_1(\theta)$ and Eq. 12.8. If the spin-down electrons are reflected, the wave function will have the form

$$\Phi = \phi_2(\theta)\, e^{-ik_\delta l} \sin(k_0 l). \tag{12.10}$$

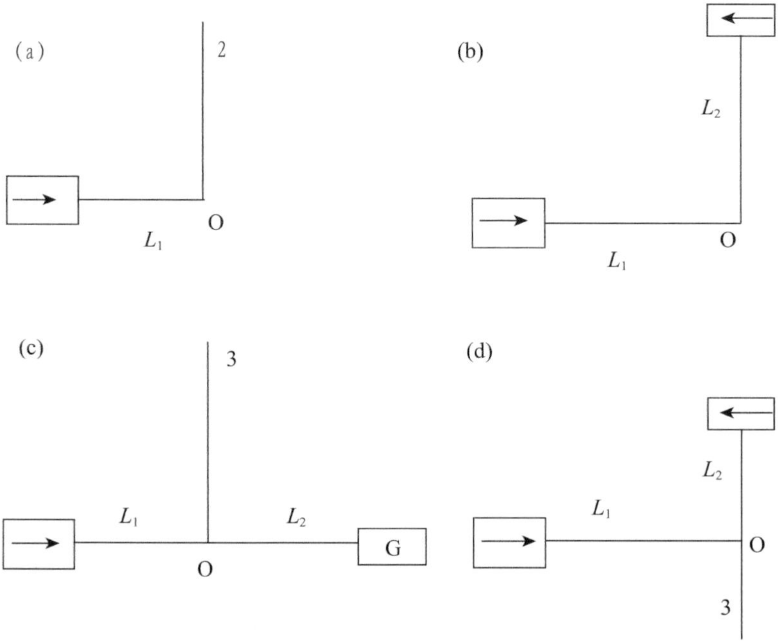

Figure 12.1 Several structures: (a) a turning structure, (b) a spin-polarized device, (c) a spin-polarized interference device, and (d) another spin-polarized interference device.

The main result of this section is that the phase of the Rashba wave function depends on the direction of the circuit θ, as shown in Eq. 12.6. The electron without considering the spin does not have this property [11], while the hole has a similar property, see Section 10.6.

12.2 Boundary Conditions of the Rashba Current

Now we know all wave functions in every circuits. Another main problem is the boundary conditions at the intersection. If the intersection is crossed by n circuits, let Φ_i be the wave function of the i-th circuit. At the intersection there are two boundary conditions: one is the continuity of the wave functions, and the other is the conservation of the current density. The first one is simple; it demands that

$$\Phi_1 = \Phi_2 = \cdots = \Phi_n. \tag{12.11}$$

The second one can be determined by the current density operator L_i, which is the operator of the derivative of the coordinate along the i-th circuit for the case of a simple electron. Now we calculate the form of the operator L_i from Eq. 12.3. Because the RSOI is related to the momentum, L_i is determined by the commutation relation as

$$L_i = -\frac{i}{\hbar}[l, H] = \frac{i}{\hbar}\begin{pmatrix} -\frac{\hbar^2}{m^*}\frac{\partial}{\partial l} & \alpha e^{-i\theta} \\ -\alpha e^{i\theta} & -\frac{\hbar^2}{m^*}\frac{\partial}{\partial l} \end{pmatrix}. \tag{12.12}$$

If we assign all positive directions of l_i to be pointing to the intersection, the second boundary condition has the form

$$\sum_{i=1}^{n} L_i \Phi_i = 0. \tag{12.13}$$

For the plane wave in Eq. 12.6, in spite of the spin-up or spin-down state, we get

$$L\Phi = \frac{\hbar k_0}{m^*}\Phi. \tag{12.14}$$

For the standing waves in Eqs. 12.9 and 12.10, with the coordinate origin at the intersection and L as the distance from the gate to the intersection, we get

$$L\Phi = -\frac{\hbar k_0}{m^*} i \cot\left[k_0 (l - L)\right]\Phi. \tag{12.15}$$

It can be expected that all $\hbar k_0/m^*$ terms in the equations of the second boundary condition (Eq. 12.13) can be counteracted. From Eq. 12.14 we see that though the Rashba states have different wave vectors of k_1 and k_2, their travel velocities are the same $\hbar k_0/m^*$. This will make the calculation more convenient.

12.3 Kinetic Property of a Rashba Wave in Branch Circuits

12.3.1 Turning Structure

First we consider the turning structure, the simplest situation in Fig. 12.1a. Suppose an incident electron with energy E enters circuit 1 (Fig. 12.1a). Then there will be transmitted waves in circuit 2 and reflected waves in circuit 1. The wave functions in circuits 1 and 2 can be written as

$$\Phi_1 = a_{10}\phi_1(0)\,e^{ik_1l_1} + a_{20}\phi_2(0)\,e^{ik_2l_1} + a_1\phi_1(\pi)\,e^{-ik_1l_1} + a_2\phi_2(\pi)\,e^{-ik_2l_1},$$

and

$$\Phi_2 = c_1\phi_1(\pi/2)\,e^{ik_1l_2} + c_2\phi_2(\pi/2)\,e^{ik_2l_2}, \quad (12.16)$$

where a_{10} and a_{20} are the coefficients of the known incident wave functions, a_1 and a_2 are the coefficients of reflected wave functions, and c_1 and c_2 are the coefficients of transmitted wave functions. From the boundary condition Eq. 12.11, we obtain

$$A_{10}\begin{pmatrix}1\\ i\end{pmatrix}+A_{20}\begin{pmatrix}1\\ -i\end{pmatrix}+A_1\begin{pmatrix}1\\ -i\end{pmatrix}+A_2\begin{pmatrix}1\\ i\end{pmatrix}=c_1\begin{pmatrix}1\\ -1\end{pmatrix}+c_2\begin{pmatrix}1\\ 1\end{pmatrix}. \quad (12.17)$$

From the boundary condition Eqs. 12.13 and 12.14, we obtain

$$A_{10}\begin{pmatrix}1\\ i\end{pmatrix}+A_{20}\begin{pmatrix}1\\ -i\end{pmatrix}-A_1\begin{pmatrix}1\\ -i\end{pmatrix}-A_2\begin{pmatrix}1\\ i\end{pmatrix}=c_1\begin{pmatrix}1\\ -1\end{pmatrix}+c_2\begin{pmatrix}1\\ 1\end{pmatrix}, \quad (12.18)$$

where

$$A_{10} = a_{10}e^{ik_1L_1}, \qquad A_{20} = a_{20}e^{ik_2L_1},$$
$$A_1 = a_1e^{-ik_1L_1}, \quad \text{and} \quad A_2 = a_2e^{-ik_2L_1}. \quad (12.19)$$

We assumed that the coordinate origin of circuit 1 is at the magnetic contact, which is at a distance L_1 from the intersection point O, and the coordinate origin of circuit 2 is at the intersection point O.

The solutions of Eqs. 12.17 and 12.18 are

$$A_1 = A_2 = 0,$$
$$c_1 = \frac{1-i}{2}A_{10} + \frac{1+i}{2}A_{20},$$

and

$$c_2 = \frac{1+i}{2}A_{10} + \frac{1-i}{2}A_{20}. \quad (12.20)$$

From Eq. 12.20, we see that in this structure there is no reflected wave. We can prove that for any simple turning structure, the reflected waves are zero, no matter what the turning angle is. If the incident wave has only one spin orientation, for example, spin-up, with the wave vector k_1, that is, $a_{20} = 0$ and $a_{10} = 1$, then from Eqs. 12.20 and 12.16 we see that it divides into two waves with the wave vectors k_1 and k_2 and the same probability 1/2. If the incident wave has the spin orientation in the positive x direction, then $a_{10} = 1/\sqrt{2}$ and $a_{20} = i/\sqrt{2}$. From Eq. 12.20 we obtain $c_1 = e^{ik_0L_1}\sin(k_\delta L_1)$ and $c_2 = e^{ik_0L_1}\cos(k_\delta L_1)$, that is, two incident waves with the wave vectors k_1 and k_2 have an interference effect as they arrive at point O. We find that the spin of the electron moving from circuit 1 to circuit 2 through the intersection has not changed. Generally, this conclusion is not related to the direction of circuit 2. So if there is no branch at the intersection like this one, the electron passes through the intersection just like in one circuit. The only difference is that the spin-up and spin-down components will be redistributed depending on the direction of circuit 2 and $k_\delta L_1$.

Figure 12.2 shows the transmission probabilities $T_{21} = |c_1|^2$ and $T_{22} = |c_2|^2$ as functions of α, when the incident electron has a spin in the positive x direction and $m^* = 0.1\ m_0$, $E = 10$ meV, and

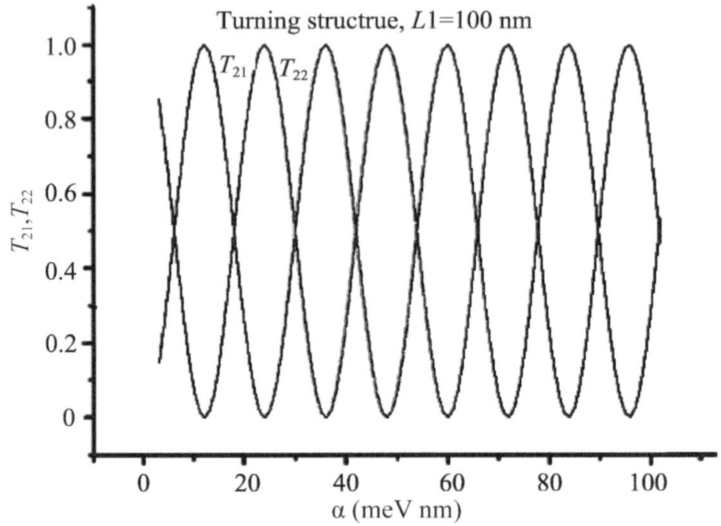

Figure 12.2 T_{21} and T_{22} of the turning structure in Fig. 12.1a as functions of the Rashba coefficient α for $m^* = 0.1\ m_0$, $E = 10$ meV, and $L_1 = 100$ nm.

$L_1 = 100$ nm. The parameters m^*, E, and L_1 are fixed in the following figures. We see that T_{21} and T_{22} change periodically with α. Due to the interference effect, when the spin at the intersection is polarized along the spin-up direction in circuit 2, $T_{21} = 1$ and $T_{22} = 0$. In the opposite case, $T_{21} = 0$, $T_{22} = 1$. It is noticed that the interference effect is only related to the phase $k_\delta L_1 \approx \alpha L_1$, independent of the electron energy E.

12.3.2 Spin-Polarized Device

Now we add a ferromagnetic contact that is magnetized in the spin-up direction in circuit 2 at the distance L_2 from the intersection O, as shown in Fig. 12.1b. According to the property of the ferromagnetic material, we expect to get a spin-polarized device. It will allow the spin-up electron to pass through completely in circuit 2 and forbid the spin-down electron from going through.

Again we assume that an incident electron with energy E and along the positive x direction enters circuit 1. The wave functions can be written as

$$\begin{aligned}\Phi_1 = a_{10}\phi_1(0)\,e^{ik_1l_1} + a_{20}\phi_2(0)\,e^{ik_2l_1} + a_1\phi_1(\pi)\,e^{-ik_1l_1}\\ + a_2\phi_2(\pi)\,e^{-ik_2l_1},\end{aligned}$$

and

$$\Phi_2 = c_1\phi_1(\pi/2)\,e^{ik_1l_2} + c_2\phi_2(\pi/2)\,e^{-ik_0l_2}\sin\left[k_0\left(l_2 - L_2\right)\right]. \quad (12.21)$$

Using the boundary conditions Eqs. 12.11 and 12.13 at the intersection O, the calculated results are

$$A_2 = -iA_1 = -\frac{A_{10} - iA_{20}}{2}e^{2ik_0L_2}$$

and

$$c_1 = \frac{1-i}{2}A_{10} + \frac{1+i}{2}A_{20}. \quad (12.22)$$

From Eq. 12.22 we see that the absolute values of c_1, A_1, and A_2 are not related to L_2, so we will neglect L_2 in the following discussion. The transmission probability $T_{21} = |c_1|^2 = \sin^2(k_\delta L_1)$ equals that in the above case, and there is no spin-down electron passing through the end of circuit 2. The reflection probability $R_{11} = R_{12} = \cos^2(k_\delta L_1)/2$, depending on the Rashba coefficient α. When

$k_\delta L_1 = m^*\alpha L_1/\hbar^2 = n\pi$, $T_{21} = 0$, and all electrons are reflected. When $k_\delta L_1 = m^*\alpha L_1/\hbar^2 = n\pi/2$, $T_{21} = 1$, and all electrons leave with spin up from circuit 2. So this structure is a polarization-modulation spin diode. If the ferromagnetic contact is magnetized in the spin-down direction in circuit 2, the reflection and transmission will be reversed.

12.3.3 Spin-Polarized Interference Device

Here we investigate the structure with a gate as shown in Fig. 12.1c. In this case, circuit 1 acts as the source, circuit 3 acts as the drain, and stub 2 acts as the gate, whose length L_2 can be controlled by a gate voltage. Again we assume that an incident electron with energy E and spin-polarized along the positive x direction enters circuit 1. The wave functions can be written as

$$\begin{aligned}\Phi_1 &= a_{10}\phi_1(0)\,e^{ik_1l_1} + a_{20}\phi_2(0)\,e^{ik_2l_1} + a_1\phi_1(\pi)\,e^{-ik_1l_1}\\ &\quad + a_2\phi_2(\pi)\,e^{-ik_2l_1},\\ \Phi_2 &= b_1\phi_1(0)\,e^{ik_\delta l_2}\sin[k_0(l_2 - L_2)]\\ &\quad + b_2\phi_2(0)\,e^{-ik_\delta l_2}\sin[k_0(l_2 - L_2)],\end{aligned}$$

and

$$\Phi_3 = c_1\phi_1(\pi/2)\,e^{ik_1l_2} + c_2\phi_2(\pi/2)\,e^{ik_2l_2}. \tag{12.23}$$

Using the boundary conditions Eqs. 12.11 and 12.13 at intersection O, we obtain

$$\begin{aligned}A_2 &= -\frac{i}{D}\cos(k_0L_2)\,A_{10},\\ A_1 &= -\frac{i}{D}\cos(k_0L_2)\,A_{20},\\ c_1 &= \frac{i}{D}\sin(k_0L_2)\left[\frac{1-i}{2}A_{10} + \frac{1+i}{2}A_{20}\right],\end{aligned}$$

and

$$c_2 = \frac{i}{D}\sin(k_0L_2)\left[\frac{1+i}{2}A_{10} + \frac{1-i}{2}A_{20}\right], \tag{12.24}$$

where

$$D = 2\sin(k_0L_2) + i\cos(k_0L_2). \tag{12.25}$$

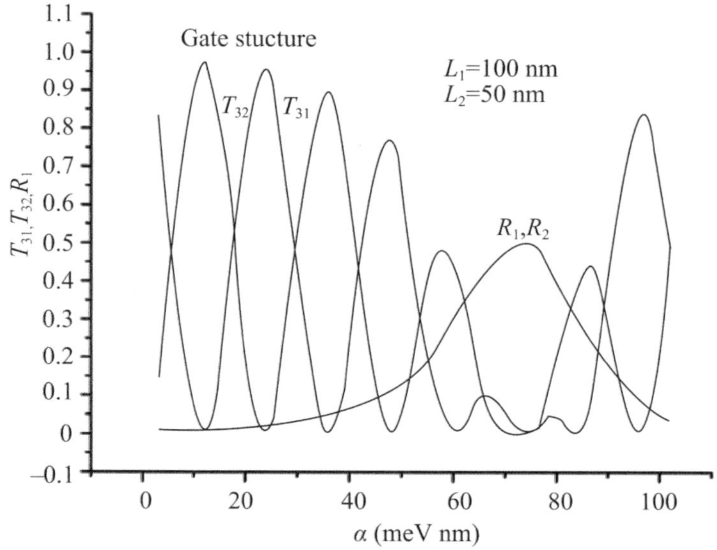

Figure 12.3 T_{31}, T_{32}, R_1, and R_2 of the spin-polarized interference device in Fig. 12.1c as functions of α for $L_2 = 50$ nm and other parameters the same as Fig. 12.2.

Comparing Eq. 12.24 to Eq. 12.20, we find that they are similar, c_1 and c_2 are multiplied by a coefficient that is only related to $|\sin(k_0L_2)/D|^2$, and the reflection is not zero now. These two effects are caused obviously by the gate. If there is no gate, the structure is the turning structure, as shown in Fig. 12.1a. The transmission probabilities T_{21} and T_{22} oscillate periodically with the Rashba coefficient α, as shown in Fig. 12.2. Now we can expect that they can be further controlled by the gate voltage.

Figure 12.3 shows the transmission and reflection probabilities as functions of α for $L_2 = 50$ nm. From Fig. 12.3 we see that T_{31} and T_{32} oscillate periodically with α, as in Fig. 12.2, but their amplitudes decrease with α, reach a minimum at $\alpha = 73$ meV nm, and then increase afterward. This modulation is caused by the interference effect related to $\cot(k_0L_2)$. Figure 12.4 shows the transmission and reflection probabilities as functions of α for $L_2 = 80$ nm. From Fig. 12.4 we see that when L_2 is changed by the gate voltage, the amplitudes of T_{31} and T_{32} increase with α, reach a maximum at

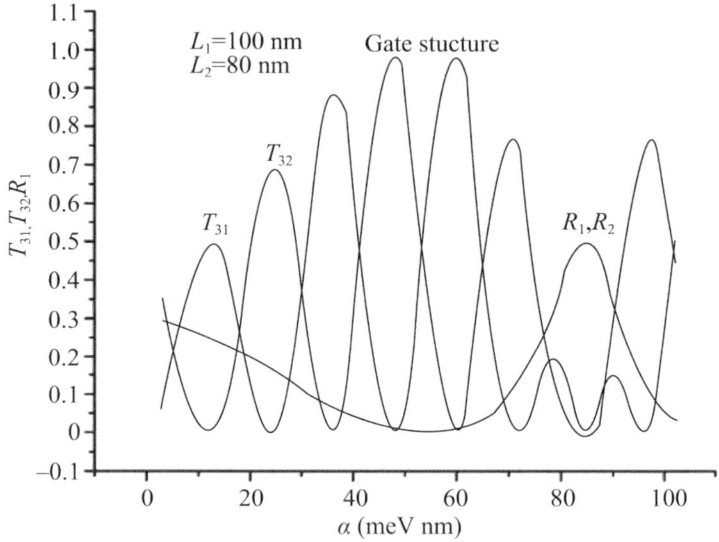

Figure 12.4 The same as Fig. 12.3 but for $L_2 = 80$ nm.

$\alpha = 50$ and 60 meV nm, then decrease to zero at $\alpha = 85$ meV nm. The interference effect of the gate is obvious, but it only changes the magnitude of T_{31} and T_{32} simultaneously, while the interference effect between electrons with spin-up and spin-down changes T_{31} and T_{32} alternatively, as shown in Fig. 12.2.

For showing the prominence of the gate, we keep the RSOI coefficient α as a constant and show the transmission probabilities T_{31}, T_{32}, R_1, and R_2 as functions of L_2 for $\alpha = 50$ meV nm in Fig. 12.5. From Fig. 12.5 we see that the amplitudes of T_{31} and T_{32} oscillate simultaneously with L_2 and their relative magnitudes are determined by the α. So making use of controlling the gate voltage, this structure is a spin-polarized interference device.

12.4 General Theory for a Structure with Multiple Branches

We assume that except the input circuit there are n_1 free output circuits ($j = 1, 2, \ldots, n_1$), n_2 circuits with gates ($k = n_1 + 1, \ldots,$

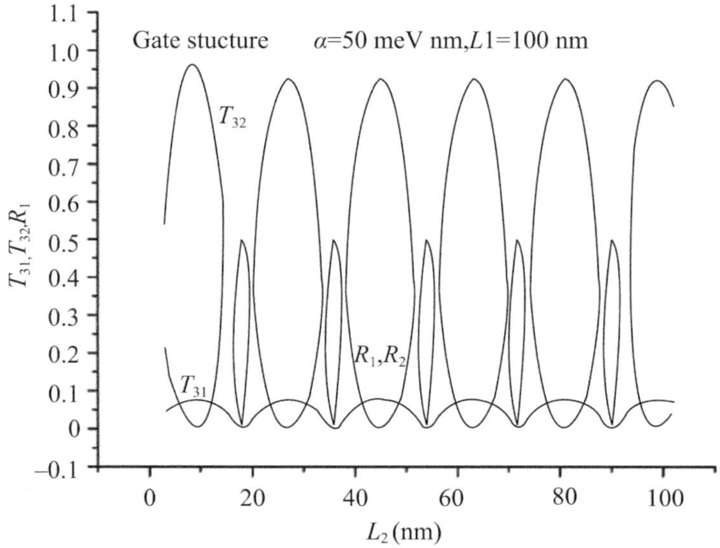

Figure 12.5 T_{31}, T_{32}, R_1, and R_2 of the spin-polarized interference device in Fig. 12.1c as functions of L_2 for $\alpha = 50$ meV nm and other parameters the same as Fig. 12.2.

$n_1 + n_2$), and n_3 circuits with ferromagnetic contacts, which allow spin-up or spin-down electrons to pass through ($m = n_1 + n_2 + 1, \ldots, n_1 + n_2 + n_3$). The orientation angle θ and the phase functions ϕ_1 and ϕ_2 (Eq. 12.6) of each circuit are denoted by the subscripts j, k, and m.

The wave functions in the input circuit are incident wave Φ_{i} and reflection wave Φ_{r},

$$\Phi_{\mathrm{i}} = a_{10}e^{ik_1l_i}\phi_{i1} + a_{20}e^{ik_2l_i}\phi_{i2},$$

and

$$\Phi_{\mathrm{r}} = a_1e^{-ik_1l_i}\phi_{i2} + a_2e^{-ik_2l_1}\phi_{i1}. \quad (12.26)$$

The output waves in the j-th circuit,

$$\Phi_j = a_{j1}e^{ik_1l_j}\phi_{j1} + a_{j2}e^{ik_2l_j}\phi_{j2}. \quad (12.27)$$

In the k-th circuit,

$$\Phi_k = a_{k1}e^{ik_\delta l_k}\sin\left[k_0\left(l_k - L_k\right)\right]\phi_{k1} + a_{k2}e^{-ik_\delta l_k}\sin\left[k_0\left(l_k - L_k\right)\right]\phi_{k2}. \quad (12.28)$$

In the m-th circuit,

$$\Phi_m = a_{m1} e^{ik_1 l_m} \phi_{m1} + a_{m2} e^{-ik_\delta l_m} \sin\left[k_0 \left(l_m - L_m\right)\right] \phi_{m2} \qquad (12.29)$$

or

$$\Phi_m = a_{m1} e^{ik_\delta l_m} \sin\left[k_0 \left(l_m - L_m\right)\right] \phi_{m1} + a_{m2} e^{ik_2 l_m} \phi_{m2} \qquad (12.30)$$

for spin-up and spin-down allowing circuits, respectively. The L_k and L_m are the distances from intersection O to the gate or contact.

The boundary conditions of the wave functions at intersection O are

$$\Phi_{\mathrm{i}} + \Phi_{\mathrm{r}} = \Phi_t, \quad t = 1, 2, \ldots n_1 + n_2 + n_3, \qquad (12.31)$$

and

$$\Phi_{\mathrm{i}} - \Phi_{\mathrm{r}} = \sum_{j=1}^{n_1} \Phi_j + i \sum_{k=n_1+1}^{n_1+n_2} \cot\left(k_0 L_k\right) \Phi_k + \sum_{m=n_1+n_2+1}^{n_1+n_2+n_3} J_m \Phi_m, \qquad (12.32)$$

where

$$J_m \Phi_m = a_{m1} \phi_{m1} + a_{m2} \left[-i \cos\left(k_0 L_m\right)\right] \phi_{m2}$$

and

$$J_m \Phi_m = a_{m1} \left[-i \cos\left(k_0 L_m\right)\right] \phi_{m1} + a_{m2} \phi_{m2} \qquad (12.33)$$

for the wave functions (Eq. 12.29 or Eq. 12.30), respectively.

Solving Eqs. 12.31 and 12.32 we can obtain the reflection coefficients a_1 and a_2 and the transmission coefficients of every circuits $a_{\mathrm{t}1}$ and $a_{\mathrm{t}2}$ for $t = 1, 2, \ldots, n_1, n_1 + n_2 + 1, \ldots, n_1 + n_2 + n_3$. For the circuits with gates $j = n_1 + 1, \ldots, n_1 + n_2$, there is no transmission wave in them, so we do not need to calculate the coefficients $a_{\mathrm{j}1}$ and $a_{\mathrm{j}2}$. For the circuits with a ferromagnetic contact $m = n_1 + n_2 + 1, \ldots, n_1 + n_2 + n_3$, one spin wave is allowed to pass, another is forbidden, so we only need to calculate the coefficient of plane wave.

In terms of the transmission probability, in the circuits $1 \le j \le n_1$, $T_j = |a_{j1}|^2 + |a_{j2}|^2$, in the circuits $n_1 + 1 \le k \le n_1 + n_2$, $T_k = 0$, and in the circuits $n_1 + n_2 + 1 \le m \le n_1 + n_2 + n_3$, $T_m = |a_{m1}|^2$ or $|a_{m2}|^2$. The reflection probability $R = |a_1|^2 + |a_2|^2$. Because the travel velocities of the electrons with spin-up and spin-down are the same $\hbar k_0 / m^*$, T and R are proportional to the square of the coefficients of the wave functions. For all the above cases we proved that $\sum T_t + R = 1$, that is, the conversation of the particle current.

For simplicity we first consider the case of $n_3 = 0$, that is, there is no circuit with a ferromagnetic contact. From the boundary condition Eqs. 12.31 and 12.32 we obtain

$$\Phi_{\mathrm{i}} - \Phi_{\mathrm{r}} = (n + iD)\,\Phi_{\mathrm{j}}, \quad D = \sum_{k=n_1+1}^{n_1+n_2} \cot(k_0 L_k). \tag{12.34}$$

From Eqs. 12.31 and 12.32 we obtain

$$\Phi_{\mathrm{i}} = \frac{1}{2}(1 + n_1 + iD)\,\Phi_{\mathrm{j}},$$

$$\Phi_{\mathrm{r}} = \frac{1}{2}(1 - n_1 - iD)\,\Phi_{\mathrm{j}},$$

$$\Phi_{\mathrm{j}} = \frac{2}{1 + n_1 + iD}\left(A_{10}\phi_{\mathrm{i}1} + A_{20}\phi_{\mathrm{i}2}\right),$$

and

$$\Phi_{\mathrm{r}} = \frac{1 - n_1 - iD}{1 + n_1 + iD}\left(A_{10}\phi_{\mathrm{i}1} + A_{20}\phi_{\mathrm{i}2}\right). \tag{12.35}$$

From Eq. 12.35 we obtain

$$a_{\mathrm{j}1} = \frac{2}{1 + n_1 + iD}\left(A_{10}\phi_{\mathrm{j}1}^{+}\phi_{\mathrm{i}1} + A_{20}\phi_{\mathrm{j}1}^{+}\phi_{\mathrm{i}2}\right),$$

$$a_{\mathrm{j}2} = \frac{2}{1 + n_1 + iD}\left(A_{10}\phi_{\mathrm{j}2}^{+}\phi_{\mathrm{i}1} + A_{20}\phi_{\mathrm{j}2}^{+}\phi_{\mathrm{i}2}\right),$$

$$A_1 = \frac{1 - n_1 - iD}{1 + n_1 + iD} A_{20},$$

and

$$A_2 = \frac{1 - n_1 - iD}{1 + n_1 + iD} A_{10}. \tag{12.36}$$

The wave function in a circuit with a gate is similarly given by Eq. 12.35. Their coefficients can also be calculated. But the current in these circuits equals zero, so the calculated coefficients are meaningless.

We apply the above result to a few special examples:

- Consider a turning structure as in Fig. 12.1a, where $n_1 = 1$, $n_2 = 0$, $\theta_{\mathrm{i}} = 0$, and $\theta_{\mathrm{j}} = \pi/2$. From Eq. 12.36 we obtain

$$a_{\mathrm{j}1} = \frac{1-i}{2} A_{10} + \frac{1+i}{2} A_{20},$$

$$a_{\mathrm{j}2} = \frac{1+i}{2} A_{10} + \frac{1-i}{2} A_{20}, \tag{12.37}$$

and

$$A_1 = A_2 = 0.$$

just as Eq. 12.20.

- Consider a spin-polarized interference device as in Fig. 12.1c, where $n_1 = 1$, $n_2 = 1$, $\theta_{\mathrm{i}} = 0$, $\theta_{\mathrm{j}} = \pi/2$, and $\theta_k = 0$. From Eq. 12.36 we obtain

$$a_{j1} = \frac{1-i}{2+i\cot(k_0 L_k)}A_{10} + \frac{1+i}{2+i\cot(k_0 L_k)}A_{20},$$
$$a_{j2} = \frac{1+i}{2+i\cot(k_0 L_k)}A_{10} + \frac{1-i}{2+i\cot(k_0 L_k)}A_{20},$$
$$A_1 = \frac{-i\cot(k_0 L_k)}{2+i\cot(k_0 L_k)}A_{20},$$

and

$$A_2 = \frac{-i\cot(k_0 L_k)}{2+i\cot(k_0 L_k)}A_{10}, \tag{12.38}$$

just as Eqs. 12.24 and 12.25.

- Another spin-polarized interference device with the free circuit and the circuit with gate exchanged, where $n_1 = 1$, $n_2 = 1, \theta_{\mathrm{i}} = 0, \theta_{\mathrm{j}} = 0$, and $\theta_k = \pi/2$. From Eq. 12.36 we obtain

$$a_{j1} = \frac{2}{2+i\cot(k_0 L_k)}A_{10},$$
$$a_{j2} = \frac{2}{2+i\cot(k_0 L_k)}A_{20},$$
$$A_1 = \frac{-i\cot(k_0 L_k)}{2+i\cot(k_0 L_k)}A_{20},$$
$$A_2 = \frac{-i\cot(k_0 L_k)}{2+i\cot(k_0 L_k)}A_{10}. \tag{12.39}$$

It means that there is no interference effect between the electrons with spin-up and spin-down and the two kinds of electrons move independently. The device behaves like an ordinary quantum interference device.

For the case that there is one circuit with a ferromagnetic contact, $n_3 = 1$, the boundary conditions Eqs. 12.31 and 12.32 become

$$\Phi_{\mathrm{i}} + \Phi_{\mathrm{r}} = \Phi_m = a_{m1}\left[-\sin(k_0 L_m)\right]\phi_{m2} \tag{12.40}$$

and

$$\Phi_\mathrm{i} - \Phi_\mathrm{r} = (1 + n_1 + i\,D)\,a_{m1}\phi_{m1} + \{(n_1 + i\,D)\,[-\sin(k_0 L_m)] + [-i\cot(k_0 L_m)]\}\,a_{m2}\phi_{m2}. \qquad (12.41)$$

From Eqs. 12.40 and 12.41 we can calculate analytically the coefficients a_{m1} and a_{m2} and other coefficients a_{j1}, a_{j2}, a_{k1}, and a_{k2}.

$$a_{m1} = \frac{2}{2 + n_1 + i\,D}\left(A_{10}\phi^+_{m1}\phi_{i1} + A_{20}\phi^+_{m1}\phi_{i2}\right),$$

$$a_{m2} = \frac{-2\left(A_{10}\phi^+_{m2}\phi_{i1} + A_{20}\phi^+_{m2}\phi_{i2}\right)}{(1 + n_1)\sin(k_0 L_m) + i\,[D\sin(k_0 L_m) + \cos(k_0 L_m)]},$$

$$a_{j1} = a_{m1}\phi^+_{j1}\phi_{m1} + a_{m2}\,[-\sin(k_0 L_m)]\,\phi^+_{j1}\phi_{m2},$$

$$a_{j2} = a_{m1}\phi^+_{j2}\phi_{m1} + a_{m2}\,[-\sin(k_0 L_m)]\,\phi^+_{j2}\phi_{m2},$$

$$a_{k1} = -\frac{1}{\sin(k_0 L_k)}\{a_{m1}\phi^+_{k1}\phi_{m1} + a_{m2}\,[-\sin(k_0 L_m)]\,\phi^+_{k1}\phi_{m2}\},$$

$$a_{k2} = -\frac{1}{\sin(k_0 L_k)}\{a_{m1}\phi^+_{k2}\phi_{m1} + a_{m2}[-\sin(k_0 L_m)]\,\phi^+_{k2}\phi_{m2}\},$$

$$A_2 = a_{m1}\phi^+_{i1}\phi_{m1} + a_{m2}\,[-\sin(k_0 L_m)]\,\phi^+_{i1}\phi_{m2} - A_{10},$$

and

$$A_1 = a_{m1}\phi^+_{i2}\phi_{m1} + a_{m2}\,[-\sin(k_0 L_m)]\,\phi^+_{i2}\phi_{m2} - A_{20}. \qquad (12.42)$$

We apply this to some special examples.

- Consider a turning structure with an output circuit with a ferromagnetic contact, as shown in Fig. 12.1b, where $n_1 = n_2 = 0$, $n_3 = 1$, $\theta_\mathrm{i} = 0$, and $\theta_m = \pi/2$. From Eq. 12.42 we obtain

$$a_{m1} = \frac{1-i}{2}A_{10} + \frac{1+i}{2}A_{20},$$

$$a_{m2} = 2i e^{i k_0 L_m}\left(\frac{1+i}{2}A_{10} + \frac{1-i}{2}A_{20}\right),$$

$$A_2 = -\frac{e^{2i k_0 L_m}}{2}(A_{10} - i\,A_{20}),$$

and

$$A_1 = -\frac{e^{2i k_0 L_m}}{2}(i\,A_{10} + A_{20}), \qquad (12.43)$$

just as Eq. 12.22.

- Consider a branch structure with two output circuits, one with a ferromagnetic contact and one free, where $n_1 = 1$, $n_2 = 0$, $n_3 = 1$, $\theta_i = 0$, $\theta_j = -\pi/2$, and $\theta_m = \pi/2$, as shown in Fig. 12.1d. From Eq. 12.42 we obtain

$$
\begin{aligned}
a_{m1} &= \frac{1-i}{3}A_{10} + \frac{1+i}{3}A_{20},\\
a_{m2} &= -\frac{2}{2\sin k_0 L_m + i\cos k_0 L_m}\left(\frac{1+i}{2}A_{10} + \frac{1-i}{2}A_{20}\right),\\
a_{j1} &= \frac{2}{2 + i\cot k_0 L_m}\left(\frac{1+i}{2}A_{10} + \frac{1-i}{2}A_{20}\right),\\
a_{j2} &= a_{m1},\\
A_2 &= \frac{1}{3}\left(-2A_{10} + iA_{20}\right) + \frac{1}{2 + i\cot k_0 L_m}\left(A_{10} - iA_{20}\right),
\end{aligned}
$$

and

$$
\begin{aligned}
A_1 = &-\frac{1}{3}\left(iA_{10} + 2A_{20}\right)\\
&+\frac{1}{2 + i\cot k_0 L_m}\left(iA_{10} + A_{20}\right). \qquad (12.44)
\end{aligned}
$$

Figure 12.6 shows the transmission probabilities T_{21}, T_{31}, T_{32}, R_1, and R_2 as functions of α for $L_2 = 50$ nm. From Fig. 12.6 we see that $T_{21} = T_{32}$ and both are not influenced by the contact in circuit 2. T_{32} oscillates with α due to the inference effect related to $\cos^2(k_\delta L_1)$, is modulated simultaneously by the interference effect related to $1/(4 + \cot^2 k_0 L_2)$, and reaches a minimum at $\alpha = 73$ meV nm. So we can modulate the electron current of one kind of spin by using a ferromagnetic contact.

12.5 Summary

We have discussed the properties of the Rashba wave function in the planar 1D waveguide and obtain the following results. Due to the Rashba effect, the plane waves of an electron with energy E divide into two kinds of waves with the wave vectors $k_1 = k_0 + k_\delta$ and $k_2 = k_0 - k_\delta$, where k_δ is proportional to the Rashba coefficient, and their spin orientations are $+\pi/2$ (spin-up) and $-\pi/2$ (spin-down) with respect to the circuit, respectively. Unlike the electron

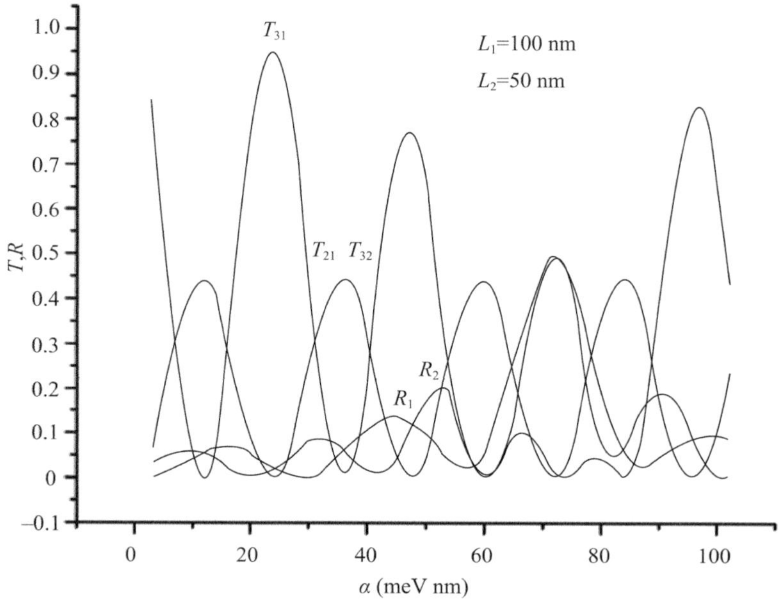

Figure 12.6 T_{21}, T_{31}, T_{32}, R_1, and R_2 of the spin-polarized interference device in Fig. 12.1d as functions of α for $L_2 = 50$ nm and other parameters the same as Fig. 12.2.

without considering the spin, the phase of the Rashba wave function depends on the direction angle θ of the circuit. The travel velocity of the Rashba waves with the wave vector k_1 or k_2 is the same, $\hbar k_0/m^*$. The boundary conditions of the Rashba wave functions at the intersection of circuits are given from the continuity of wave functions and the conservation of current density. We considered three cases: a free circuit, a circuit with a gate, and a circuit with a ferromagnetic contact. In the case of gate or ferromagnetic contact in the circuit, the Rashba wave function becomes a standing wave form $e^{\pm ik_\delta l}\sin\left[k_0\left(l - L\right)\right]$, where L is the position coordinate of the gate or contact. Using the boundary conditions of Rashba wave functions we studied the transmission and reflection probabilities of a Rashba electron moving in several structures and found the interference effects of the two Rashba waves with different wave vectors and the Rashba waves with the gate and the ferromagnetic contact. Finally, we derived a general theory for a structure with multiple branches.

The theory of Rashba 1D waveguide, including the wave functions and the boundary conditions at the intersection, can be used to design various spin-polarized devices.

References

1. Liu, D. Y., Xia, J. B., and Chang, Y. C. (2010). *Sci. China: Phys. Mech. Astron.*, **53**, 1.
2. van Wees, B. J., van Houten, H., Beenakker, C. W. J., et al. (1998). *Phys. Rev. Lett.*, **60**, 848.
3. Wharam, D. A., Thornton, T. J., Newbury, R., et al. (1998). *J. Phys. C*, **21**, L209.
4. Fiederling, R., Keim, M., Reuscher, G., et al. (1999). *Nature*, **402**, 787.
5. Gruber, T., Keim, M., Fiederling, R., et al. (2001). *Appl. Phys. Lett.*, **78**, 1101.
6. Matsukura, F., Ohno, H., Shen, A., et al. (1998). *Phys. Rev. B*, **57**, R2037.
7. Datta, S., and Das, B. (1990). *Appl. Phys. Lett.*, **56**, 665.
8. Rashba, E. I. (1960). *Sov. Phys. Solid State*, **2**, 1109.
9. Dresselhaus, G. (1955). *Phys. Rev.*, **100**, 580.
10. Sheng, J. S., and Chang, K. (2006). *Phys. Rev. B*, **74**, 235315.
11. Xia, J. B. (1992). *Phys. Rev. B*, **45**, 3593.
12. Andrada, E. A., La Rocca, G. C., and Bassani, F. (1994). *Phys. Rev. B*, **50**, 8523.

Chapter 13

1D Quantum Waveguide Theory of Rashba Electrons in Curved Circuits

Since the Datta–Das spin field-effect transistor [1] was proposed, the Rashba spin-orbit interaction (RSOI) [2] has attracted considerable attention on account of its conceivable applications in spintronics. A great deal of spintronic devices based on RSOI have been proposed [3–6], most of which focus on ballistic straight waveguides [7–9] and rings or analogous structures [10–12]. They are expected to control spin-polarized transport on the basis of the interferences of different paths [13]. Conductance properties of rings or analogous structures with several diametrically connected leads have been discussed [5, 14]; some researchers consider similar structures as a spin filter [6, 12]. Földi et al. [10] accounted for the spin transformation properties of a quantum ring with RSOI and indicated that it can serve as a one-qubit quantum gate for electron spin, and Shelykh et al. [6] invested a gated ballistic AB ring with three asymmetrically situated electrodes and showed that it was able to demonstrate the properties of both the quantum splitter and spin filter. In recent years, there have also been experiments about the spin transport in these structures [15–17], where the AB effect plays an important role.

Quantum Waveguide in Microcircuits (Second Edition)
Jian-Bai Xia, Duan-Yang Liu, and Wei-Dong Sheng

ISBN 978-981-4968-30-0 (Hardcover), 978-1-003-65146-8 (eBook)
www.jennystanford.com

But the AB rings are not made of straight circuits such as structures discussed in Chapter 12, so transports in these rings cannot be studied using the theory in the previous chapter. Therefore we divide the curved part into N segments, where N is a big number so that every segment can be regarded as a small straight line. Then we investigate the electron's movement in every straight segment and every vertex. When every straight segment tends to be infinitely small, the wave function of the Rashba electron passing through a curve can be obtained.

13.1 Transfer Matrix of a Rashba Electron in a 1D Two-Terminal Structure [18]

Firstly, we apply boundary conditions in Chapter 12 to a turning structure. An incident electron with energy E enters circuit 1 at angle θ, passes through a corner, and then enters circuit 2 at angle ϕ. The wave functions in the circuits can be written as

$$\Phi_1 = a_{10}\phi_1(\theta) e^{ik_1 l_1} + a_{20}\phi_2(\theta) e^{ik_2 l_1} + a_1\phi_1(\theta+\pi) e^{-ik_1 l_1} + a_2\phi_2(\theta+\pi) e^{-ik_2 l_1}$$

and

$$\Phi_2 = c_1\phi_1(\varphi) e^{ik_1 l_2} + c_2\phi_2(\varphi) e^{ik_2 l_2}, \qquad (13.1)$$

where a_{10} and a_{20} are amplitudes of the inject waves, a_1 and a_2 are those of the reflecting waves, and c_1 and c_2 are those of the output waves. l_1 and l_2 are the coordinates of circuits 1 and 2, respectively. ϕ_1 and ϕ_2 are the phase factors of spin-up and spin-down electrons, given in Eq. 12.6, k_1 and k_2 are the wave vectors, given in Eq. 12.8, respectively. Applying Eq. 13.11 and Eq. 13.13, we can find that there is no reflected wave function in circuit 1, at the intersection $a_1 = 0$, $a_2 = 0$.

In a circuit with angle θ, we can write the wave function as

$$\Phi = \{[\phi_1(\theta), \phi_2(\theta)]\}\Psi = [\phi_1(\theta), \phi_2(\theta)]\begin{pmatrix} a_1 \\ a_2 \end{pmatrix}. \qquad (13.2)$$

In the following part of the chapter, we will adopt Ψ to describe the wave function in most time. If we take the origins of the coordinates

l_1 and l_2 all at the intersection, then Eq. 13.1 can be written as

$$[\phi_1(\varphi)\ \phi_2(\varphi)]\begin{pmatrix} c_1 \\ c_2 \end{pmatrix} = [\phi_1(\theta)\ \phi_2(\theta)]\begin{pmatrix} a_{10} \\ a_{20} \end{pmatrix}. \tag{13.3}$$

We can use a simple transfer matrix

$$\mathbf{M}_{\theta\to\varphi} = \begin{pmatrix} \phi_1^+(\varphi) \\ \phi_2^+(\varphi) \end{pmatrix}[\phi_1(\theta)\ \phi_2(\theta)] = \frac{1}{2}\begin{pmatrix} 1+e^{-i(\varphi-\theta)} & 1-e^{-i(\varphi-\theta)} \\ 1-e^{-i(\varphi-\theta)} & 1+e^{-i(\varphi-\theta)} \end{pmatrix} \tag{13.4}$$

to describe the transform of the wave function Ψ at the intersection of the turning structure. Accordingly the transfer matrix of one circuit with length L can be written as

$$\mathbf{M}_{\mathrm{L}} = \begin{pmatrix} e^{ik_1L} & 0 \\ 0 & e^{ik_2L} \end{pmatrix}. \tag{13.5}$$

In the following, we will calculate the spin transport on the basis of Eq. 13.4 and Eq. 13.5 when there is no branch. If at the intersection there are branches, we can calculate that by utilizing the boundary conditions Eq. 12.11 and Eq. 12.13.

All the results above are in the condition of no magnetic field. Now we consider the two-dimensional (2D) system in the external magnetic field perpendicular to the circuit plane. In this case, for a close ring the wave vector k in Hamiltonian will be replaced by

$$k_{\mathrm{eff}} = k \pm eA/\hbar, \tag{13.6}$$

where the $\pm$ sign depends on the relative orientations of the magnetic field and the k, $|A| = \Phi_m/L$ is the vector potential, and Φ_m and L are the magnetic flux through the closed loop and its perimeter, respectively. So if we substitute k_{eff} for k, the k_1 and k_2 in Eq. 13.1 and Eq. 13.5 will be changed to

$$k_1 = k_0 + k_\delta - k_A, \quad k_1 = k_0 - k_\delta - k_A,$$
$$k_0 = \frac{m^*}{\hbar^2}\sqrt{\alpha^2 + \frac{2\hbar^2}{m^*}E},\ k_\delta = \frac{m^*}{\hbar^2}\alpha, \quad \text{and} \quad k_A = \pm\frac{eA}{\hbar}. \tag{13.7}$$

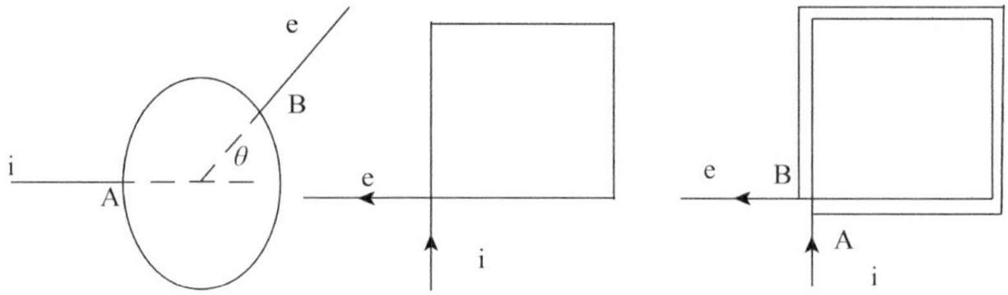

Figure 13.1 Several structures: (a) an AB ring, (b) an AB SL, and (c) a AB double SL.

13.2 Electron Structure of a Closed Circle [1]

The structures studied in this section are shown in Fig. 13.1. Figure 13.1a is an Aharonov–Bohm (AB) ring, and the angle between the inject circuit i and the output circuit e is θ. Figure 13.1b is an AB square loop structure, and Fig. 13.1c is an AB double square loop. First we calculate the electron structure of a closed circular ring (as Fig. 13.1a without inject and output circuits) by the transfer matrix method. According to Eq. 13.4 and Eq. 13.5, we can get the transfer matrix of arbitrary structure with no branches. For a closed circle, we divide the circle into N segments, each segment can be seen as a straight line segment, and we calculate the transfer matrices in every segments and every intersections. When the N is large enough, the product of all matrices according to the order of transport is the total transfer matrix M.

It can be inferred from Eq. 13.4 and Eq. 13.5 that the total transfer matrix **M** is just concerned with the electron energy E, the Rashba coefficient α, the external magnetic B, and the radius of the circle R. Now we fix up the last three coefficients, so the **M** can be written as $\mathbf{M}(E)$. It is noticed that the eigenstates of the circle exist just when $\mathbf{M}(E)$ has an eigen value of 1, that is, when

$$\det\left[\mathbf{M}\left(E\right)-I\right]=0 \tag{13.8}$$

and the corresponding eigen vector describes the eigenstate in the form of Ψ.

Next we will give the analytical solution of this problem. We consider one segment and the intersection at its heel. If the N is large enough, the length of the segment can be written as $R\delta\theta$, and

$\delta\theta = (\phi - \theta)$ is the turning angle at the intersection in Eq. 13.4. If the wave vector is in the counter-clockwise (CCW) direction, $\delta\theta > 0$, and for the wave vector in the clockwise (CW) direction, $\delta\theta < 0$. $\delta\theta$ is an infinitesimal quantity, and $N\delta\theta = 2\pi$. For one segment, the transfer matrix is

$$\mathbf{M}_{\delta\theta} = \frac{1}{2}\begin{pmatrix} 1+e^{-i\delta\theta} & 1-e^{-i\delta\theta} \\ 1-e^{-i\delta\theta} & 1+e^{-i\delta\theta} \end{pmatrix}\begin{pmatrix} e^{ik_1 R\delta\theta} & 0 \\ 0 & e^{ik_1 R\delta\theta} \end{pmatrix}. \tag{13.9}$$

Because $\mathbf{M}_{\delta\theta}$ is very close to I, so we assume that its eigen value is $e^{im\delta\theta}$, where m is an integral. Ignoring twice and more power terms of $\delta\theta$ and solving the secular equation

$$\det\left(\mathbf{M}_{\delta\theta} - e^{im\delta\theta}\right) = 0, \tag{13.10}$$

we obtain the eigenenergy for $B = 0$,

$$\varepsilon = \left(m + \frac{1}{2}\right)^2 + \frac{1}{4} \pm \left(m + \frac{1}{2}\right)\sqrt{1+\bar{\alpha}^2}, \tag{13.11}$$

where m is the quantum number of the quantum energy level. In this paper, we use the dimensionless physical quantities. If the energy unit $E_0 = \hbar^2/(2m^* R^2)$, then the energy $\varepsilon = E/E_0$, the magnetic field $b = (\hbar eB/m^* c)/E_0$, and the Rashba coefficient $\bar{\alpha} = (\alpha/R)/E_0 = \alpha(2m^* R/\hbar^2)$.

We can also get the eigen vectors of the transfer matrix $\mathbf{M}_{\delta\theta}$,

$$\mathbf{M}_{\delta\theta}\begin{pmatrix} \cos\varphi' \\ \sin\varphi' \end{pmatrix} = e^{im\delta\theta}\begin{pmatrix} \cos\varphi' \\ \sin\varphi' \end{pmatrix}, \tag{13.12}$$

where

$$\tan\varphi' = \bar{\alpha} \pm \sqrt{1+\bar{\alpha}^2}.$$

Because the total transfer matrix $\mathbf{M} = (\mathbf{M}_{\delta\theta})^N$, these vectors are also the eigen vectors of $\mathbf{M}$, and the eigen value becomes $e^{im2\pi}$. It is obvious that when m is an integer, these states describe eigen Rashba states of the circle. In the presence of a magnetic field, the corresponding eigenenergy is

$$\varepsilon = \left(m + \frac{b}{4} + \frac{1}{2}\right)^2 + \frac{1}{4} \pm \left(m + \frac{b}{4} + \frac{1}{2}\right)\sqrt{1+\bar{\alpha}^2}. \tag{13.13}$$

According to Eq. 13.2 and Eq. 13.12, the eigen Rashba states in the circle with a suffix m are

$$\Psi_{\mathrm{m}}(\theta) = [\phi_1(\theta)\ \phi_2(\theta)]\begin{pmatrix} \cos\varphi' \\ \sin\varphi' \end{pmatrix} = \frac{1}{\sqrt{2\pi}}e^{im\theta}\begin{pmatrix} \cos\varphi \\ \sin\varphi e^{i\theta} \end{pmatrix}, \tag{13.14}$$

where $\tan\varphi = (1+\sqrt{1+\bar{\alpha}^2})/\bar{\alpha}$ is for the + sign in Eq. 13.13 and $\tan\varphi = (1-\sqrt{1+\bar{\alpha}^2})/\bar{\alpha}$ is for the – sign, which are defined as spin-up and spin-down states, respectively. The spin orientation of each state can be described as a function of θ,

$$S(\theta) = \sin(2\varphi)\cos\theta\hat{e}_x + \sin(2\varphi)\sin\theta\hat{e}_y + \cos(2\varphi)\hat{e}_z. \quad (13.15)$$

We find that the spin-up and spin-down states have the opposite local spin orientation, and everywhere their components in the x–y plane point to the center along the radius, or in the opposite direction, respectively. These results are identical to those obtained from solving the Schrödinger equation with the Hamiltonian in the polar coordinates [19, 20]

$$H_{\text{po}} = \begin{pmatrix} \left(-i\frac{\partial}{\partial\theta}+\frac{b}{4}\right)^2 & \frac{1}{2}\left[\alpha e^{-i\theta}\left(-i\frac{\partial}{\partial\theta}+\frac{b}{4}\right)+\alpha\left(-i\frac{\partial}{\partial\theta}+\frac{b}{4}\right)e^{-i\theta}\right] \\ \frac{1}{2}\left[\alpha e^{i\theta}\left(-i\frac{\partial}{\partial\theta}+\frac{b}{4}\right)+\alpha\left(-i\frac{\partial}{\partial\theta}+\frac{b}{4}\right)e^{i\theta}\right] & \left(-i\frac{\partial}{\partial\theta}+\frac{b}{4}\right)^2 \end{pmatrix} \quad (13.16)$$

Notice that the nondiagonal elements are written to ensure the Hermitian of the Hamiltonian. On the basis of Eq. 13.16, we can also obtain Eq. 13.13 to Eq. 13.15.

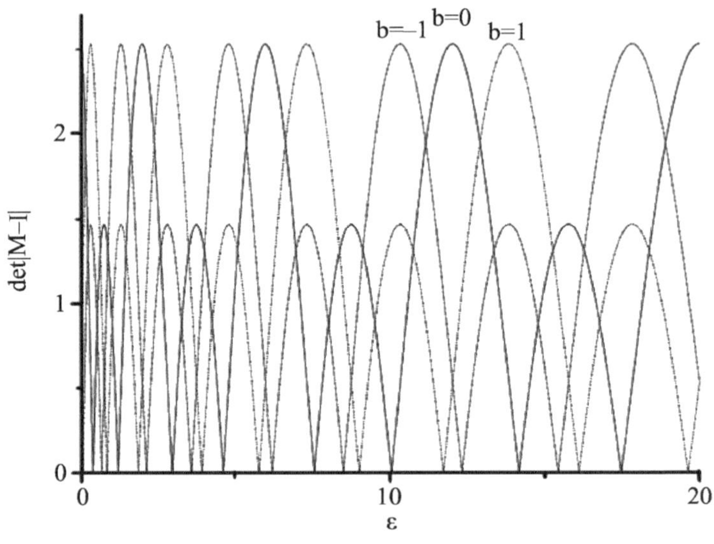

Figure 13.2 $|\det[\mathbf{M}(E)-I]|$ of the closed circular ring as functions of the electron energy ε for $\bar{\alpha}=1$ and $b=0$, 1, and -1.

We calculate the eigenenergies numerically. Figure 13.2 shows the numerical results of the transfer matrix method, the $|\det(\mathbf{M}(E)-I)|$ as functions of ε for $\alpha = 1$ and $b = 0, 1$, and -1. From the figure we see that the zero points correspond to the energies of eigenstates, for example, in the case of $b = 0$, $\varepsilon = 0.38, 1.2, 3.0, 4.6, 7.6, 10.0, 14.1, 17.4$, etc., which are in agreement with the results of Eq. 13.13. In the calculation, we take $N = 1000$.

13.3 Electron Structure of a Closed Square Loop [18]

Although the spin states and electron structure in a close circle have been investigated in some literature [19, 20] by other methods, our method can be utilized to calculate more anomalistic geometry, such as a closed square loop (as in Fig. 13.1b without inject and output circuits). Assume that the side length of the square loop is L. We can get the transfer matrix for a CCW circuit as

$$\mathbf{M}_{\text{SL}} = -\frac{1}{4}\begin{pmatrix} A^4 - 3A^3B - A^2B^2 - AB^3 & i\left(B^4 - B^3A - B^2A^2 + BA^3\right) \\ i\left(A^4 - 3A^3B - A^2B^2 + AB^3\right) & B^4 - B^3A - B^2A^2 - BA^3 \end{pmatrix}, \tag{13.17}$$

where $A = e^{ik_1^L}$, $B = e^{ik_2L}$, k_1, and k_2 are defined in Eq. 13.8. The matrix in CW is also easy to obtain. Similarly, we use the dimensionless physical quantities, taking the energy unit $E_0 = \hbar^2(2m^*L^2)$. Then the energy $\varepsilon = E/E_0$, the magnetic field $b = (\hbar eB/m^*c)/E_0 = 2eBL^2/\hbar c$, and the Rashba coefficient $\alpha = (\alpha/L)/E_0 = \alpha(2m^*L/\hbar^2)$. We can also get the eigenenergies and vectors on the basis of Eq. 13.8. From Eq. 13.8 and Eq. 13.17 we obtain the secular equation,

$$\begin{pmatrix} C - y & iD \\ iD^* & C^* - y \end{pmatrix} = 0, \tag{13.18}$$

where

$$C = -e^{2i\bar{\alpha}} + 3e^{i\bar{\alpha}} + 1 + e^{-i\bar{\alpha}},$$
$$D = -e^{i\bar{\alpha}} + 1 + e^{-i\bar{\alpha}} - e^{-2i\bar{\alpha}},$$
$$y = 4e^{-4ix},$$

and

$$x = k_0 L - k_A L = \sqrt{\left(\frac{\bar{\alpha}}{2}\right)^2 + \varepsilon} - \frac{b}{8}. \tag{13.19}$$

From Eq. 13.18 we obtain the eigenenergies of a closed square loop,

$$\varepsilon = \left(\pm\frac{\varphi}{4} + \frac{\pi}{2}m + \frac{b}{8}\right)^2 - \left(\frac{\bar{\alpha}}{2}\right)^2$$

and

$$\varphi = \arccos\left(\frac{1 + 4\cos\bar{\alpha} - \cos 2\bar{\alpha}}{4}\right), \tag{13.20}$$

where m is an integer and $\pm(\phi/4) + (\pi/2)m + (b/8) > 0$. As we know, this is the first result of eigenenergies of Rashba states in a square loop obtained by the transfer matrix method, which cannot be obtained by solving the Schrödinger equation. With the transfer matrix method we can calculate the eigenstates of a closed loop of any shape, for example, triangle, pentagon, hexagon, rectangle, and ellipse.

About the eigenstates, in the straight-line section, the states are just denoted by Eq. 13.4 and Eq. 13.6; the spin orientation of the Rashba states with the wave vectors k_1 and k_2 are all perpendicular to the circuit; the angles between the spin and the circuit are $+\pi/2$ and $-\pi/2$; respectively; and the coefficients of the spin-up and spin-down states, which describe eigenstates in the form of Ψ in Eq. 13.2, can be determined by $M_{\mathrm{SL}}\Psi = \Psi$.

13.4 Spin Interference in an AB Ring [18]

Now we consider an interference structure, the AB ring as in Fig. 13.1a. We use the transfer matrices Eq. 13.4 and Eq. 13.5 to describe the spin transport in a circular circuit, and the boundary conditions Eq. 13.11 and Eq. 13.13 to describe the transport through the intersection. We obtained the reflection and transmission coefficients in AB rings as functions of the electron energy ε, the Rashba coefficient α, and the external magnetic field B.

We assume that an electron with energy ε enters circuit i of the AB ring from left to intersection A, then two transmitted waves depart from intersection A, meet at intersection B, and depart from circuit e. At the same time, two reflected waves depart from circuit i. Write the wave functions in the four circuits of the AB ring and the boundary conditions at the intersections A and B. In the two arms of the AB ring, the wave functions at point B are related with those at point A by the transfer matrices. There will be 20 coefficients for all wave functions in all circuits. We have 12 equations from the boundary conditions at two intersections and 8 equations from the transfer matrices in two arms of the geometries. So we can solve this problem and get coefficients of reflection and transmission waves uniquely.

The transmission and reflection coefficients are $T_1 = |t_{\text{up}}|^2$, $T_2 = |t_{\text{down}}|^2$, and $R_1 = |r_{\text{up}}|^2$, $R_2 = |r_{\text{down}}|^2$. Figure 13.3 shows T_1, T_2, R_1, and R_2 as functions of the electron energy ε for the magnetic field $b = 0$, $\theta = 0$, $\bar{\alpha} = 1$, and coefficients of injected waves are $a_{10} = 1$ and $a_{20} = 0$. In this case, $R_1 = 0$ and $T_2 = 0$. From Fig. 13.3 we see that the zero points of the R_2 curve are $\varepsilon = 0.3$, 1.3, 2.8, 4.8, 7.3, 10.3, 13.8, 17.8, etc., which just correspond to the eigenenergies of eigenstates in the closed AB ring, calculated from Eq. 13.13. It

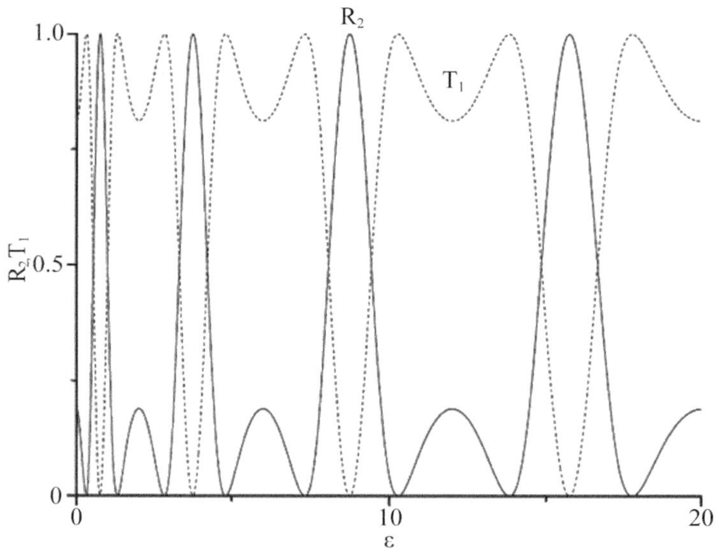

Figure 13.3 T_1 and R_2 of the AB ring in Fig. 13.1a as functions of ε for $\theta = 0, \bar{\alpha} = 1, b = 0, a_{10} = 1$, and $a_{20} = 0$.

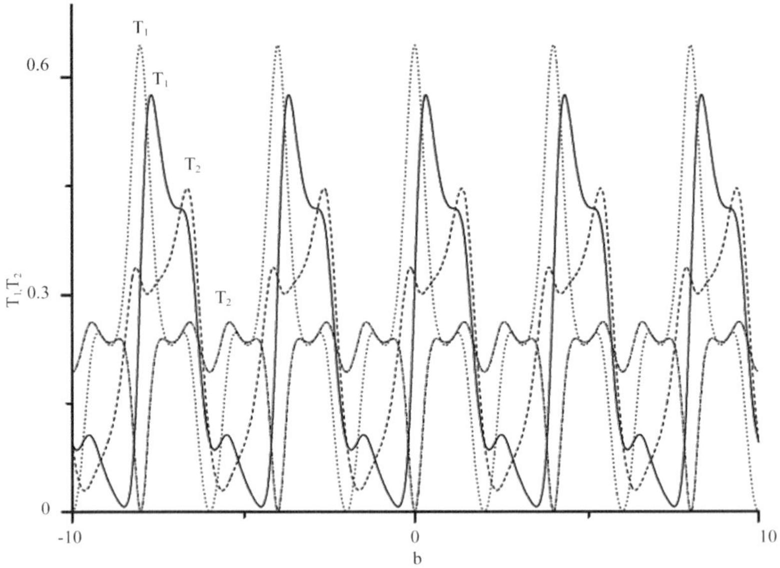

Figure 13.4 T_1 and T_2 of the AB ring in Fig. 13.1a as functions of **b** for $\theta = 0$, $\bar{\alpha} = 1$, $\varepsilon = 0.5$, and $a_{10} = 1$, $a_{20} = 0$ (T1,T2) and $a_{10} = 0.707$, $a_{20} = 0.707i$(T1',T2').

means that when the energy of the injected electron equals the energy of the eigenstate in the closed ring, there will be no reflection completely and there will be resonance.

Figure 13.4 shows the T_1 and T_2 as functions of the magnetic field b for the fixed electron energy $\varepsilon = 0.5$, $\theta = 0$, $\bar{\alpha} = 1$, and the coefficients of injected waves $a_{10} = 1$, $a_{20} = 0$, and $a_{10} = 0.7071$, $a_{20} = 0.7071i$. From Fig. 13.4 we see that for both cases, T_1 and T_2 curves all oscillate with the magnetic field b and the oscillating period is $b = 4$, though their zero points are different. From the definition of $b = 2eBR^2/c\hbar$, we obtain the period of the magnetic flux is $\Phi_m = hc/e$, independent of the magnitude of the ring. This is the first proof of the AB effect of the Rashba spin current.

13.5 Spin Interference in an AB Square Loop [18]

For comparing our results with the experiment in Ref. [15] and analytical account in Ref. [21] of Koga, we consider the spin interference in a square loop as shown in Fig. 13.1b, similar to the geometry of sample 2 in Koga's paper. We illustrate our method in detail. Using Eqs. 13.4 and 13.6, the wave functions in the input circuit i and the output circuit e can be written as

$$\Psi_{\mathrm{i}} = a_{10}\begin{pmatrix}1\\-1\end{pmatrix}e^{ik_1l_1} + a_{20}\begin{pmatrix}1\\1\end{pmatrix}e^{ik_2l_1} + a_1\begin{pmatrix}1\\1\end{pmatrix}e^{-ik_1l_1} + a_2\begin{pmatrix}1\\-1\end{pmatrix}e^{-ik_2l_1}$$

and

$$\Psi_{\mathrm{e}} = d_1\begin{pmatrix}1\\-i\end{pmatrix}e^{ik_1l_2} + d_2\begin{pmatrix}1\\i\end{pmatrix}e^{ik_2l_2}. \qquad (13.21)$$

The wave functions in the closed square loop at points A and B can be written as

$$\Psi_{\mathrm{c}}(A) = b_1\begin{pmatrix}1\\i\end{pmatrix}e^{ik_1l_2} + b_2\begin{pmatrix}1\\-i\end{pmatrix}e^{ik_2l_2} + b_3\begin{pmatrix}1\\-i\end{pmatrix}e^{-ik_1l_2} + b_4\begin{pmatrix}1\\i\end{pmatrix}e^{-ik_2l_2}$$

and

$$\Psi_{\mathrm{c}}(B) = c_1\begin{pmatrix}1\\1\end{pmatrix}e^{ik_1l_2'} + c_2\begin{pmatrix}1\\-1\end{pmatrix}e^{ik_2l_2'} + c_3\begin{pmatrix}1\\-1\end{pmatrix}e^{-ik_1l_2'} + c_4\begin{pmatrix}1\\1\end{pmatrix}e^{-ik_2l_2'}, \qquad (13.22)$$

where the origins of the coordinates l_2 and l_2' are at points A and B, respectively.

Using the boundary conditions Eqs. 13.11 and 13.13 we obtain the following group of equations:

$$a_{10} + a_{20} + a_1 + a_2 = b_1 + b_2 + b_3 + b_4,$$
$$-a_{10} + a_{20} + a_1 - a_2 = i\left(b_1 - b_2 - b_3 + b_4\right),$$
$$b_1 + b_2 + b_3 + b_4 = d_1 + d_2,$$
$$b_1 - b_2 - b_3 + b_4 = -d_1 + d_2,$$
$$c_1 + c_2 + c_3 + c_4 = d_1 + d_2,$$
$$c_1 - c_2 - c_3 + c_4 = i\left(-d_1 + d_2\right),$$
$$a_{10} + a_{20} - a_1 - a_2 + c_1 + c_2 - c_3 - c_4 = b_1 + b_2 - b_3 - b_4 + d_1 + d_2,$$

and

$$-a_{10} + a_{20} - a_1 + a_2 + c_1 - c_2 + c_3 - c_4 = i\left(b_1 - b_2 + b_3 - b_4 - d_1 + d_2\right). \quad (13.23)$$

The coefficients c_1, c_2, c_3, and c_4 are related to the coefficients b_1, b_2, b_3, and b_4 by the transfer matrices,

$$\begin{pmatrix} c_1 \\ c_2 \end{pmatrix} = \mathbf{M}_1 \begin{pmatrix} b_1 \\ b_2 \end{pmatrix}, \begin{pmatrix} c_3 \\ c_4 \end{pmatrix} = \mathbf{M}_2 \begin{pmatrix} b_3 \\ b_4 \end{pmatrix}. \quad (13.24)$$

The transfer matrices $\mathbf{M}_1$ and $\mathbf{M}_2$ are calculated in a way similar to Eq. 13.17.

$$\mathbf{M}_1 = \frac{1}{4}\begin{pmatrix} (1+i)\left(-A^4 + 2A^3B + A^2B^2\right) & (1-i)\left(A^3B + AB^3\right) \\ (1-i)\left(A^3B + AB^3\right) & (1+i)\left(A^2B^2 + 2AB^3 - B^4\right) \end{pmatrix}$$

and

$$\mathbf{M}_2 = \frac{1}{4}\begin{pmatrix} (1+i)\left(-A^{*4} + 2A^{*3}B^* + A^{*2}B^{*2}\right) & (1-i)\left(A^{*3}B^* + A^*B^{*3}\right) \\ (1-i)\left(A^{*3}B^* + A^*B^{*3}\right) & (1+i)\left(A^{*2}B^{*2} + 2A^*B^{*3} - B^{*4}\right) \end{pmatrix}, \quad (13.25)$$

where $A = e^{ik_1^L}$ and $B = e^{ik_2L}$.

Equations 13.23 to 13.25 are for the case of no magnetic field. In the case of a magnetic field perpendicular to the loop plane, the k_1 and k_2 in Ψ_i and Ψ_e in Eq. 13.21 are $k_0 \pm k_\delta$, while those in Ψ_c in Eq. 13.22 are $k_0 \pm k_\delta + k_A$ and $k_A = \pm\, eA/\hbar = \pm\, e\Phi_m/\hbar L$, where the $\pm$ sign of k_A depends on the direction of the wave, CW or CCW (see Eq. 13.6).

Solving Eqs. 13.23 and 13.24 we obtain the T_1, T_2, R_1, and R_2 of the square loop as functions of electron energy ε for $b = 0, \bar{\alpha} = 1, a_{10} = 1$, and $a_{20} = 0$, as shown in Fig. 13.5. From Fig. 13.5 we see that $R_1 = 0$, $T_1 = T_2$, and the zero points of R_2 are $\varepsilon = 1.9, 2.6, 8.9$, and 10.4, which are just the energies of eigenstates in the closed square loop given by Eq. 13.20. Figure 13.6 shows the T_1, T_2, R_1, and R_2 of the square loop as functions of the magnetic field b for $\varepsilon = 0.5, \bar{\alpha} = 1$, $a_{10} = 1$, and $a_{20} = 0$. From Fig. 13.6 we see that the T_1, T_2, and R_2 all oscillate with the magnetic field and the oscillating period is $b = 12.5$. According to the definition of the b in the square loop $b = 2eBL^2/\hbar c$, the oscillating period corresponds to $\Phi_m = hc/e$, the same as the AB ring.

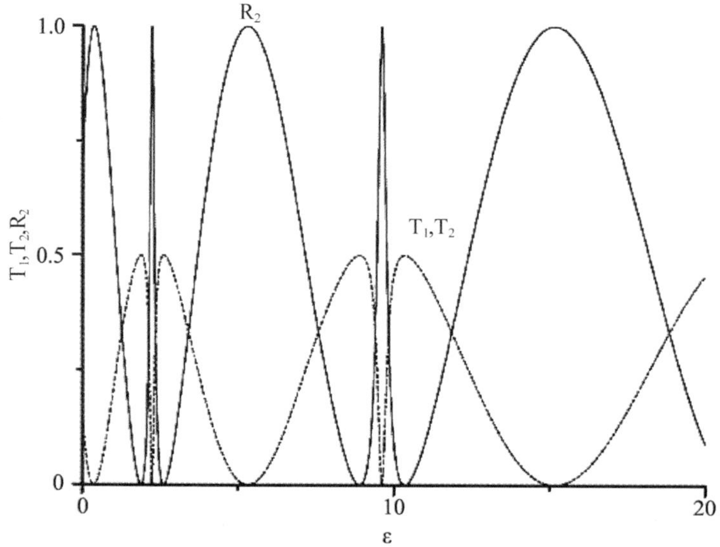

Figure 13.5 T_1, T_2, and R_2 of the AB SL in Fig. 13.1b as functions of ε for $b = 0, \bar{\alpha} = 1, a_{10} = 1$, and $a_{20} = 0$.

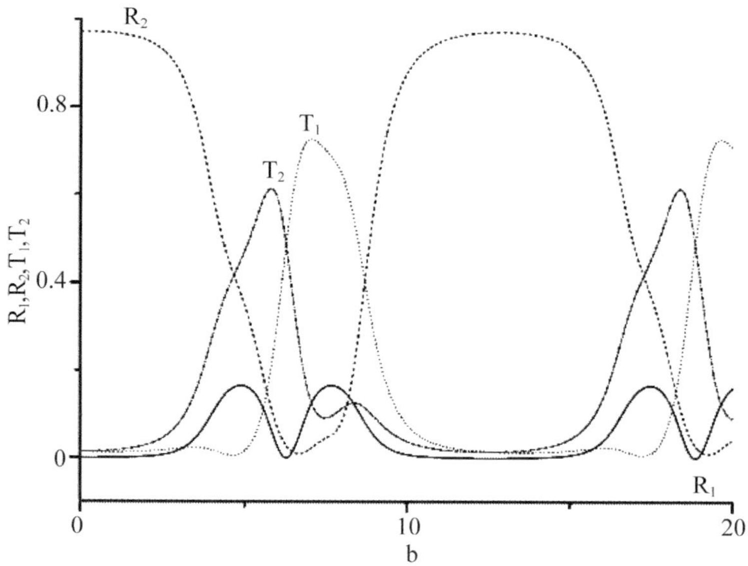

Figure 13.6 T_1, T_2, R_1, and R_2 of the AB SL in Fig. 13.1b as functions of b for $\varepsilon = 0.5, \bar{\alpha} = 1, a_{10} = 1$, and $a_{20} = 0$.

13.6 Spin Interference in an AB Double Square Loop [18]

In the experiment in Ref. [15] Koga et al. found that the period of the AB oscillation is $\Phi_m = hc/2e$, half of the theoretical prediction. We think that their circuit is actually an AB double square loop, as shown in Fig. 13.1c, and the Rashba electron travels twice in the loop. Similarly we write the wave functions in each circuit, and use the boundary conditions and the transfer matrices, and obtain the set of equations of wave function coefficients.

Solving the set of equations we obtain the T_1, T_2, R_1, and R_2 of the double square loop as functions of the electron energy ε for $b = 0$, $\bar{\alpha} = 1$, $a_{10} = 1$, and $a_{20} = 0$, as shown in Fig. 13.7. From Fig. 13.7 we see that $R_1 = 0$, $T_1 = T_2$, and the zero points of R_2 are $\varepsilon = 0.6$, 1.9, 2.6, 4.7, 6.0, 8.9, 10.4, 14.1, and 16.3, which are the energies of eigenstates in the closed double square loop. In there $\varepsilon = 1.9$, 2.6, 8.9, and 10.4 are the energies of eigenstates in the closed single square loop. So if the closed loop has n circles, there are n times of

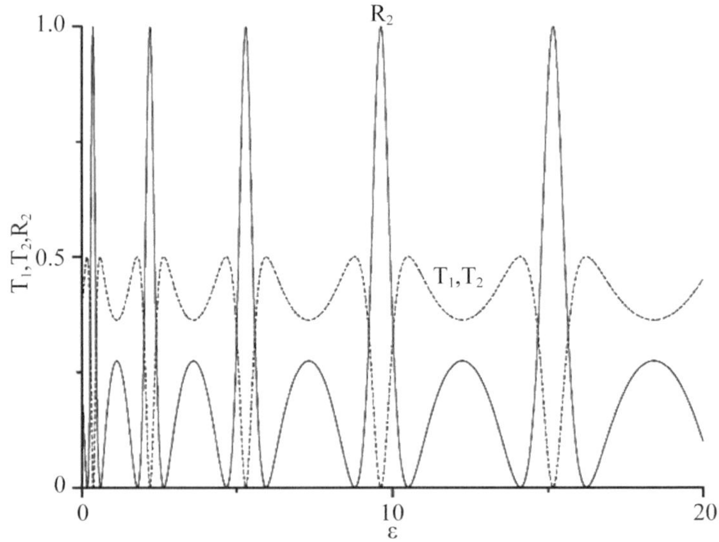

Figure 13.7 T_1, T_2, and R_2 of the AB double SL in Fig. 13.1c as functions of ε for $b = 0$, $\bar{\alpha} = 1$, $a_{10} = 1$, and $a_{20} = 0$.

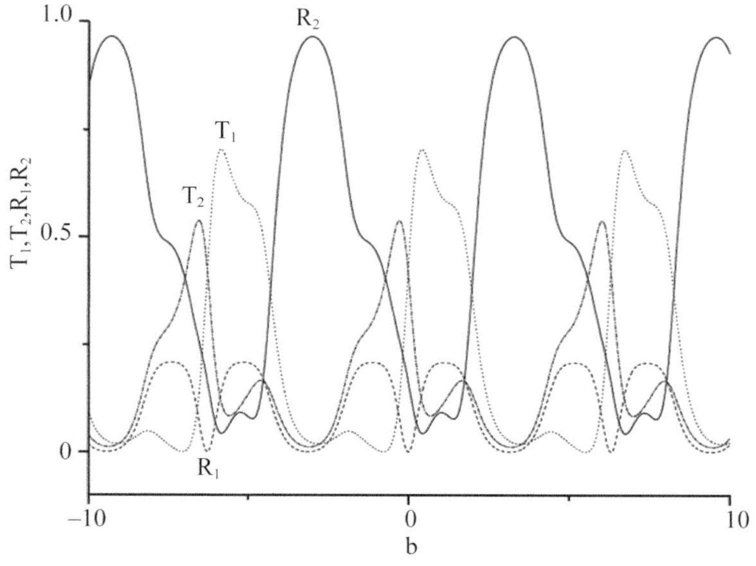

Figure 13.8 T_1, T_2, R_1, and R_2 of the AB double SL in Fig. 13.1c as functions of b for $\varepsilon = 0.5$, $\bar{\alpha} = 1$, $a_{10} = 1$, and $a_{20} = 0$.

eigenstates in the loop. Figure 13.8 shows the T_1, T_2, R_1, and R_2 of the double square loop as functions of the magnetic field b for $\varepsilon = 0.5$, $\bar{\alpha} = 1$, $a_{10} = 1$, and $a_{20} = 0$. From Fig. 13.8 we see that the T_1, T_2, and R_2 all oscillate with the magnetic field and the oscillating period is $b = 6.28$. According to the definition of the b in the square loop, the oscillating period corresponds to $\Phi_m = 3.14\hbar c/e = hc/2e$. It is half of that of the single square loop, in agreement with the experimental results [15].

Finally, we consider that electrons in Koga's experiment [15] are not spin-polarized, so we average our results for all spin orientation and get the total transmission coefficient T. Figure 13.9 shows T as a function of α (θ in Koga's papers) for $b = 0$ and $\varepsilon = 0.5$ and 50, corresponding to $E = 0.01$ meV and 1.0 meV for $L = 1200$ nm and $m^* = 0.047\ m_e$, as in Ref. 15. We note that the period is small, just about 1/4 of that in Koga's papers, and the curve is not very regular. So we consider that though Koga et al. got the biggest change of σ, they didn't control $\Delta\theta$ from 0 to 0.75 π. Maybe the variety range of α is just about 1/4 of what they expected.

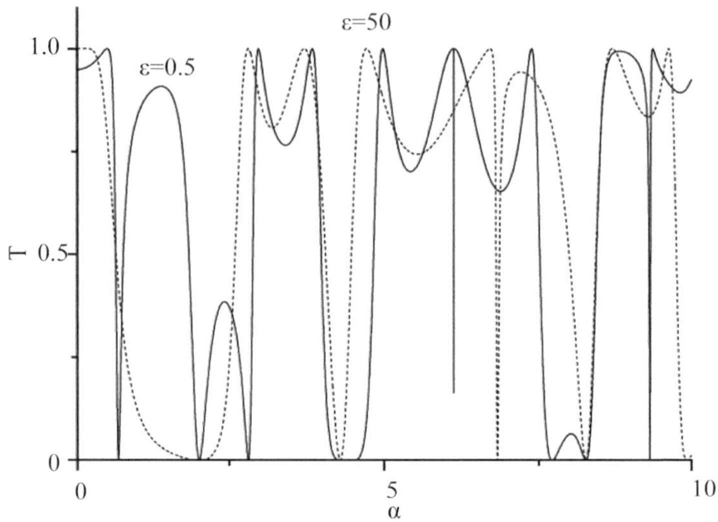

Figure 13.9 T as functions of α (θ in Koga's papers) for $b = 0$ and $\varepsilon = 0.5$ and 50.

13.7 Summary

We studied the spin waveguide transport of electrons with Rashba spin-orbit interaction (RSOI) in the curved circuits. The eigenstates of the circular and square loops were studied by using the transfer matrix method. the transfer matrix $\mathbf{M}(E)$ of a circular arc is obtained by dividing the circular arc into N segments and multiplying the transfer matrix of each straight segment. The energies of eigenstates in a closed loop are obtained by solving the equation $\det[\mathbf{M}(E) - I] = 0$. For the circular ring, the eigenenergies obtained with this method are in agreement with those obtained by solving the Schrödinger equation. For the SL, the analytic formula of the eigenenergies is obtained first. The transport properties of the AB ring, AB SL, and double SL are studied using the boundary conditions and the transfer matrix method. In the case of no magnetic field, the zero points of the reflection coefficients are just the energies of eigenstates in closed loops. In the case of a magnetic field, the transmission and reflection coefficients all oscillate with the magnetic field. The oscillating period is $\Phi_m = hc/e$, independent of the shape of the loop, where Φ_m is the magnetic flux through the loop. Finally, we compared our results with Koga's experiment.

References

1. Datta, S., and Das, B. (1990). *Appl. Phys. Lett.*, **56**, 665.
2. Rashba, E. I. (1960). *Sov. Phys. Solid State*, **2**, 1109.
3. Belluci, S., Carillo, E., and Onorato, P. (2008). *Eur. Phys. J. B*, **66**, 509.
4. Zhang, R. L., Zhang, Z. I., Peng, R. W., et al. (2008). *Appl. Phys.*, **103**, 07B727.
5. Souma, S., and Nikolić, B. K. (2005). *Phys. Rev. Lett.*, **94**, 106602.
6. Shelykh, L. A., Galkin, N. G., and Bagraev, N. T. (2005). *Phys. Rev. B*, **72**, 235316.
7. Yao, J., and Yang, Z. Q. (2006). *Phys. Rev. B*, **73**, 033314.
8. Wang, X. F., Vasilopoulos, P., and Peters, F. M. (2002). *Appl. Phys. Lett.*, **80**, 1400.
9. Wang, L. G., Chang, K., and Chan, K. S. (2006). *J. Appl. Phys.*, **99**, 043701.
10. Földi, P., Molnár, B., Benedict, M. G., and Peters, F. M. (2005). *Phys. Rev. B*, **71**, 033309.
11. Belluci, S., and Onorato, P. (2008). *Phys. Rev. B*, **78**, 235312.
12. Kálmán, O., Földi, P., Benedict, M. G., and Peters, F. M. (2008). *Physica E*, **40**, 567.
13. Aronov, A. G., and Lyanda-Geller, Y. B. (1993). *Phys. Rev. Lett.*, **70**, 343.
14. Nitta, J., Meijer, F. E., and Takayanagi, H. (1999). *Appl. Phys. Lett.*, **75**, 695.
15. Koga, T., Sekine, Y., and Nitta, J. (2006). *Phys. Rev. B*, **74**, 041302(R).
16. Bergsten, T., Kobayashi, T., Sekine, Y., and Nitta, J. (2006). *Phys. Rev. Lett.*, **97**, 196803.
17. Konig, M., Tschetschetkin, A., Hankiewiez, E. M., et al. (2006). *Phys. Rev. Lett.*, **96**, 076804.
18. Liu, D. Y., Xia, J. B., and Chang, Y. C. (2009). *J. Appl. Phys.*, **106**, 093705.
19. Sheng, J. S., and Chang, K. (2006). *Phys. Rev. B*, **74**, 235315.
20. Bolgakov, E. N., and Sadreev, A. F. (2002). *Phys. Rev. B*, **66**, 075331.
21. Koga, T., Nitta, J., and van Veenhuizen, M. (2004). *Phys. Rev. B*, **70**, 161302(R).

Chapter 14

Spin Polarization of a Rashba Electron with a Mixed State

In recent years, much attention has been paid to the field of RSOI [1] in low-dimensional semiconductor structures due to its potential applications in spintronic devices [2–4], which is based on the idea of the possible manipulation of the electron spin by an electric field [5]. Although spin transistors have yet to be achieved, the fundamental interest in understanding the effect and application of RSOI on macroscopic low-dimensional semiconductor structures continues [6, 7]. Many investigations focus on ballistic macroscopic rings [8–11] because a quantum ring exhibits an intriguing spin interference phenomenon [12, 13]. For instance, Chang et al. have studied persistent current in a quantum ring with two kinds of SOIs [9] and have shown that effective periodic potential caused by SOIs results in the weakening of the spin current and the localization of electrons. Ballistic electron transport of a Rashba electron through a chain of quantum circular rings has been investigated by Molnar et al. [10]. They have shown a periodic dependence on the incident electron's energy, through the parameter ka, the magnetic field B, and the strength of the RSOI α. Recently Naeimi et al. have shown that a double quantum ring in the presence of RSOI and a

Quantum Waveguide in Microcircuits (Second Edition)
Jian-Bai Xia, Duan-Yang Liu, and Wei-Dong Sheng

ISBN 978-981-4968-30-0 (Hardcover), 978-1-003-65146-8 (eBook)
www.jennystanford.com

magnetic flux can work as a spin-inverter [11]. Most investigations on quantum rings with RSOI focus on the circular ring because its Hamiltonian can be written as a one-dimensional model [14]. Without this advantage, spin transport in other shapes of rings is not convenient to calculate.

In previous work the injected electron is in the pure state, but for many cases, we expect a spin polarizer to get a spin polarization current from a nonpolarized state. To achieve this objective, in addition to an electric field, a magnetic field is also required. In this chapter, we study in detail the spin transport of Rashba electrons in an AB square ring with a magnetic flux. In addition, we study the spin polarization and its manipulation by α or B.

14.1 Transfer Matrix of a Rashba Electron in an AB Ring with a Magnetic Flux

In the presence of Rashba spin-orbital interaction (RSOI) and of a magnetic field B perpendicular to the $x-y$ plane, the appropriate Hamiltonian of an electron in the two-dimensional (2D) system in the $x-y$ plane is

$$H = \begin{pmatrix} \frac{\hbar^2}{2m^*}(\mathbf{p}+e\mathbf{A})^2 + V(x,y) & \frac{\alpha}{\hbar}(ip_x+p_y) \\ \frac{\alpha}{\hbar}(-ip_x+p_y) & \frac{\hbar^2}{2m^*}(\mathbf{p}+e\mathbf{A})^2 + V(x,y) \end{pmatrix}, \tag{14.1}$$

where m^* is the electron effective mass, α is the Rashba coefficient, and the Zeeman energy has been neglected [13]. For a one-dimensional (1D) circular ring structure this Hamiltonian can be rewritten as

$$H = \frac{\hbar^2}{2m^*R^2}\left[\left(-i\frac{\partial}{\partial\varphi} + \frac{\bar{\alpha}}{2}\sigma_r - \frac{\phi}{\phi_0}\right)^2 - \frac{\bar{\alpha}^2}{4}\right], \tag{14.2}$$

where ϕ is the azimuthal angle, R is the radius, $\bar{\alpha} = \alpha\left(\frac{2m^*R}{\hbar^2}\right)$ is the normalized Rashba constant, $\sigma_r = \cos\varphi\sigma_x + \sin\varphi\sigma_y$, ϕ is the magnetic flux through the ring, and $\phi_0 = h/e$. We can introduce the dimensionless Hamiltonian. Then the eigenenergy and eigenstates in this system can be given as

$$E^{\mu} = \left(m + \frac{b}{4} + \frac{1}{2} + \frac{(-1)^{\mu}}{2}\sqrt{1+\bar{\alpha}^2}\right)^2 - \frac{\bar{\alpha}^2}{4} \tag{14.3}$$

and

$$\Phi^{\mu} = e^{im\varphi}\chi^{\mu}(\varphi), \tag{14.4}$$

where $\mu = 1$ and 2, which refer to the up- and down-spin states, respectively, $b = 2eBR^2/\hbar$, and the orthogonal spinors $\chi^{\mu}(\varphi)$ can be expressed in the terms of the eigenvectors $(1{,}0)^{\mathrm{T}}$ and $(0{,}1)^{\mathrm{T}}$ of the Pauli matrix σ_z as

$$\chi^1(\varphi) = \begin{pmatrix} \cos\dfrac{\beta}{2} \\ e^{i\varphi}\sin\dfrac{\beta}{2} \end{pmatrix} \tag{14.5}$$

and

$$\chi^2(\varphi) = \begin{pmatrix} \sin\dfrac{\beta}{2} \\ -e^{i\varphi}\cos\dfrac{\beta}{2} \end{pmatrix}, \tag{14.6}$$

where $\beta \equiv \arctan(-\bar{\alpha})$. These results are obtained from the 1D Hamiltonian Eq. 14.2, but for a variety of 1D ring, such as a square ring or an elliptical ring, we can't find the 1D Hamiltonian. To study a Rashba electron's transport in these structures, we divide a curved line into N segments [15]. For a curved line, such as an elliptical ring or a circular ring, N is large enough and every segment is very small, and then every segment can be approximated to be a line segment along the tangential direction. For every segment we can obtain the eigenstates and we can write the wavefunction in the curve as

$$\Phi = (\psi_1(\theta), \psi_2(\theta))\Psi = (\psi_1(\theta), \psi_2(\theta))\begin{pmatrix} a(l) \\ b(l) \end{pmatrix}, \tag{14.7}$$

where

$$\psi_1(\theta) = \frac{1}{\sqrt{2}}\begin{pmatrix} 1 \\ ie^{i\theta} \end{pmatrix}, \quad \psi_2(\theta) = \frac{1}{\sqrt{2}}\begin{pmatrix} 1 \\ -ie^{i\theta} \end{pmatrix},$$

l denotes the coordinates on the curve line, and θ is the azimuthal angle of the tangent line of the curve. Adopt Ψ to describe the wave function and by using Griffith's boundary conditions [16–18] in each vertex, we can relate the wave function at the two endpoints by

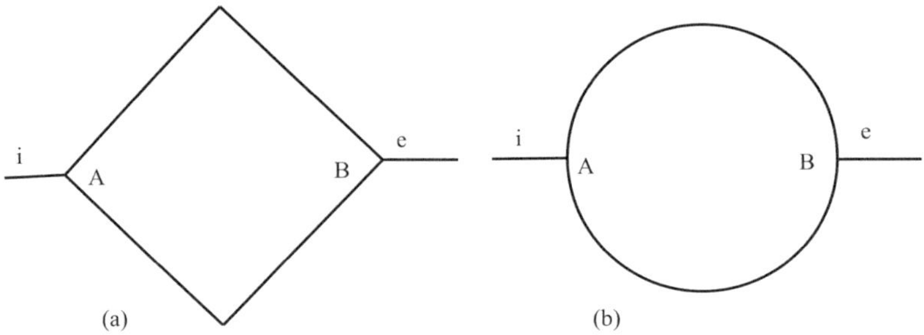

Figure 14.1 Two structures: (a) an AB square ring and (b) an AB circular ring.

a transfer matrix. For a polyline structure, such as a square ring, every lead is a natural line segment, so N is the number of leads. All analyses on the curved line remain correct, and actually a curved line is approximated to be a multistage polyline in our method. In a circular ring this method gave results that are identical to those obtained from 1D Hamiltonian [15], so this method is reasonable.

Consider a 1D square ring as shown in Fig. 14.1a. Here the electron current is injected from circuit i and the output circuit e is on the right side. We assume that at point A the down arm is with the polar angle θ_1 and the up arm is with the polar angle θ_2. Therefore in Eq. 14.7 θ_1 and θ_2 correspond to the corresponding side of the square ring. The side length of the square is denoted by L, and the strength of RSOI in the square is denoted by α_s. In this chapter, we use the dimensionless physical quantities. Similar to R in the circular ring, L can be treated as the size of the square ring. If the energy unit $E_0 = \hbar^2/(2m^* L^2)$, then the energy $\varepsilon = E/E_0$, the magnetic field $b = (\hbar e B/m^*)/E_0$, and the Rashba coefficient $\bar{\alpha}_s = (\alpha_s/L)/E_0 = \alpha_s(2m^* L/\hbar^2)$. Similar to Ref. [15], we can write the state of the electron in every part of the structure. We adopt Ψ to describe the wave function in the two arms of the square ring:

$$\Psi^j_{\text{up}} = \begin{pmatrix} a^j_{\text{up}} & (l^j_{\text{up}}) \\ b^j_{\text{up}} & (l^j_{\text{up}}) \end{pmatrix}, \tag{14.8}$$

$$\Psi^j_{\text{down}} = \begin{pmatrix} a^j_{\text{down}} & (l^j_{\text{down}}) \\ b^j_{\text{down}} & (l^j_{\text{down}}) \end{pmatrix}, \tag{14.9}$$

$$\Phi_i = \begin{pmatrix} a_i \\ b_i \end{pmatrix} e^{ik_0 l_i} + \begin{pmatrix} a_r \\ b_r \end{pmatrix} e^{-ik_0 l_i}, \tag{14.10}$$

and

$$\Phi_0 = \begin{pmatrix} a_t \\ b_t \end{pmatrix} e^{ik_0 l_o}, \tag{14.11}$$

where up and down denote the upper and lower arms of the square ring and $j = 1$ or 2, which correspond to the clockwise and counterclockwise motions of electron through the square ring, respectively. In particular, $l^1_{\rm up}$ and $l^2_{\rm up}$ or $l^1_{\rm down}$ and $l^2_{\rm down}$ have opposite directions. According to the transfer matrix method, we can obtain the Ψ of each arm if we know its value at any one point, so there are 12 unknown coefficients in $\Psi^j_{\rm up}$, $\Psi^j_{\rm down}$, $\Phi_{\rm i}$, and Φ_0. Using Griffith's boundary conditions [16–18], we can determine all unknown coefficients. We assume that the original points of $l_{\rm i}$ and $l_{\rm o}$ are point A and point B, respectively, then when $a_{\rm i} = 1$ and $b_{\rm i} = 0$, we have $a_{\rm r} = r_{11}$, $b_{\rm r} = r_{12}$, $a_{\rm t} = t_{11}$, and $b_{\rm t} = t_{12}$; similarly, when $a_{\rm i} = 0$ and $b_{\rm i} = 1$, we have $a_{\rm r} = r_{21}$, $b_{\rm r} = r_{22}$, $a_{\rm t} = t_{21}$, and $b_{\rm t} = t_{22}$, where $\sigma = 1$ or 2, which denote the spin-up and spin-down states in the z direction. The spin-dependent transmission coefficient of an electron with incoming spin σ and outgoing spin σ' can be written as $T_{\sigma\sigma'} = |t_{\sigma\sigma'}|^2$. For comparison the spin transport in an AB circular ring as shown in Fig. 14.1b is also calculated.

14.2 Description of Spin Polarization of a Rashba Electron

In the above discussion, the injected electron is in the pure state, but for many cases, this condition can't be satisfied. We assume that the rate for spin-up is P_1 and the rate for spin-down is P_2. Then it is convenient to introduce a spin polarization P to describe the difference between the spin-up and spin-down electron transmissions:

$$P = \frac{P_1 - P_2}{P_1 + P_2}. \tag{14.12}$$

Obviously, $-1 \leq P \leq 1$, $P = -1$, and $P = +1$ represent complete spin polarization in the z direction. Now we define

$$Q \equiv P_1/P_2. \tag{14.13}$$

Then $P = (Q-1)/(Q+1)$. There is a one-to-one correspondence between Q and P, and Q and P have the same changing trend, so we can study the variation of Q instead of the variation of P. We can obtain the relationship of Q of the injected electron Q^{i} and outgoing electron Q^{o}:

$$Q^{\mathrm{o}} = \frac{T_{11}Q^{\mathrm{i}} + T_{21}}{T_{21}Q^{\mathrm{i}} + T_{22}}. \tag{14.14}$$

First we consider the change of Q though a structure ΔQ. In the condition $Q^{\mathrm{i}} < Q^{\mathrm{fix}}$, we have $\Delta Q < 0$; in the condition $Q^{\mathrm{i}} > Q^{\mathrm{fix}}$, $\Delta Q > 0$, and in the condition $Q^{\mathrm{i}} = Q^{\mathrm{fix}}$, $\Delta Q = 0$, where

$$Q^{\mathrm{fix}} = \frac{-(T_{11} - T_{22}) + \sqrt{(T_{11} - T_{22})^2 + 4T_{12}T_{21}}}{2T_{12}}. \tag{14.15}$$

We can determine that the spin polarization can be changed though a structure, and $Q^{\mathrm{fix}}(P^{\mathrm{fix}})$ is the sign value of $Q(P)$ of the outgoing electron. Because $T_{12} = T_{21}$, Q^{fix} is determined by $(T_{11} - T_{22})/T_{12}$. Another condition we are interested in is the injected electron is totally nonpolarized. In this case $Q^{\mathrm{i}} = 1$, so $Q^{\mathrm{o}} = (T_{11} + T_{21})/(T_{12} + T_{22}) \equiv Q^{\mathrm{sp}}$. In this condition, $P^{\mathrm{sp}} = (Q^{\mathrm{sp}} - 1)/(Q^{\mathrm{sp}} + 1)$ can describe the polarization of the spin of the outgoing electron when the injected electron is nonpolarized.

14.3 Spin Transport in a Square Ring and a Circular Ring with a Magnetic Flux

In this section we present the numerical result for the spin-dependent electron transport in a square ring and a circular ring. The effect of the RSOI strength α, the magnetic field B, and the incident electron energy E are investigated for different conditions, and two structures are compared with each other. Further, we use the dimensionless physical quantities as in section 14.2, but for B, we measure the magnetic flux ϕ in units of $\phi_0 = h/e$.

In Fig. 14.2 we show the spin-dependent electron transmission coefficients T_{11}, $T_{12}(T_{21})$, and T_{22} and the electron spin polarizations

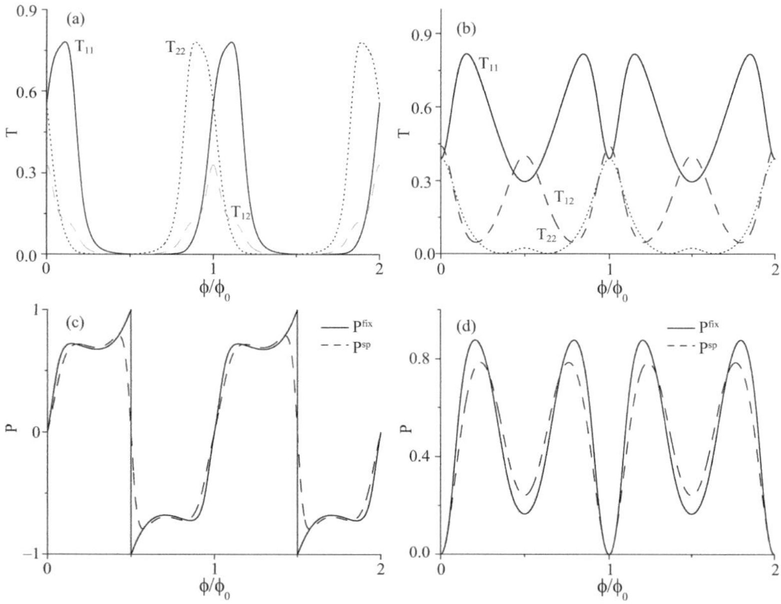

Figure 14.2 (a, c) The spin-dependent electron transmission coefficients T_{11}, $T_{12}(T_{21})$, and T_{22} and (b, d) the electron spin polarizations P^{fix} and P^{sp} of the emergent electron as a function of the magnetic flux ϕ. (a, c) for the AB square ring, and (b, d) for the circular ring.

P^{fix} and P^{sp} of the emergent electron as a function of the magnetic flux ϕ. The relevant parameters in our calculation are that $E = 10$ and $\bar{\alpha} = 1.0$. Figure 14.2a shows the electron transmission coefficients for the square ring in Fig. 14.1a, and Fig. 14.2b shows the electron transmission coefficients for the circular ring in Fig. 14.1b. We can see that for both cases, the electron transmission coefficient curves all oscillate with the magnetic flux ϕ and the oscillating period is $\phi/\phi_0 = 1$. It proves that the AB effect exists in two kinds of AB rings. In these figures, $T_{11} \neq T_{22}$ for any regions in which ϕ/ϕ_0 is not an integer number. According to our theory in Section 14.2, this means that the AB rings can act as spin polarizers in the presence of both the RSOI and the perpendicular magnetic field. In Fig. 14.2c and Fig. 14.2d we show the P^{fix} and P^{sp} as a function of the magnetic flux ϕ for the square ring and the circular ring, respectively. High polarization of spin can be reached in two structures, but there are some differences. In our calculation, in the circular ring, the peak

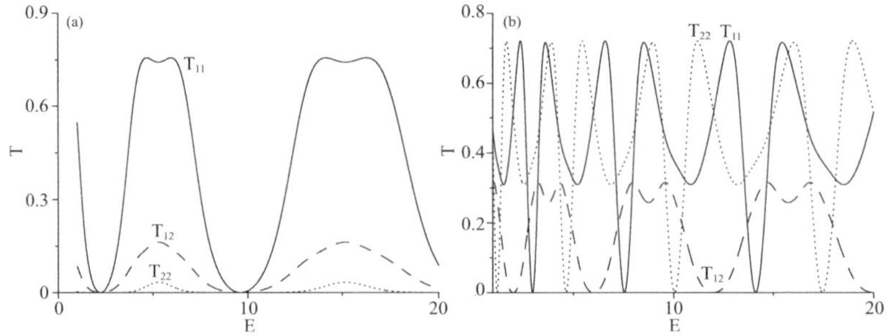

Figure 14.3 T_{11}, $T_{12}(T_{21})$, and T_{22} as a function of the injected electron's energy (a) for the AB square ring and (b) for the AB circular ring.

of P^{fix} or P^{sp} is sharp, while in the square ring, the peak area and the valley area of P is wide. This means that the square ring has a higher stability as a spin polarizer. This fact shows that the geometry of the structure is an important parameter for modulation of the spin transport if other parameters are fixed.

Now we investigate the effects of the energy of the injected electron on the spin transport. Figures 14.3a and 14.3b show T_{11}, $T_{12}(T_{21})$, and T_{22} as a function of the injected electron's energy E for $\bar{\alpha} = 1.0$ and $\phi/\phi_0 = 0.4$ in the square and circular rings, respectively. From Fig. 14.3 we see that the electron transmission coefficients oscillate with the energy, but "the period" increases with increasing E. This result is due to the expression of the eigenenergy of corresponding close rings in Ref. [15]. Comparing the electron transmission coefficients in the two rings, we find that the electron transmission coefficients in the circular ring oscillate more rapidly because the energy-level spacing is smaller in the circular ring. This result shows again that the square ring has a higher stability as a spin device.

Figure 14.4a and Fig. 14.4b show T_{11}, $T_{12}(T_{21})$, and T_{22} as a function of the Rashba strength $\bar{\alpha}$ in the square ring and in the circular ring, respectively. The relevant parameters are $E = 10$ and $\phi/\phi_0 = 0.4$. Again we see that the electron transmission coefficients in the circular ring oscillate much more rapidly, so it means that the stability of the circular ring is worse than the square ring as a spin device.

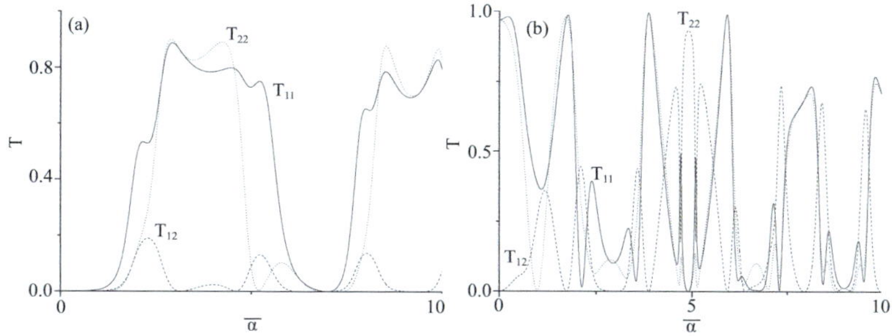

Figure 14.4 T_{11}, $T_{12}(T_{21})$, and T_{22} as a function of the Rashba strength $\bar{\alpha}$ (a) for the AB square ring and (b) for the AB circular ring.

14.4 Spin Polarization of a Rashba Electron in a Quantum Ring

To obtain a higher spin polarization and more effective modulation, we should take account of the effect of more parameters together. In Fig. 14.5 we show the effect of the Rashba strength and the magnetic flux on the spin polarization transport. We focus on the realization of spin polarizer, so the spin current density is as important as the spin polarization. We define the effective P

$$P^{\mathrm{ef}} = \frac{1}{2} P^{\mathrm{sp}} \sum_{\sigma,\sigma'=1,2} T_{\sigma\sigma'}. \tag{14.16}$$

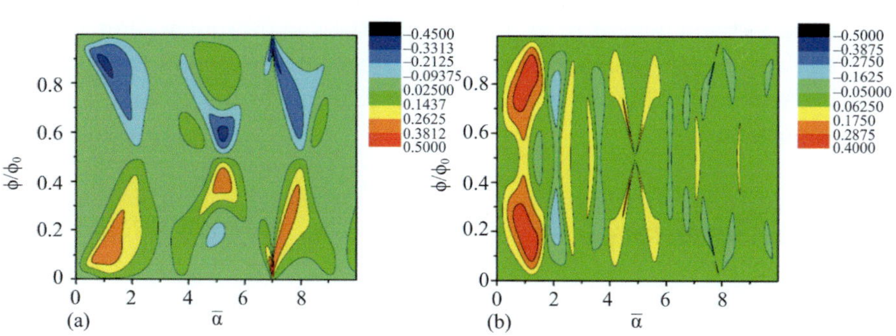

Figure 14.5 Contour map of effective spin polarization P^{ef} as a function of the Rashba strength $\bar{\alpha}$ and the magnetic flux $\phi(a)$ for the AB square ring and (b) for the AB circular ring.

If the injected electron is nonpolarized, the sign of P^{ef} denotes the spin-polarization direction on the z axis and its absolute value contains information of both electron current density and spin polarization of the outgoing electron. Only when both of them reach a higher value, P^{ef} have a considerable value. The contour maps of P^{ef} as a function of $\bar{\alpha}$ and ϕ in the square ring and in the circular ring are shown in Fig 14.5a and Fig. 14.5b, respectively. In our calculation the incident electron energy $E = 10$. As shown in Fig. 14.5, $|P^{\text{ef}}|$ is very small for most of the region of $(\bar{\alpha}, \phi/\phi_0)$, so we must choose suitable parameters for a spin polarizer. For instance, we can set $\bar{\alpha} = 1.0$ and modulate ϕ in both structures. This condition is discussed in Fig. 14.2. Another result is that we don't need large Rashba strength and magnetic field. If the scale of the structure is about 100 nm and $m^*/m_e \simeq 0.1$, then we get $\alpha = \bar{\alpha}\hbar^2/(2m^*L) \simeq 4.0$ meV nm and $B \simeq 0.1$T. Comparing Fig. 14.5a with Fig. 14.5b, we can clearly find that P^{ef} changes more rapidly in the circular ring and there is a wider region of $(\bar{\alpha}, \phi/\phi_0)$ in the square ring, which is useful for modulating the spin polarization. These results are reasonable. Many research studies have indicated that the transport in an AB ring is related to the energy band structures of a corresponding closed ring. Eigenenergies of Rashba states in a square loop are given in Ref. [15], and we can find the energy-level spacing is much larger than that in Eq. 14.3. As a result the spin transport of a Rashba electron in the square ring is not so sensitive to relevant parameters as that in the circular ring. In addition, when the $\bar{\alpha}$ and ϕ/ϕ_0 are the same in two kinds of rings, the arms of the square ring are longer and a spin-dependent quantum interference phenomenon is more likely to occur. This fact may be another reason why the geometry is important.

14.5 Summary

In this chapter, we studied the spin polarization in a square ring and in a circular ring with the RSOI and the magnetic flux. We developed the 1D quantum waveguide theory in straight waveguide systems and applied it in the 1D curve systems. The effects of Rashba strength, magnetic field, and incident electron energy were taken into account. We found that for appropriate values of parameters, both kinds of AB rings can work as spin polarizers and modulation by some parameters is realizable. Comparing the results in two structures, it was found that the spin polarization in the square ring has a wider peak (valley) area and it means that the square ring has higher stability as a spin polarizer.

References

1. Rashba, E. I. (1960). *Sov. Phys. Solid State*, **2**, 1109.
2. Pareek, T. P. (2004). *Phys. Rev. Lett.*, **92**, 76601.
3. Koga, T., Nitta, J., and van Veenhuizen, M. (2004). *Phys. Rev. B*, **70**, 161302(R).
4. Koga, T., Sekine, Y., and Nitta, J. (2006). *Phys. Rev. B*, **74**, 041302(R).
5. Datta, S., and Das, B. (1990). *Appl. Phys. Lett.*, **58**, 665.
6. Simchi, H., Esmaeilzadeh, M., and Mazidabadi, H. (2013). *Physica E*, **54**, 220.
7. Fallah, F., and Esmaeilzadeh, M. (2011). *AIP Adv.*, **1**, 032113.
8. Splettstoesser, J., Governale, M., and Zulicke, U. (2003). *Phys. Rev. B*, **68**, 165341.
9. Sheng, J. S., and Chang, K. (2006). *Phys. Rev. B*, **74**, 235315.
10. Molnar, B., Vasilopoulos, P., and Peeters, F. M. (2005). *Phys. Rev. B*, **72**, 075330.
11. Naeimi, A. S., Eslami, L., Esmaeilzadeh, M., et al. (2013). *J. Appl. Phys.*, **113**, 014303.
12. Mathur, H., and Stone, A. D. (1992). *Phys. Rev. Lett.*, **68**, 2964.
13. Molnar, B., Peeters, F. M., and Vasilopoulos, P. (2004). *Phys. Rev. B*, **69**, 155335.
14. Meijer, F. E., Morpurgo, A. F., and Klapwijk, T. M. (2005). *Phys. Rev. B*, **66**, 033107.
15. Liu, D.-Y., Xia, J.-B., and Chang, Y.-C. (2009). *J. Appl. Phys.*, **106**, 093705.
16. Liu, D.-Y., Xia, J.-B., and Chang, Y.-C. (2010). *Sci. China Ser. G*, **53**, 16.
17. Griffith, S. (1953). *Trans. Faraday Soc.*, **49**, 345.
18. Xia, J. B. (1992). *Phys. Rev. B*, **45**, 3593.

Chapter 15

Two-Dimensional Quantum Waveguide Theory of Rashba Electrons

The first model of spin transistor proposed by Datta [1] working by controlling the strength of Rashba spin-orbit interaction (RSOI) [2] was demonstrated to be feasible in narrow-gap semiconductors [3]. Then many device ideas were brought forward [4–10]. Recently, Koo et al. [11] reported the demonstration of the spin-injected field-effect transistor in a high-mobility InAs heterostructure. They confirmed the Datta's theory [1]—an oscillatory channel conductance as a function of monotonically increasing gate voltage. Besides, a great deal of one-dimensional (1D) devices were proposed [12, 13]. The 1D quantum waveguide theory have been introduced in the previous three chapters. When the width of the waveguide is not narrow enough, the 1D approach cannot work; the transport of the electrons will be in some transverse excited states synchronously, not only the ground state; and the transmission will be concerned with the structure shape.

The two-dimensional (2D) quantum waveguide transport for a spinless electron has been introduced in Chapter 11, but that for a Rashba electron has rarely been studied [5]. In our work for this chapter we applied the 1D quantum waveguide theory of

Quantum Waveguide in Microcircuits (Second Edition)
Jian-Bai Xia, Duan-Yang Liu, and Wei-Dong Sheng

ISBN 978-981-4968-30-0 (Hardcover), 978-1-003-65146-8 (eBook)
www.jennystanford.com

a Rashba electron to the 2D cases, which can operate the spin flip and the electron block more effectively. We used the transfer matrix method [16, 17] with the boundary conditions derived in Refs. [14] and [15] to study the 2D ballistic transport of Rashba electrons in straight structures of various shapes, including square stub, triangular stub, and periodic structures [19]. We found that the boundary condition for the derivative of the wave functions at the discontinuous interface used by Wang et al. [5] is not reasonable, so it is necessary to formulate the theory and calculate the results correctly. And the 2D quantum waveguide theory for a Rashba electron may be useful for designing the spin-injected field-effect transistor.

15.1 Transfer Matrix Method Considering Spin [19]

The transfer matrix method's general formalism of a spinless situation has been described in Chapter 11 and many papers [16–18]. Now we apply it to Rasbha electrons. If ballistic electrons move along a straight channel in two-dimensional electron gasses (2DEGs) that work in the presence of the Rasbha spin-orbit interaction (RSOI), we can divide the channel into some segments (see Chapter 11) and in each segment the width is constant. If we assume that 2DEGs are in the $x - y$ plane and the channel is in the x direction, in each segment the wave function can be written as

$$\Psi^m(x, y) = \sum_{n=1}^{N} \left[a_{1n}^m \phi_1(0) e^{ik_{1n}^m x} + a_{2n}^m \phi_2(0) e^{ik_{2n}^m x} + b_{1n}^m \phi_1(0) e^{-ik_{2n}^m x} + b_{2n}^m \phi_2(0) e^{-ik_{1n}^m x}\right] \varphi_n^m(y), \quad (15.1)$$

where m denotes the m-th segment and ϕ_1 and ϕ_2 are the phases of the Rashba wave functions in Eq. 12.6 [14]. The argument θ in ϕ_1 and ϕ_2 is the azimuth angle of the circuit; in our case $\theta = 0$ or π. k_{1n}^m and k_{2n}^m are the wave vectors [14] depending on the transverse energy of the electron in the m-th segment,

$$k_{1n}^m = k_{0n}^m + k_\delta, \quad k_{2n}^m = k_{0n}^m - k_\delta$$

and

$$k_{0n}^{m} = \frac{m^*}{\hbar^2}\sqrt{\alpha^2 + \frac{2\hbar^2}{m^*}E_{\|}^{mn}}, \quad k_\delta = \frac{m^*}{\hbar^2}\alpha, \tag{15.2}$$

where α is the Rashba coefficient, $E_{\|}$ is the longitudinal kinetic energy

$$E_{\|}^{mn} = E - \frac{\hbar^2}{2m^*}\left(\frac{n\pi}{W_m}\right)^2, \tag{15.3}$$

W_m denotes the width of the m-th segment, and n denotes the n-th transverse sublevel. The sum over n includes evanescent modes, that is, k_{0n}^{m} is imaginary.

In this chapter we use dimensionless units of physical quantities. The length of the unit is the width of the input channel W_0, the wave vector unit is π/W_0, and the units of the energy and the Rashba coefficient are

$$\varepsilon_0 = \frac{\hbar^2}{2m^*}\left(\frac{\pi}{W_0}\right)^2 \quad \text{and} \quad \alpha_0 = \frac{\hbar^2}{2m^*}\left(\frac{\pi}{W_0}\right), \tag{15.4}$$

respectively. With the dimensionless units, k_{0n}^{m} and k_δ in Eq. 15.2 can be written as

$$k_{0n}^{m} = \sqrt{\left(\frac{\alpha}{2}\right)^2 + E - \left(\frac{nW_0}{W_n}\right)^2} \quad \text{and} \quad k_\delta = \frac{\alpha}{2}. \tag{15.5}$$

Then we match the wave function and the current density, J_x left-multiplying the wave function, at each interface of segments, with different widths. J_x is the current density operator (4.12) [14]

$$J_x = -\frac{i}{\hbar}[l, H] = \frac{i}{\hbar}\begin{pmatrix} -\frac{\hbar^2}{m^*}\frac{\partial}{\partial l} & \alpha e^{-i\theta} \\ -\alpha e^{i\theta} & -\frac{\hbar^2}{m^*}\frac{\partial}{\partial l} \end{pmatrix}. \tag{15.6}$$

It is noticed that the result of the operator J_x left-multiplying the wave function is rather simple [14].

$$J_x\Psi_{\sigma n} = J_x\phi_\sigma(\theta)\,e^{ik_{\sigma n}x} = \frac{\hbar k_{0n}}{m^*}\Psi_{\sigma n}. \tag{15.7}$$

This means that the current density is independent of spin σ, which is in contrast to the result of Ref. [5], where the authors used the derivative operator instead of the current density operator J_x.

Considering the orthogonality of ϕ_1 and ϕ_2, we divide the Rashba electron wave functions into two parts, spin-up and spin-down. We write the expression of Ψ^m in Eq. 15.1 in the matrix form and adopt a 4N dimensional column vector to describe the wave function in the m-th segment as the following form:

$$\Psi^m = \phi_1(0) \otimes \Psi_1^m + \phi_2(0) \otimes \Psi_2^m \tag{15.8}$$

and

$$\begin{aligned}\Psi_\sigma^m &= \left(a_\sigma 1 e^{ik_{\sigma 1}^m x}, a_{\sigma 2} e^{ik_{\sigma 2}^m x}, \cdots a_{\sigma N} e^{ik_{\sigma N}^m x}, b_{\sigma 1} e^{-ik_{3-\sigma 1}^m x},\right.\\ &\quad \left. b_{\sigma 2} e^{-ik_{3-\sigma 2}^m x}, \cdots b_{\sigma N} e^{-ik_{3-\sigma N}^m x}\right)^T\\ &= \left(A_\sigma^m, B_\sigma^m\right)^T,\end{aligned} \tag{15.9}$$

where $\sigma = 1$ or 2. We assume the hard wall approximation at the edge of channel, hence the transverse wave functions

$$\varphi_n^m(y) = \sqrt{\frac{2}{W_m}} \sin\left[\frac{n\pi}{W_m}\left(y - y_0^m\right)\right], \tag{15.10}$$

where y_0^m is the coordinate of the down edge of the m-th segment. This assumption is adopted in the whole chapter. The boundary conditions of the wave function and the current density at the interface of the m-th and the $(m+1)$th segment are

$$\Psi^m = \Psi^{m+1}$$

and

$$J_x \Psi^m = J_x \Psi^{m+1}. \tag{15.11}$$

Then left-multiplying Eq. 15.11 by $\phi_1(0)$ and $\phi_2(0)$, respectively, we get

$$\Psi_\sigma^m = \Psi_\sigma^{m+1} \tag{15.12}$$

and

$$J_x \Psi_\sigma^m = J_x \Psi_\sigma^{m+1}, \tag{15.13}$$

where $\sigma = 1$ or 2. So it is derived from Eqs. 15.12 and 15.13 that for movement in the straight channel, there is no mixing between the wave functions of spin-up ($\sigma = 1$) and spin-down ($\sigma = 2$) and the transfer matrix, which relates Ψ_σ on two sides of the interface, is independent of σ. In the following we will omit the index σ.

There are two cases: the transverse width of the m-th segment W_m is smaller than W_{m+1} of the $(m+1)$th segment, or vice versa. For the former, we multiply both sides of Eq. 15.12 by $\varphi_n^{m+1}(y)$ and both sides of Eq. 15.13 by $\varphi_l^m(y)$ and integrate over the transverse coordinate range of the $(m+1)$th and m-th segments, respectively. We get

$$\sum_{l=1}^{N}\left(a_l^m + b_l^m\right) S_{l,n} = a_n^{m+1} + b_n^{m+1} \tag{15.14}$$

and

$$k_{0l}^m\left(a_l^m - b_l^m\right) = \sum_{n=1}^{N} k_{0n}^{m+1}\left(a_n^{m+1} - b_n^{m+1}\right) S_{l,n}, \tag{15.15}$$

where $S_{l,n}$ is defined as

$$S_{l,n} = \int_{W_n} \varphi_l^m(y)\,\varphi_n^{m+1}(y)\,dy. \tag{15.16}$$

Now if we adopt the expression of Ψ_m in Eq. 15.8 and Eq. 15.9, we can obtain the transfer matrix connecting the Ψ_m on the left side of the interface and the Ψ_{m+1} on the right side [16]

$$\begin{aligned}\Psi^m &= M_{m,m+1}\Psi^{m+1},\\ M_{m,m+1} &= \frac{1}{2}\begin{pmatrix} M^+ + M^- & M^+ - M^- \\ M^+ - M^- & M^+ + M^- \end{pmatrix},\end{aligned} \tag{15.17}$$

$$M^+ = \left(\mathbf{S}^T\right)^{-1}, \text{ and } \quad M^- = \left(K^m\right)^{-1}\mathbf{S}\mathbf{K}^{m+1}. \tag{15.18}$$

K^m is the diagonal matrix with elements k_{0n}^m. For the case of $W_m > W_{m+1}$, we can similarly obtain the transfer matrix.

In the straight m-th segment, the transfer matrix connecting the Ψ_m on the left end and the right end of the segment is the diagonal matrix, dependent on σ.

$$M_{m,\sigma} = \begin{pmatrix} P_\sigma^- & 0 \\ 0 & P_\sigma^+ \end{pmatrix} \tag{15.19}$$

and

$$\left(P_\sigma^-\right)_{nl} = e^{-ik_{\sigma n}^m L_m}\delta_{nl}, \ \left(P_\sigma^+\right)_{nl} = e^{ik_{3-\sigma n}^m L_m}\delta_{nl},$$

where L_m is the length of the m-th segment.

In this chapter, we assume that an incident electron with energy E_0 enters the channel and the state of the incident electron is known.

So the coefficients of the wave functions at the input end and the output end A_σ^{in} are certain and $B_\sigma^{\text{out}} = 0$. The total transfer matrix $\mathbf{M}_\sigma$ is the multiplication of all partial transfer matrices in due order, which is just related with E, σ, and the channel's structure.

$$\mathbf{M}_\sigma = \prod_m \mathbf{M}_{m,\sigma}\mathbf{M}_{m,m+1} = \begin{pmatrix} \mathbf{M}_{\sigma,1} & \mathbf{M}_{\sigma,2} \\ \mathbf{M}_{\sigma,3} & \mathbf{M}_{\sigma,4} \end{pmatrix}. \tag{15.20}$$

From the transfer matrix Eq. 15.20 we can obtain A_σ^{out} and B_σ^{in}.

$$A_\sigma^{\text{out}} = (M_{\sigma,1})^{-1} A_\sigma^{\text{in}} \text{ and } B_\sigma^{\text{in}} = M_{\sigma,3}(M_{\sigma,1})^{-1} A_\sigma^{\text{in}}. \tag{15.21}$$

The transmission and reflection coefficients t_{ij} and r_{ij} are related to the wave function coefficients. Supposing that the incident electron propagates in the ground transverse mode of the input channel [16]

$$A^{\text{in}} = \begin{pmatrix} k_{01}^{-1/2} \\ 0 \\ \vdots \\ 0 \end{pmatrix}, \; B^{\text{in}} = \begin{pmatrix} r_{11}k_{01}^{-1/2} \\ r_{21}k_{02}^{-1/2} \\ \vdots \\ r_{N1}k_{0N}^{-1/2} \end{pmatrix}, \text{ and } A^{\text{out}} = \begin{pmatrix} t_{11}k_{01}^{-1/2} \\ t_{21}k_{02}^{-1/2} \\ \vdots \\ t_{N1}k_{0N}^{-1/2} \end{pmatrix}. \tag{15.22}$$

The total transmission and reflection probabilities T and R equal

$$T_i = \sum_{j=1}^{N} \left|t_{ji}\right|^2 \quad \text{and} \quad R_i = \sum_{j=1}^{N} \left|r_{ji}\right|^2, \tag{15.23}$$

for one electron incident in the i-th transverse mode. The summation is over all the traveling waves in the longitudinal direction. It can be proved that the current density is conserved.

$$T_i + R_i = 1. \tag{15.24}$$

15.2 Spin Interference in Two Kinds of 2D Waveguides [19]

First we studied the transmission probabilities in the stub structure as shown in Figs. 15.1a and 15.1b. We take the dimensionless parameters as $L_1 = L_2 = L_3 = 1$, $W_1 = W_3 = 1$, $W_2 = 1.8$, and $\alpha = 0.5$, and define the effective wave vector of the incident electron,

$$k_{\text{eff}} = \sqrt{\left(\frac{\alpha}{2}\right)^2 + E_0}, \tag{15.25}$$

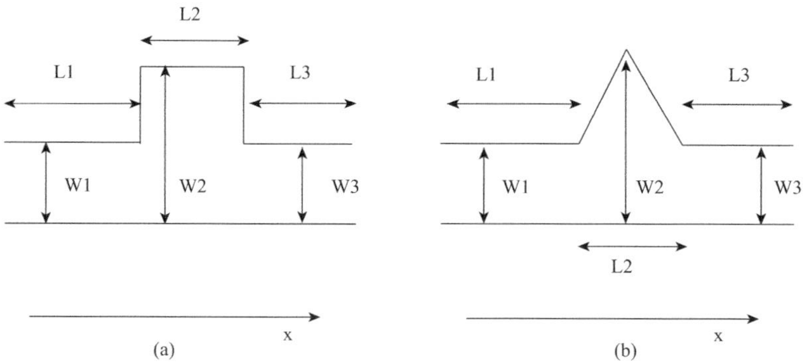

Figure 15.1 Two kinds of waveguides with different stubs: (a) a square stub and (b) a triangle stub.

where E_0 is the energy of the incident electron. For $1 < k_{\text{eff}} < 2$, the transmission probabilities as functions of k_{eff} for two structures are shown in Fig. 15.2. From Fig. 15.2 we see that the T versus k_{eff} relation is independent of σ (=1 or 2, corresponding to $\pm\alpha$). Furthermore, it is also independent of the magnitude of α if we take k_{eff} as the variable. That means the calculated results of the two-dimensional (2D) quantum waveguide transport for the spinless electron published in pervious papers [16–18] are also valid for the Rashba electrons; only the variable k_{eff} is changed from $\sqrt{E_0}$ to Eq. 15.25. In Fig. 15.2 the T curve has a sharp dip at $k_{\text{eff}} = 1.85$ and the position of the dip depends on the length of the stub W_2; here we take $W_2 = 1.8$. If we look at the stub as a gate, over which we make a metal electrode, then the gate length W_2 can be controlled by the gate voltage. So the transmission probability of the electron with a fixed energy E_0 can be controlled by the gate voltage. For square stubs (solid line), all vales have those energies where k_{0n} in Eq. 15.2 equals 1. If we assumed stubs are totally closed, these points correspond with eigenenergies. We can infer that this rule holds true for triangular stubs.

For $2 < k_{\text{eff}} < 3$, the transmission probabilities as functions of k_{eff} are shown in Fig. 15.3. There are two modes of incident electrons: one is at the first sub-band and the other is at the second sub-band. Similarly, there are two modes of outgoing electrons, so the transmission probabilities have four qualities, T_{ij} (i, j = 1 or 2),

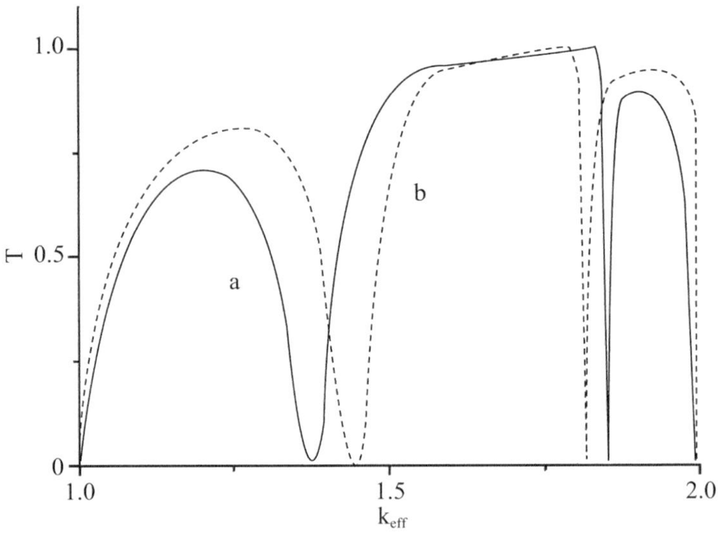

Figure 15.2 T as a function of E in the structure in Figs. 15.1a and 15.1b, when $1 < k_{\text{eff}} < 2$ and $N = 5$.

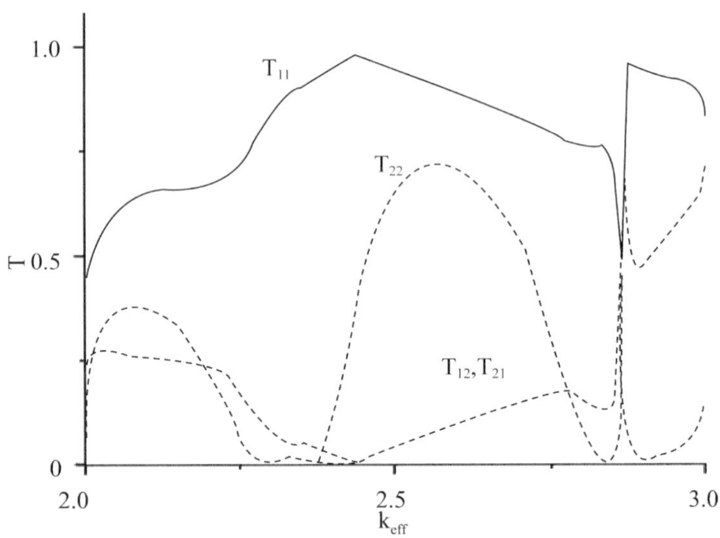

Figure 15.3 T_{ij} $(i, j = 1 \text{ or } 2)$ as a function of E in the structure in Fig. 15.1a when $2 < k_{\text{eff}} < 3$ and $N = 5$.

which denotes the transmission probability from the j-th mode to the i-th mode. From Fig. 15.3 we see that the T_{11} and T_{21} are larger and the T_{12} and T_{21} are smaller, but not zero. This means that in the transport process there is coupling between different modes. We also found that $T_{12} = T_{21}$.

Though the transmission probabilities are independent of the spin orientation (σ), if we put a ferromagnetic contact oriented in the x direction at the output end, which permits only electrons with spin σ_x to pass through, then the spin polarization along the x direction will depend on the Rashba coefficient α and the total length of the structure

$$P = \frac{J_+ - J_-}{J_+ + J_-}, \tag{15.26}$$

where J_+ and J_- are current densities of spin along the $+x$ and $-x$ directions, respectively. Figure 15.4 shows the spin polarizations as functions of α for $k_{\text{eff}} = 1.7$. The coefficients of incident wave function are $A_1^{\text{in}} = 1/\sqrt{2}\,(1, 0, 0\ldots)^T$ and $A_2^{\text{in}} = 1/\sqrt{2}\,(i, 0, 0\ldots)^T$. From Fig. 15.4 we see that the spin polarization of the outgoing Rashba electron can be modulated by the Rashba coefficient and the relation of P and α is immovable even when E or the structure of the waveguide changes.

Figure 15.5 shows the transmission probabilities as functions of the width of the stubs for the structure of Fig. 15.1a with parameters $L_1 = L_3 = 1$, $L_2 = 0.5$, $W_1 = W_3 = 1$, $N = 3$, $\alpha = 0.889$, and incident electron energies $E_0 = 3.4$, 3.5, and 3.6. We choose this value of α for comparing with results of Fig 15.7. From Fig. 15.5 we see that the transmission probability equals zero at a definite width of stub W_{stub}, which is different for different incident electron energies. Figure 15.6 shows the transmission probabilities as functions of the width of the stubs for the same parameters as Fig. 15.5, but $\alpha = 0.5$. Comparing Figs. 15.5 and 15.6, we found that the positions of the transmission probability valleys change with the Rashba coefficient α; when α increases, the W_{stub} of the transmission probability valley decreases. This means that we can control the transmission probability by changing the W_{stub}, that is, changing the gate voltage over the stub, or changing the Rashba coefficient α, that is, changing the perpendicular electric field to the 2DEG.

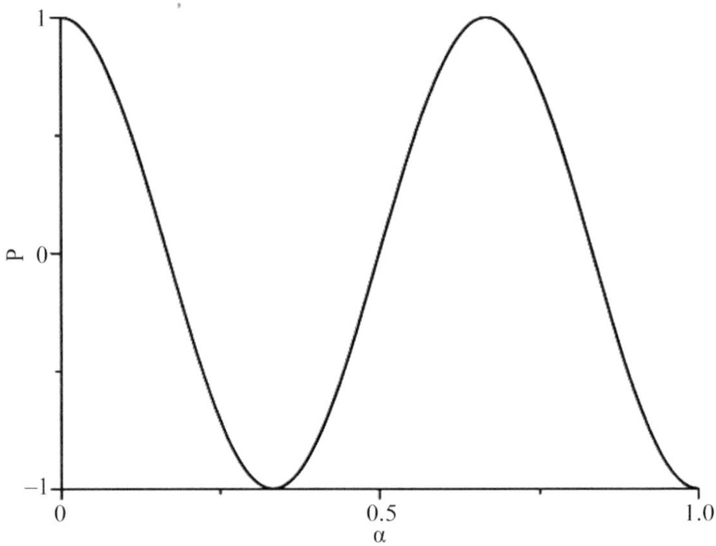

Figure 15.4 P as a function of α; this function is the same as the 1D case and is not related with the structure.

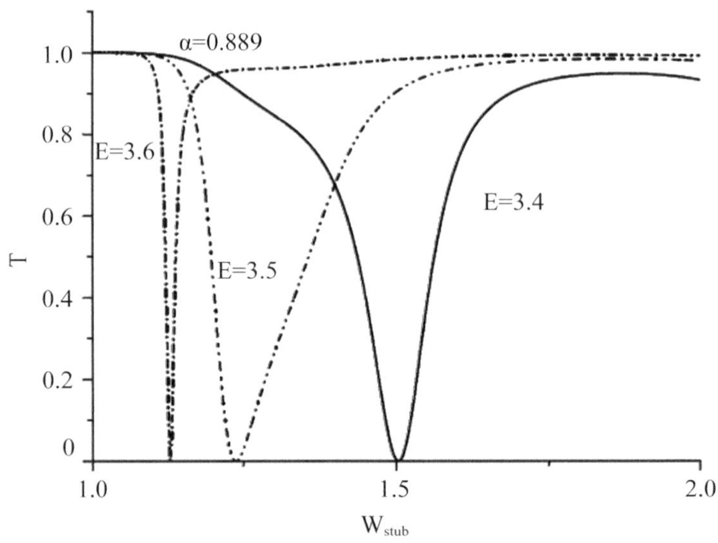

Figure 15.5 T as a function of W_{stub} in the waveguide with a single square stub; the length of the stub is 0.5, $\alpha = 0.889$, and $N = 3$ for $E_0 = 3.4$, 3.5, and 3.6.

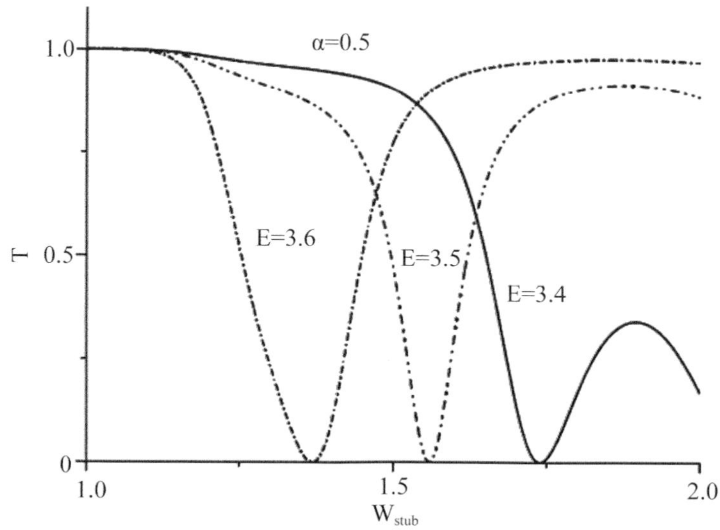

Figure 15.6 T as a function of W_{stub} in the waveguide with a single square stub; the length of the stub is 0.5, $\alpha = 0.5$, and $N = 3$ for $E_0 = 3.4$, 3.5, and 3.6.

Then we investigate the influence of the number of stubs. We investigate the structure where there are three configurations that are the same as the structure of Fig. 15.5, with the parameters $N = 3$ and $\alpha = 0.889$. We select this α value so that $P = 1$. Figure 15.7 shows T as a function of W_{stub} when $E_0 = 3.4$ and 3.8. From Fig. 15.7 we see that the transmission probabilities equal zero at the same W_{stub} as the single stub structure but the half-widths of the valleys decrease greatly. Trends in two curves are similar: $T \approx 1$ in most range of W_{stub}, and T decreases to zero in a very small range. In these regions in which $T \approx 0$, we can completely block the electron. Figure 15.8 shows T as a function of W_{stub} when square stubs are replaced by triangular ones, and $E = 3.4$ and 3.8. From Figs. 15.7 and 15.8 we see that these regions in which $T \approx 0$ depend on the shape of the structure. With the multistub structure we can control the T more effectively. In Fig. 15.8 the dip in the solid line looks strange because it is too sharp to be expressed. And there are some very sharp dips, which can be explained by the (nW_0/W_n) in Eq. 15.5. It make the condition of resonance become unsatisfied rapidly.

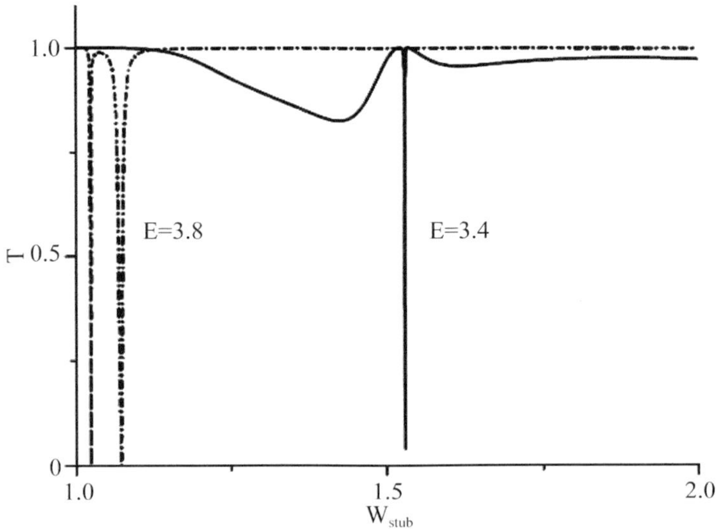

Figure 15.7 T as a function of W_{stub} in waveguides with three square stubs that are the same as Fig. 15.5, for $E_0 = 3.4$ and 3.8.

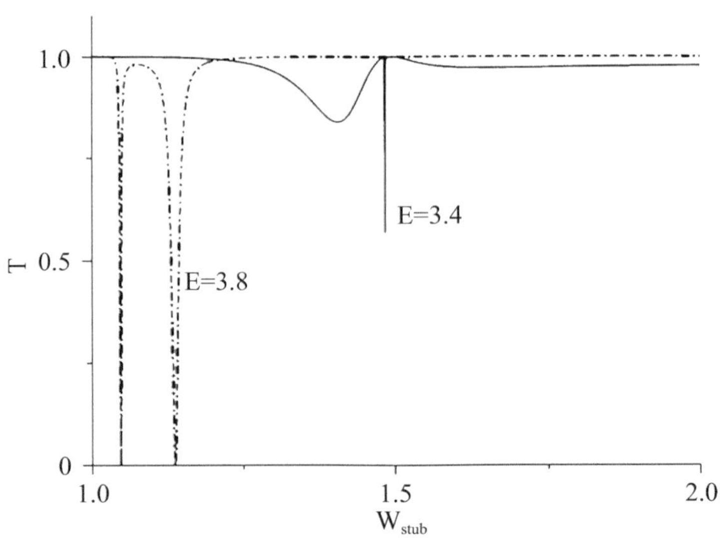

Figure 15.8 T as a function of W_{stub} in waveguides with three triangular stubs that are the same, other parameters are the same as

Physically, in the case of $T = 0$ there is an eigen standing wave state in the stub region between the input and output waveguides for each structure. Figure 15.2 shows that when E is in a small region T decreases rapidly. It is because for these E values there is almost a complete standing wave in the mid-region, so at the interfaces of input and output waveguides between the mid-region, $\Psi \approx 0$, so $R = 1$ and $T = 0$. And the eigenenergy E of these standing waves depends on the E, W_{stub}, and the shape of the structure. So we can select a structure of a suitable shape and parameter and apply gate voltage and perpendicular electric field to block the electron.

15.3 The Unitary Condition [19]

Now we prove that the unitary condition Eq. 15.24 is always tenable. Considering that $\mathbf{M}_{m\sigma}$ are unitary matrices, we just need to prove that currents on the two sides of the interface of the m-th and the $(m+1)$th segment are equal. For every wave function Ψ in the waveguide, the current in the x direction is $\text{Re}(\Psi^{+} J_x \Psi)$. Considering the orthogonality of $\varphi_n^m(y)$ and the express of Ψ^m in Eq. 15.1, $J^m = J_1^m + J_2^m$ and

$$J_\sigma^m = \text{Re}\left[\sum_{n=1}^{N} k_{0n}\left(a_{\sigma n}^m e^{ik_{\sigma n}^m x} + b_{\sigma n}^m e^{-ik_{3-\sigma n}^m x}\right)^* \left(a_{\sigma n}^m e^{ik_{\sigma n}^m x} - b_{\sigma n}^m e^{-ik_{3-\sigma n}^m x}\right)\right]. \quad (15.27)$$

So if we define a new Ψ^m and $J_x\Psi^m$ as Eq. 15.8 and

$$\Psi_\sigma^m = \left(a_{\sigma 1} e^{ik_{\sigma 1}^m x} + b_{\sigma 1} e^{-ik_{3-\sigma 1}^m x},\ a_{\sigma 2} e^{ik_{\sigma 2}^m x} + b_{\sigma 2} e^{-ik_{3-\sigma 2}^m x},\ \cdots a_{\sigma N} e^{ik_{\sigma N}^m x} + b_{\sigma N} e^{-ik_{3-\sigma N}^m x}\right)^T,$$

and

$$J_x \Psi_\sigma^m = K^m \left(A_\sigma^m - B_\sigma^m\right)^T. \quad (15.28)$$

Then formula $J_\sigma^m = \text{Re}\left[\left(\Psi_\sigma^m\right)^* J_x \Psi_\sigma^m\right]$ remains effective and Eq. 15.14 and Eq. 15.15 can be written as

$$\Psi_\sigma^m = \left(S^T\right)^{-1} \Psi_\sigma^{m+1} \quad (15.29)$$

and

$$J_x \Psi_\sigma^m = S\left(J_x \Psi_\sigma^{m+1}\right). \tag{15.30}$$

Left-multiplying Eq. 15.30 by Hermitian conjugation of Eq. 15.29 and taking the real component

$$J_\sigma^m = \mathrm{Re}\left[\left(\Psi_\sigma^m\right)^+ J_x \Psi_\sigma^m\right] = \mathrm{Re}\left[\left(\Psi_\sigma^{m+1}\right)^+ J_x \Psi_\sigma^{m+1}\right] = J_\sigma^{m+1}. \tag{15.31}$$

So the unitary condition is satisfied in our method.

15.4 Summary

We studied the 2D ballistic transport of Rashba electrons in a straight structure in 2DEGs. We derived the transfer matrix relating the wave functions at the input end and at the output end of the structure and found that for movement in a straight structure, there is no mixing between the wave functions of spin-up ($\sigma = 1$) and spin-down ($\sigma = 2$) and the transfer matrix is independent of σ in the interface. We investigated the influence of the structure on electron transport. It is found that the transmission probabilities as functions of the effective wave vector k_{eff} are independent of the sign and magnitude of the Rashba coefficient α. The transmission probability depends on the shape of the structure, especially the stub width. If we change the stub width by the gate voltage, we can control the transmission probability of the electron with a fixed energy. The transmission probabilities of the electron with an energy higher than the second transverse sub-band include contributions from different sub-bands, which means in the transport process there is coupling between different sub-bands. The spin polarization along the x direction depends on the Rashba coefficient, so we can control the spin polarization of the outgoing Rashba electron by changing the Rashba coefficient, for example, changing the applied perpendicular electric field. For the multistub structure, T decreases rapidly in a certain small region of W_{stub} and T keeps high at the other W_{stub} for a certain Rashba coefficient α. And the anti-resonance is found, where the quasi-confined state is formed in the center part of the structure. So, we can modulate a spin-polarized current by controlling the incident electron energy E_0 and by changing the shape, parameters of the stub, the applied gate voltage, and the perpendicular electric field. With the 2D stub-like structures we can design the spin-injected field-effect transistor. We proved the unitary condition of the transfer matrix, that is, the conservation condition of the current density.

References

1. Datta, S., and Das, B. (1990). *Appl. Phys. Lett.*, **56**, 665.
2. Rashba, E. I. (1960). *Sov. Phys. Solid State*, **2**, 1109.
3. Nitta, J., Meijer, F. E., and Takayanagi, H. (1999). *Appl. Phys. Lett.*, **75**, 695.
4. Kiselev, A. A., and Kim, K. W. (2001). *Appl. Phys. Lett.*, **78**, 775.
5. Wang, X. F., Vasilopoulos, P., and Peeters, F. M. (2002). *Appl. Phys. Lett.*, **80**, 1400.
6. Bulgakov, E. N., and Sadreev, A. F. (2002). *Phys. Rev. B*, **66**, 075331.
7. Sheng, J. S., and Chang, K. (2006). *Phys. Rev. B*, **74**, 235315.
8. Bellucci, S., Carillo, F., and Onorato, P. (2008). *Eur. Phys. J. B*, **66**, 509.
9. Zhang, R. L., Zhang, Z. J., Peng, P. W., et al. (2008). *J. Appl. Phys.*, **103**, 07B727.
10. Souma, S., and Nikolić, B. K. (2005). *Phys. Rev. Lett.*, **94**, 106602.
11. Koo, H. C., Kwon, J. H., Eom, J., et al. (2009). *Science*, **325**, 1515.
12. Bellucci, S., and Onorato, P. (2008). *Phys. Rev. B*, **77**, 075303.
13. Wang, L. G., Chang, K., and Chan, K. C. (2006). *J. Appl. Phys.*, **99**, 043701.
14. Liu, D. Y., Xia, J. B., and Chang, Y. C. (2010). *Sci. China: Phys. Mech. Astron.*, **53**, 1.
15. Liu, D. Y., Xia, J. B., and Chang, Y. C. (2009). *J. Appl. Phys.*, **106**, 093705.
16. Sheng, W.-D., and Xia, J.-B. (1996). *J. Phys. C*, **8**, 3635.
17. Weisshaar, A., Lary, J., Goodnick, S. M., and Tripathi, V. K. (1991). *J. Appl. Phys.*, **70**, 355.
18. Akis, R., Vasilopoulos, P., and Debray, P. (1995). *Phys. Rev. B*, **52**, 2805.
19. Liu, D. Y., and Xia, J. B. (2010). *J. Appl. Phys.*, **108**, 053717.

Chapter 16

Conductance of Rashba Electron in a Quantum Waveguide with Smooth Boundary

Duan-Yang Liu,[a] Yi-Zhu Lin,[b] and Jian-Bai Xia[c]

[a] *College of Science, Beijing University of Chemical Technology, Beijing 100029, China*

[b] *School of Electronic Information Engineering, China West Normal University, Sichuan 637002, China*

[c] *State Key Laboratory for Superlattice and Microstructures, Institute of Semiconductors, Beijing 100083, China*

liudy@mail.buct.edu.cn, Linyz@cwnu.edu.cn

By using the transfer matrix method, we theoretically investigated Rashba electron's spin transport in a straight waveguide with a stub which has a smooth boundary. It is proved that the two spin states propagate independently. The conductance of the waveguide has been calculated. Conductance quantization is common in this structure when we change the Fermi energy and the width of the stub. If the shape of the stub converges toward the limit–a quadrate stub, the conductance will reduce distinctly due to reflection.

In 1990, Datta [1] proposed an innovative device in which spin current can be modulated by changing Rashba spin-orbit interaction (RSOI) [2] strength via gate voltage. This proposal which

Quantum Waveguide in Microcircuits (Second Edition)
Jian-Bai Xia, Duan-Yang Liu, and Wei-Dong Sheng

ISBN 978-981-4968-30-0 (Hardcover), 978-1-003-65146-8 (eBook)
www.jennystanford.com

was called spin field effect transistor (SFET) has been expected to work in quantum computer based on spintronics. Since then researchers have been trying to experimentally demonstrate it in laboratory, for example, Koo et al. reported that in a high-mobility InAs hetrostructure they observed oscillatory channel conductance which was modulated by gate voltage and exactly fit to Datta's theory. On the other hand, a series of similar devices in theory have been proposed [4–10], most of which are one dimensional (1D) devices [11, 12], i.e. the lateral width is so narrow that there is only one lateral model occupied by electrons. 1D devices feature manifold advantages, like simple and easy to calculate, yet it is extremely demanding with regard to fabrication. However, if multiple lateral models in waveguides were involved, shape factor of the device would have been used to control the spin current [13, 14]. Therefore, in recent studies, two-dimensional (2D) waveguides with straight [15], nonuniform [16], and random configurations [17] in presence of RSOI have been investigated by using scattering matrix method.

In previous work, we had developed 1D quantum waveguide theory of Rashba electron [18, 19], which in our opinion is reasonable to be applied to 2D cases. In this work, we applied this theory to a 2D waveguide with smoothly fluctuant boundaries. In contrast to other published results, we used the transfer matrix method [20, 21] with the boundary condition derived in Ref.18 to study the 2D ballistic transport Rashba electrons in this structure. We expected to investigate the influence of the shape of stub and Fermi energy on the conductance.

16.1 How to Calculate Conductance of Rashba Electron

We consider a waveguide in two-dimensional electron gas (2DEG) with a RSOI strength α in x–y plane, the spin-dependent Hamiltonian of a charged particle of effective mass m thus is [23]

$$\begin{pmatrix} \mathbf{P}^2/2m^* + V(x,y) & \alpha/\hbar(ip_x + p_y) \\ \alpha/\hbar(-ip_x + p_y) & \mathbf{P}^2/2m^* + V(x,y) \end{pmatrix}, \tag{16.1}$$

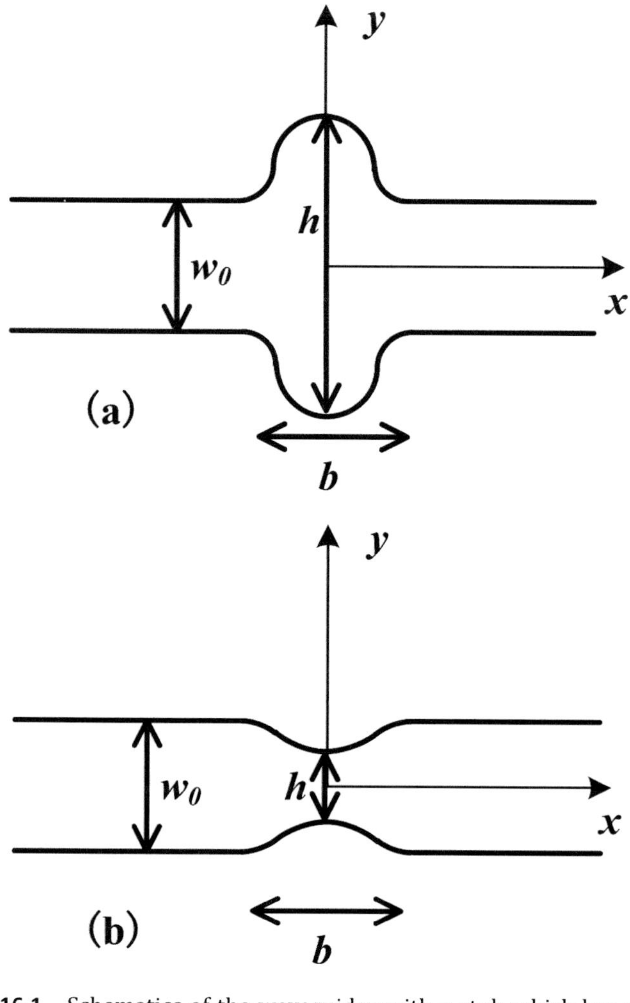

Figure 16.1 Schematics of the waveguides with a stub which has smooth boundaries: (a) the stub is wider than tunnel, (b) the stub is narrower than the tunnel.

where $V(x, y)$ is assumed to be a hard wall approximation at the boundaries of the waveguide in this work. Waveguides under investigation have transverse symmetric and smooth boundaries as shown in Fig. 16.1. The $x-$axis along the waveguide is the axis of symmetry. The width of the waveguide is described by the function

$w(x) = w0 + \delta w(x)$. We assume that $\delta w(x) \neq 0$ in a limited area $-b/2 < x < b/2$, where a stub can be created. To get a smooth stub, we take

$$\delta w(x) = \begin{cases} (h - w_0)\exp\left(\beta\left(\dfrac{b^2}{4(x+b/2)(x-b/2)} + 1\right)\right), & -b/2 < x < b/2 \\ 0, & |x| \geq b/2 \end{cases}. \quad (16.2)$$

In this equation, h is the width of the stub, and β is a parameter which indicates the rate of change of the waveguide's width near $x = 0$. If $h > w$, the width of the stub is greater than that of the incident waveguide as shown in Fig. 16.1a; and if $h < w$, there is a narrow stub which is shown in Fig. 16.1b. If β increases, $\delta w(x)$ will decrease faster near $x = 0$. If $\beta \to +\infty$, $\delta w(x) \to \delta-$ function. If $\beta \to 0$, $\delta w(x)$ will approach a rectangular function, i.e. the stub will become a quadrate stub in the area $|x| < b/2$. On the other hand, we can say when β is small enough, the width will change mainly and rapidly near $|x| = b/2$, and the limit is a rectangular stub.

In this work, we calculate the conductance by using transfer matrix method which has been proved to be an effective method for electron transport in 2DEG [20–22] in presence of RSOI [5, 19, 23]. If ballistic electrons with energy E move along the waveguide in Fig. 16.1, we can divide the waveguide into several segments, and in each segment the width is constant. In each segment the wave function can be written as

$$\Psi^m(x, y) = \frac{1}{\sqrt{2}} \sum_{n=1}^{N} \begin{bmatrix} a_{1,n}^m \begin{pmatrix} 1 \\ i \end{pmatrix} e^{ik_{1,n}^m x} + a_{2,n}^m \begin{pmatrix} 1 \\ -i \end{pmatrix} e^{ik_{2,n}^m x} \\ +b_{1,n}^m \begin{pmatrix} 1 \\ i \end{pmatrix} e^{-ik_{2,n}^m x} + b_{2,n}^m \begin{pmatrix} 1 \\ -i \end{pmatrix} e^{-ik_{1,n}^m x} \end{bmatrix} \varphi_n^m(y), \quad (16.3)$$

where N is the number of transverse modes occupied in the incident circuit or in the stub, $\sigma = 1, 2$ denotes the two spin states, and m denotes the m-th segment. $\varphi_n^m(y) = \sqrt{2/W_m}\sin[(n\pi/W_m)(y - y_0^m)]$, and $k_{1,n}^m$, $k_{2,n}^m$ are the wave vectors [19],

$$k_{1,n}^m = k_{0,n}^m + k_\delta, \; k_{2,n}^m = k_{0,n}^m - k_\delta, \quad (16.4)$$

$$k_{0,n}^m = \frac{m^*}{\hbar^2}\sqrt{\alpha^2 + \frac{2\hbar^2}{m^*}E_{||}}, \; k_\delta = \frac{m^*}{\hbar^2}\alpha, \quad (16.5)$$

where $E_{||}$ is the longitudinal kinetic energy, $E_{||} = E - (\hbar^2/2m^*)$ $(n\pi/W_m)^2$, W_m denotes the width of the m -th segment. The sum over n may include evanescent modes, i.e. $k_{0,n}^m$ is imaginary.

By considering the form of $\Psi^m(x, y)$, it can be represented by

$$\begin{aligned}\Psi_\sigma^m &= (a_{\sigma,1}e^{ik_{\sigma,1}^m x}, a_{\sigma,2}e^{ik_{\sigma,2}^m x} \ldots a_{\sigma,N}e^{ik_{\sigma,n}^m x},\\ &\quad b_{\sigma,1}e^{-ik_{3-\sigma,1}^m x}, b_{\sigma,2}e^{-ik_{3-\sigma,2}^m x} \ldots b_{\sigma,N}e^{-ik_{3-\sigma,n}^m x})^T\\ &= (A_\sigma^m, B_\sigma^m)^T, \end{aligned} \tag{16.6}$$

Then we can match the wavefunction in the incident point Ψ^{in} and the wavefunction in the outside segment Ψ^{out} by adopting boundary conditions on every dividing line between two segments[18, 23]. With the transfer matrix method, we obtain [24]

$$\Psi_\sigma^{in} = M_\sigma \Psi_\sigma^{out}, \tag{16.7}$$

Based on this equation, we can calculate all spin current in each transverse mode in incident wavefunction and output wavefunction according to current operator in Ref. 23, then all transmission and reflection coefficients $t_{\sigma,ij}$ and $r_{\sigma,ij}$ can be extracted [24]. The total transmission and reflection probabilities of certain spin T_σ and R_σ equal

$$T_{\sigma,i} = \sum_{j=1}^{N} \left|t_{\sigma,ji}\right|^2, R_{\sigma,i} = \sum_{j=1}^{N} \left|r_{\sigma,ji}\right|^2. \tag{16.8}$$

It is worth noting that two parts of the wave, spin up and spin down, propagate independently. Actually it can be proved that

$$T_{1,ij} = \left|t_{1,ij}\right|^2 = \left|t_{2,ij}\right|^2 = T_{2,ij} = T_{ij}. \tag{16.9}$$

and the current density is conserved [23].

$$T_{\sigma,i} + R_{\sigma,i} = 1. \tag{16.10}$$

so the transmission probabilities are independent of the spin orientation σ.

As all transmission and reflection probabilities can be obtained, if the temperature is low enough, we can obtain the total conductance of the quantum waveguide from the Landauer–Büttiker formula,

$$G(E_F) = G_0 \sum_{i,j=1}^{N} T_{ij}, \tag{16.11}$$

where E_F is the Fermi energy of the 2DEG, and $G_0 = e^2/\pi\hbar$. Considering the form of Eq. (16.11), we evaluate conductance G in units of G_0 in this work.

16.2 Conductance and Boundary Shape

In this section, we present the numerical results for conductances of structures in Fig. 16.1 and discussions. Hereafter, unless explicitly specified, $m^* = 0.07m_e$, $\alpha = 10$ meV.nm, $w_0 = 40$ nm, $b = 40$ nm. Mainly we focus on the influence of the Fermi energy E_F of the 2DEG and the shape of the stub.

Fig. 16.2 presents the conductance as a function of the Fermi energy of waveguide in Fig. 16.1(a) with $h = 80$ nm. We choose two values of β for comparison, i.e. (1) $\beta = 1$, (2) $\beta = 0.01$. For the case $\beta = 1$, the boundary of the stub has more gentle gradient. As shown in the figure, in the area $E_F < 40$ meV, the conductances in two cases are nearly same and increase in a step-like manner. In the area $E_F > 40$ meV, this characteristic remains in the waveguide with $\beta = 1$; for the waveguide with $\beta = 0.01$, the conductance oscillates when E_F rises. The phenomenon that the conductance increases by the same amount step by step is so called 'conductance quantization'.

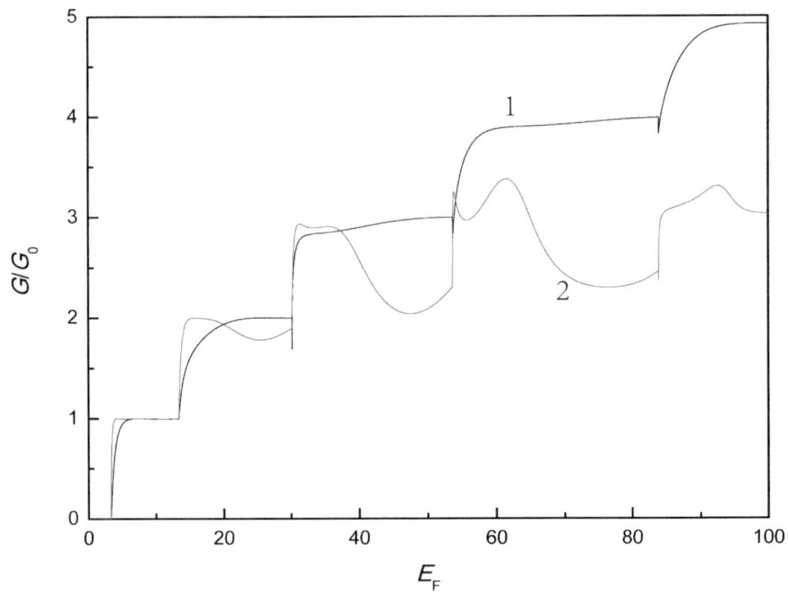

Figure 16.2 Conductance G of waveguide in Fig. 16.1(a) vs the Fermi energy E_F with different stub's shape parameter β: (1) $\beta = 1$, (2) $\beta = 0.01$.

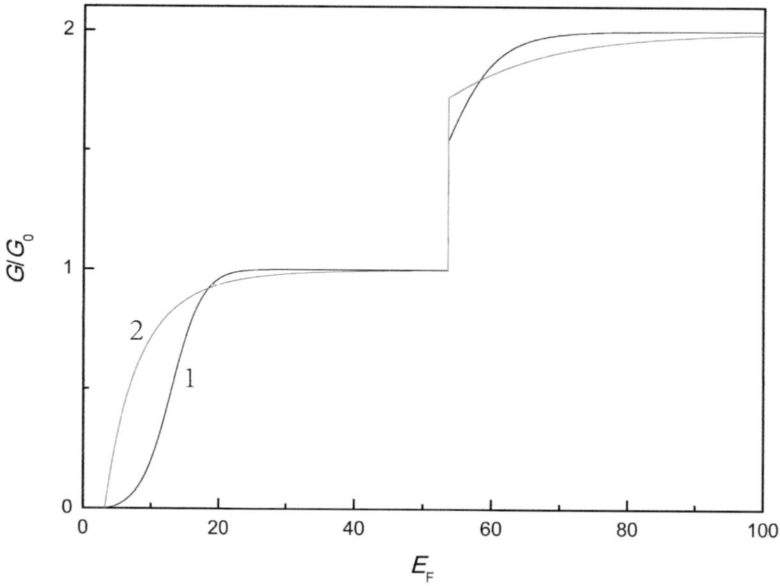

Figure 16.3 Conductance G of waveguide in Fig. 1(b) vs the Fermi energy E_F with different stub's shape parameter β: (1) $\beta = 1$, (2) $\beta = 0.01$.

The reason is that when a new transverse mode in the waveguide has been occupied by electrons, according to Eq. (16.11), the new mode will contribute to the conductance with a unit G_0. However, T_{ij} is not always equal to or approximately equal to 1; when there are several occupied transverse modes and the width of the waveguide declines rapidly, T_{ij} would be far from 1 due to the significant reflection. In Fig. 16.2 it is the exactly the same for when $E_F > 40$ meV and $\beta = 0.01$, so the conductance is much smaller than that in the case $\beta = 1$.

Fig. 16.3 also shows G as a function of E_F, but for the case $h = 20$ nm, i.e. the stub is narrower than the waveguide. We choose two values of β for comparison, i.e. (1) $\beta = 1$, (2) $\beta = 0.01$. again. Unlike Fig. 16.2, the curve of G looks like two stairs for both two cases. This means the shape of the stub has little influence on the conductance. This is because there are only two transverse modes in the stub, so it is close to the case of one-dimensional approach, the shape of the waveguide can be ignored. Or we can say, the significant reflection

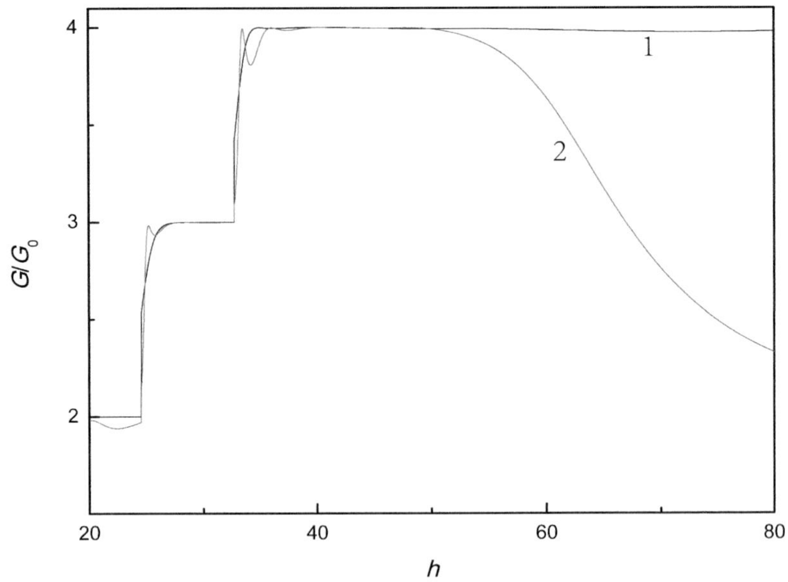

Figure 16.4 Conductance G vs the stub's width h with different stub's shape parameter β: (1) $\beta = 1$, (2) $\beta = 0.01$.

is more likely to occur when there are more occupied transverse modes.

To further investigate the influence of the width of the stub, we study the relationship between the conductance and the width of the stub directly. In Fig. 16.4, G as a function of h has been shown. Two values of β are chosen for comparison: (1) $\beta = 1$, (2) $\beta = 0.01$. To observe the tendency of the conductance when more transverse modes are occupied, we choose a relatively large Fermi energy $E_F =$ 80 meV. Similar to Fig. 16.3, when $h < 40$ nm, the curve of G looks like stairs for both two cases. Conductance quantization appears when h increases and a higher transverse mode becomes occupied. For waveguide with $\beta = 1$ this phenomenon remains the same for all regions of h. However, for the waveguide with $\beta = 0.01$ the conductance G reduces significantly when $h > 50$ nm. Because in this case the boundary is much sharper, the reflection is much more remarkable.

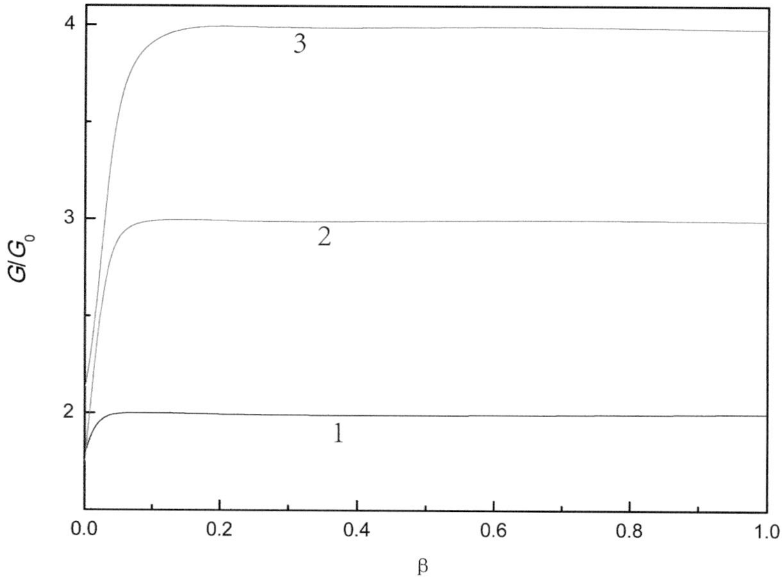

Figure 16.5 Conductance G vs stub's shape parameter β with different Fermi energy: (1) $E_F = 30$ meV, (2) $E_F = 50$ meV, (3) $E_F = 80$ meV.

Finally we discuss the influence of β directly. Fig. 16.5 shows G as a function of β in the waveguide with $h = 80$ nm, we choose this value of h because there are relatively more transverse modes occupied by electrons. And there are three values of E_F that have been involved: (1) $E_F = 30$ meV, (2) $E_F = 50$ meV, (3) $E_F = 80$ meV. As shown in Fig. 16.5, It is worth noting that G is almost independent of β if β exceeds a certain value for all cases; and this time the value of G in unit of G_0 is close to the number of the transverse modes occupied by electrons in the incident waveguide. It means that once β exceeds a certain small value (≈ 0.1), the electron can freely penetrate the waveguide in every transverse mode occupied. On the other hand, if β is very small, such as $\beta < 0.05$, the conductance increases rapidly when β increases. Because when β is small enough, the width of the waveguide drops abruptly near $|x| = b/2$, and the steep decline of the width of the waveguide along the propagate direction results in significant reflection of electrons.

16.3 Summary

We studied the two-dimensional ballistic transport of Rashba electrons in a straight structure with a stub which has a smooth boundary in 2DEG by using transfer matrix method, and the conductance of the waveguide has been calculated. We found that for the movement in a straight structure, there is no mixing between the wave functions of spin up ($\sigma = 1$) and spin down ($\sigma = 2$). We investigated the influence of the shape of the structure and Fermi energy on the conductance. Because the number of transverse modes occupied is determined by Fermi energy E_F, the width of the stub h and another shape parameter β which indicates the rate of change of the waveguide's width in the stub, the phenomenon called 'conductance quantization' is common for many cases. For the case β is not very small, the waveguide has gentle boundary and G in unit of G_0 is nearly equal to the transverse modes occupied in the waveguide. If β is small enough, the stub is close to a rectangular stub; and then the conductance will reduce significantly. This is because there is a significant reflection on the right boundary of the stub.

Acknowledgments

This work was supported by The National Natural Science Foundation of China No. 11504016, the Fundamental Research Funds of China West Normal University under Grand 15E001, the Meritocracy Research Funds of China West Normal University under Grand 17YC048, and Scientific Research Fund of Sichuan Provincial Education Department under Grand 17ZB0424.

References

1. S. Datta and B. Das. *Appl. Phys. Lett.* **58**, 665 (1990).
2. E. I. Rashba. *Sov. Phys. Solid. State*, **2**, 1109 (1960).
3. H. C. Koo, J. H. Kwon, J. Epm, J. Chang, S. H. Han, and M. Johnson, *Science*, **325**, 1515 (2009).

4. A. A. Kiselev and K. W. Kim, *Appl. Phys. Lett.* **78**, 775 (2001).
5. X. F. Wang, P. Vasilopoulos, and F. M. Peeters, *Appl. Phys. Lett.* **80**, 1400 (2002).
6. E. N. Bulgakov and A. F. Sadreev, *Phys. Rev. B* **66**, 075331 (2002).
7. J. S. Sheng and K. Chang, *Phys. Rev. B* **74**, 235315 (2006).
8. S. Bellucci, F. Carillo, and P. Onorato, *Eur. Phys. J. B* **66**, 509 (2008).
9. R. L. Zhang, Z. J. Zhang, R. W. Peng, X. Wu, De Li, Jia. Li, and L. S. Cao, *J. Appl. Phys.* **103**, 07B727 (2008).
10. S. Souma and B. K. Nikolić, *Phys. Rev. Lett.* **94**, 106602 (2005).
11. S. Bellucci and P. Onorato, *Phys. Rev. B* **77**, 075303 (2008).
12. L. G. Wang, K. Chang and K. S. Chan, *J. Appl. Phys.* **99**, 043701 (2006).
13. L. He, G. Bester, and A. Zunger, *Phys. Rev. Lett.* **94**, 016801 (2005).
14. Y. Ban and E. Ya. Sherman, *Appl. Phys. Lett.* **99**, 112101 (2011).
15. L. Zhang, P. Brusheim, and H. Q. Xu, *Phys. Rev. B* **72**, 045347 (2005).
16. G. B. Akguc and J. Gong, *Phys. Rev. B* **77**, 205302 (2008).
17. X. B. Xiao, H. L. Li, G. H. Zhou, and N. H. Liu, *Eur. Phys. J. B* **85**, 305 (2012).
18. D. Y. Liu Duan Yang, J.-B. Xia and Y. C. Chang, *Sci. China. Ser G* **53**, 16 (2010).
19. D.-Y. Liu, J.-B. Xia and Y.-C. Chang, *J. Appl. Phys.* **106**, 093705 (2009).
20. W.-D. Sheng and J.-B. Xia, *J. Phys. C* **8**, 3635 (1996).
21. A. Weisshaar, J. Lary, S. M. Goodnick, and V. K. Tripathi, *J. Appl. Phys.* **70**, 355 (1991).
22. R. Akis, P. Vasilopoulos, and P. Debray, *Phys. Rev. B* **52**, 2805(1995).
23. D.-Y. Liu, J.-B. Xia, *J. Appl. Phys.* **108**, 053717 (2010).
24. D.-Y. Liu, J.-B. Xia, *Physica E* **98**, 4 (2018).

Chapter 17

Spin Flip in a Quantum Ring

Nowadays much of research in semiconductor spintronics has been shifting toward the field of the Rashba spin-orbit interaction (RSOI) [1] in low dimensional semiconductor structures [2–7]. Many low dimensional semiconductor structures with RSOI have been intensively studied [8–11], and these structures are expected to be potential spin devices, such as spin-inverter or spin-polarizer. In these quantum structures, spin transport is a basic problem, so many investigations focus on ballistic macroscopic circular rings [12–18] as its one dimensional Hamiltonian [19]. In macroscopic circular rings, Many intriguing spin interference phenomena [20, 21] have been found. For instance, Ballistic electron transport of Rashba electron through a chain of quantum circular rings has been investigated by Molnar et al [14]. They have shown that a periodic of spin transport is determined by the incident electron's energy E, the magnetic field B, and the strength of the RSOI α. Recently Naeimi et al. have shown that a double quantum rings in the presence of RSOI and a magnetic flux can work as a spin-inverter [18].

In this paper, we study the spin transport of electrons in an elliptical ring with the RSOI, the same case in an circular ring is studied for comparison. We focus on the spin flip in two kinds of

Quantum Waveguide in Microcircuits (Second Edition)
Jian-Bai Xia, Duan-Yang Liu, and Wei-Dong Sheng

ISBN 978-981-4968-30-0 (Hardcover), 978-1-003-65146-8 (eBook)
www.jennystanford.com

rings, and expect to find a suitable structure and conditions for a spin inverter.

17.1 Spin Transport in Two Kinds of Quantum Ring

For a ring in $x - y$ plane, in the presence of Rashba spin-orbital interaction, the effective Hamiltonian of an electron in this two dimensional system could be written as:

$$H = \begin{pmatrix} \mathbf{p}^2/2m^* + V(x,y) & \alpha/\hbar\,(ip_x + p_y) \\ \alpha/\hbar\,(-ip_x + p_y) & \mathbf{p}^2/2m^* + V(x,y) \end{pmatrix}, \tag{17.1}$$

where m^* is the electron effective mass, α is the Rashba coefficient. For a one-dimensional (1D) circular ring, Eq. (17.1) can be rewritten into a 1D form [19]. With this 1D Hamiltonian, the eigenvalues, eigenstates and spin transport problems could be easily solved. However, for other 1D rings, such as hexagonal or elliptical rings, we cannot find 1D Hamiltonian. To solve problems of spin transport in these structures, we adopted the method of dividing a curved line into N segments [22]. For a curved line, such as an elliptical ring or a circular ring, N is large enough and every segment is very small, so each segment can be approximated to be a line segment along the tangential direction. For every linear segment, we could easily obtain its eigenstates [23]:

$$\phi_1(\theta) = \frac{1}{\sqrt{2}}\begin{pmatrix} 1 \\ ie^{i\theta} \end{pmatrix}, \quad \phi_2(\theta) = \frac{1}{\sqrt{2}}\begin{pmatrix} 1 \\ -ie^{i\theta} \end{pmatrix}, \tag{17.2}$$

where θ is the azimuthal angle of the segment. Based on Eq. (17.2), we can write the wavefunction in the curve as

$$\Phi = \left(\left(\phi_1(\theta)\ \phi_2(\theta)\right)\right)\Psi = \left(\left(\phi_1(\theta)\ \phi_2(\theta)\right)\right)\begin{pmatrix} a(l) \\ b(l) \end{pmatrix}, \tag{17.3}$$

where l denotes the coordinates on the curved line, θ is the azimuthal angle of the tangent line of the curve. By Adopting Ψ to describe the wave function and by using the Griffith's boundary conditions [22–25] in each vertex, we can relate the wave function at the two endpoints by a transfer matrix. For a polyline structure, such as a hexagonal ring, every lead is a natural line segment, so N is the number of the leads. In circular ring this method gave results which are identical with those obtained from one-dimensional Hamiltonian [22], so this method is reasonable.

In each arm of an Aharonov-Bohm (AB) ring, the transfer matrix connect the wave functions at a junction with the wave function at the other junction. According to the wave function must be continuous and the spin current must be conserved at the junctions, so we can just determine all unknown coefficients when an electron beam is injected into the AB ring. In our method, rings with any shape could be easily dealt with. For instance, spin transport in a polyline structure is more convenient to be calculated in our method.

Consider an one dimensional AB ring, for example, an elliptical ring as show in Fig. 17.1a, the electron current is injected from circuit i at point A, and output circuit e is at the right side with intersection point B. We assume the semimajor axis and the semiminor axis of the ellipse is denoted by a_e and b_e, respectively, and the strength of RSOI in the ring is denoted by α. The size of the AB ring can be denoted by the distance between point A and point B, so the characteristic length of the elliptical ring can be denoted by a_e. In this paper, we adopt the dimensionless physical quantities. Taking the energy unit $E_0 = \hbar^2/(2m^* a_e^2)$, then the energy $\epsilon = E/E_0$, and the Rashba coefficient $\overline{\alpha} = (\alpha/a_e)/E_0 = \alpha\left(2m^* a_e/\hbar^2\right)$. For a circular ring as shown in Fig. 17.1b, their dimensionless physical quantities are similar, just the a_e is replaced by the radius of the circle R. For any one dimensional AB ring, the wave function in the upper and lower arms can be divided into two components: one in clockwise direction and another in counterclockwise direction, so they can be written as:

$$\Psi_{up}^{j} = \begin{pmatrix} a_{up}^{j}(l_{up}^{j}) \\ b_{up}^{j}(l_{up}^{j}) \end{pmatrix}, \tag{17.4}$$

$$\Psi_{down}^{j} = \begin{pmatrix} a_{down}^{j}(l_{down}^{j}) \\ b_{down}^{j}(l_{down}^{j}) \end{pmatrix}, \tag{17.5}$$

where up and down denote the upper and lower arms of the square ring, $j = 1, 2$ correspond to the clockwise and counterclockwise directions, respectively. In particular, l_{up}^1 and l_{up}^2, or l_{down}^1 and l_{down}^2 have opposite positive direction. The wavefunction of the electron in

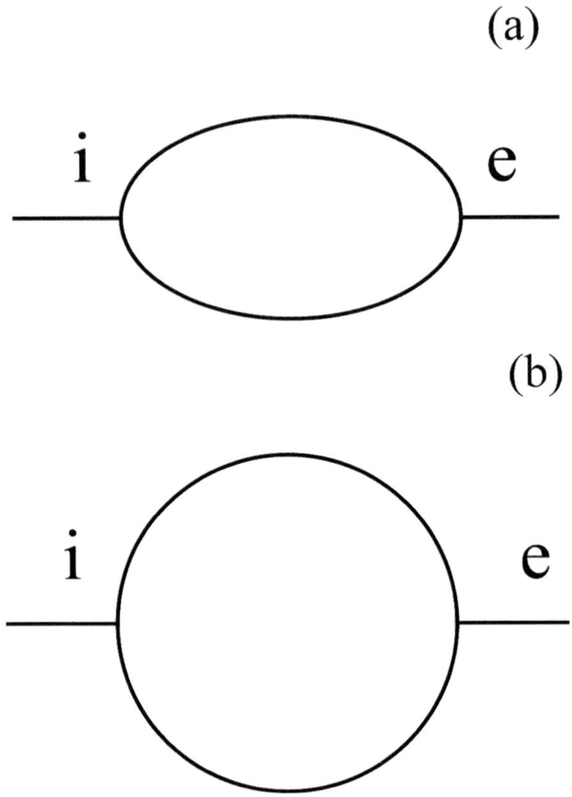

Figure 17.1 Two structures: (a) AB elliptical ring, (b) AB circular ring.

the input and output lead can be written as:

$$\Phi_i = \begin{pmatrix} a_i \\ b_i \end{pmatrix} e^{ik_0 l_i} + \begin{pmatrix} a_r \\ b_r \end{pmatrix} e^{-ik_0 l_i}, \tag{17.6}$$

$$\Phi_o = \begin{pmatrix} a_t \\ b_t \end{pmatrix} e^{ik_0 l_o}, \tag{17.7}$$

According to the transfer matrix method, we can obtain Ψ of each arm if we know its value at any one point, so there are 12 unknown coefficients in Ψ^j_{up}, Ψ^j_{down}, Φ_i, Φ_o. Using Griffith's boundary conditions [23–25], we can determine all unknown coefficients. We assume the original point of l_i and l_o are point A and point B, respectively, then when $a_i = 1$, $b_i = 0$, we have $a_r = r_{11}$, $b_r = r_{12}$,

$a_t = t_{11}, b_t = t_{12}$. And similarly, for $a_i = 0, b_i = 1$, we have $a_r = r_{21}$, $b_r = r_{22}$, $a_t = t_{21}$, $b_t = t_{22}$, where $\sigma = 1, 2$ denote spin up and down states that are quantized in z direction. The spin dependent transmission coefficient of an electron with incoming spin σ and outgoing spin σ' can be written as $T_{\sigma\sigma'} = |t_{\sigma\sigma'}|^2$.

For a spin-inverter, we expect that T_{12} and T_{21} are as large as possible, and for T_{11}, T_{22} to be the opposite, therefore the spin flip degree could be given as Ref [18]:

$$P = (T_{\sigma\sigma} - T_{\sigma\sigma'})/(T_{\sigma\sigma} + T_{\sigma\sigma'}), \tag{17.8}$$

where $\sigma \neq \sigma'$. For $P = 1$, it means that the conservation of the spin degree of freedom during the transition of the spin current. And $P = -1$ means the spin flip during the spin injection and transmission. Now there is a problem that for $\sigma = 1, 2$, P not always has same value, and for a spin-inverter, $T_{\sigma\sigma}$ and $T_{\sigma\sigma'}$ should be identical for spin up or spin down injection. We have calculated the spin transport in many configurations, and find out that $T_{12} = T_{21}$ is satisfied in any case, and $T_{11} = T_{22}$ is satisfied in the absence of the perpendicular magnetic field. Considering this fact, we can expect an AB ring as a spin-inverter just in the presence of RSOI.

17.2 Spin Flip in Quantum Rings

In previous work, we found that P is associated with $\overline{\alpha}$ only in a certain AB ring, which possesses the same behavior as the Datta spin filed effect transistor [26]. In this paper, we intent to compare spin flip in different AB rings, especially in the elliptical ring and in the circular ring.

In Fig. 17.2 we show the P as a function of the RSOI strength $\overline{\alpha}$ in the elliptical ring, and in the circular ring. The relevant parameters in our calculation are as follows: $\varepsilon = 5$, and $b_e/a_e = 0.6$ for the elliptical ring. We can see that the P in the elliptical ring oscillates between -1 and 1 when the $\overline{\alpha}$ changes. This means that as a spin-inverter, the elliptical ring is sensitive and can work in a reasonable range of $\overline{\alpha}$, i.e. $\overline{\alpha} < 2$. The P in the circular ring has a different expression because it decreases monotonously when the $\overline{\alpha}$ increases, and it has asymptotic value -1. Therefore, in a single

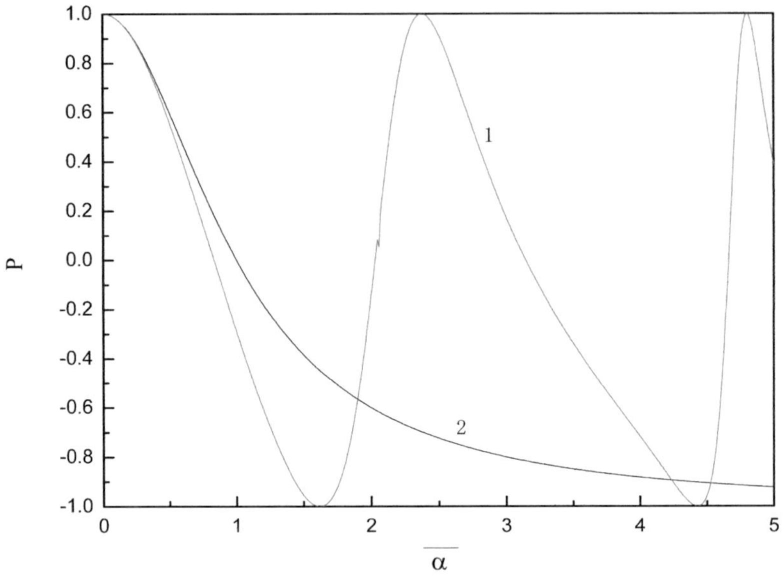

Figure 17.2 P as a function of the RSOI strength $\overline{\alpha}$: (1) in the elliptical ring, and (2) in the circular ring.

circular ring it is unrealistic to modulate the spin of the emergent electron completely. And only when $\overline{\alpha}$ is very large (> 5), the P can reach a value less than -0.9. For an AB ring with $a_e(/R) = 50$ nm, $m^*/m_e = 0.1$, $\alpha \simeq \overline{\alpha} * 7.62$ meV.nm, so to make the AB ring work as a spin-inverter well, we need a RSOI strength $\alpha > 40$ meV.nm in a circular ring, but in a elliptical ring we just need $\alpha \simeq 12$ meV.nm. On the other hand, if a large α can be reached, a spin-inverter based on a circular ring can work in a vast range of α, so it has a high stability.

To study the influence of the geometry on the spin transport ulteriorly, we compare P in elliptical rings with different eccentricities. The contour maps of P as a function of the RSOI strength $\overline{\alpha}$ and the semiminor axis b_e in a elliptical ring is shown in Fig. 17.3. Here the energy of the electron $\varepsilon = 5$. We find that if b_e/a_e is small (as $b_e/a_e < 30$), spin inversion can occur for some regions of $\overline{\alpha}$, and these regions are nearly same for all b_e, such as $\overline{\alpha} \simeq 1.62$. For these cases, P oscillates between 1 and -1 when the $\overline{\alpha}$ increases, but if b_e/a_e is close to 1, i.e. the elliptical ring is close to a circular

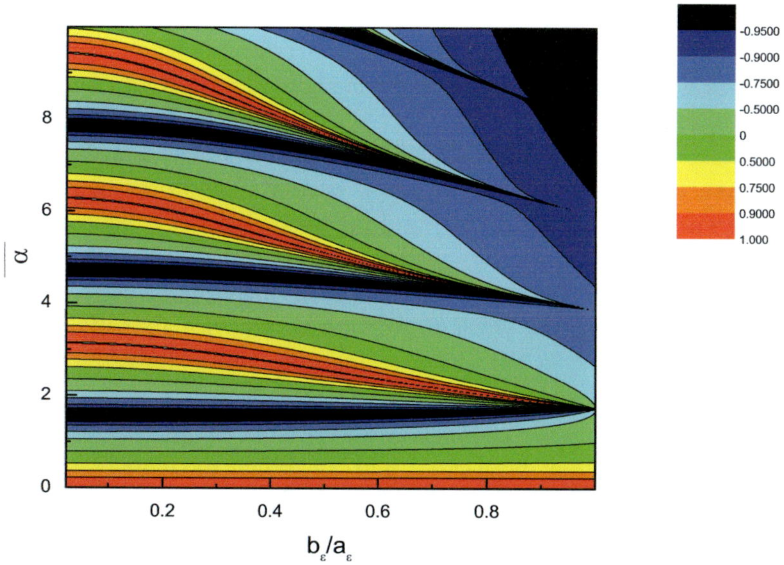

Figure 17.3 Contour map of P as a function of the Rashba strength $\overline{\alpha}$ and the ratio of the semiminor axis b_e and the semimajor axis a_e in the elliptical ring.

one, P decreases monotonically and have a asymptotic value -1 when the $\overline{\alpha}$ increases. As b_e/a_e increases, P as a function of the RSOI strength $\overline{\alpha}$ changes more and more gently. We infer that the increase of the transverse size of the ring weakens the quantum confinement effect, and then undermines the interference of two spin states in the longitudinal dimension.

To produce ideal ellipses is virtually impossible, and to produce an inscribed polygon of the ellipse is a concise and effective method. To consider this possible geometry, we calculated P in the inscribed polygon of the elliptical ring and the circular ring. We assume one focus of the elliptical ring and the the centre of the circular ring are poles of polar coordinates in each system, respectively. And every edge of the inscribed polygon have same polar angle. Because two arms of AB rings in this paper are symmetrical, so we can assume each arm of the ring is replaced by a M section polyline. Obviously the inscribed polygon of the circular ring will be a regular polygon of $2M$ sides.

The contour maps of P as a function of the RSOI strength $\overline{\alpha}$ and half of the edges' number of the inscribed polygon M in the elliptical ring and the circular ring is shown in Fig. 17.4a and Fig. 17.4b, respectively. Here the energy of the electron $\varepsilon = 5$, and $b_e/a_e = 0.2$ in the elliptical ring. In Fig. 17.4a, we find that the influence of the edges' number on P is very small in the inscribed polygon of the elliptical ring, and for every inscribed polygon, P has nearly same curve followed the RSOI strength $\overline{\alpha}$. This result shows that an AB ring similar to a elliptical ring in shape can replace the elliptical ring, and work as a spin-inverter with a small α. The result is different in the circular ring. Fig. 17.4b shows that the curve of P as $\overline{\alpha}$ changes has changed greatly when M increases. If M is small, such as the inscribed polygon is square, hexagon, or octagon, P oscillates between 1 and -1. When the $\overline{\alpha}$ increases, P can reach -1 at a low value of $\overline{\alpha}$ (about 2~3). If M is relatively large, such as $M > 8$, P decreases monotonically and have a asymptotic value -1 when the $\overline{\alpha}$ increases, and to reduce P to the same value, the required value of $\overline{\alpha}$ increases when M increases. This result is similar to the result in Fig. 17.3, and shows that as the limit case of elliptical rings and of inscribed polygons, the circular ring has much different spin transport character from elliptical rings and inscribed polygons. In addition, Fig. 17.4 shows that there is a limit result of spin transport in the inscribed polygon of the elliptical ring and the circular ring when M becomes very large, so our method in which a curve is divided into many segments is self-consistent.

The elliptical ring or the regular polygon with a small number of edges(such as a regular hexagon) can work as a spin-inverter with the normalized Rashba constant about 2 which corresponds to RSOI strength $\alpha \simeq 16$ meV.nm. This value is in the range of that obtained experimentally in InGaAs [27, 28], so these spin-inverters are realizable. A spin-inverter of the circular ring requires $\alpha > 40$meV.nm, which is hard to achieve in traditional III-V semiconductor. However, recently giant RSOI with $\alpha > 400$meV.nm has been obtained in the bulk Rashba semiconductor BiTeI [29]. Although the requirement is harsh, the value of RSOI strength required in the circular ring will be possible to reach in more common situation in the near future.

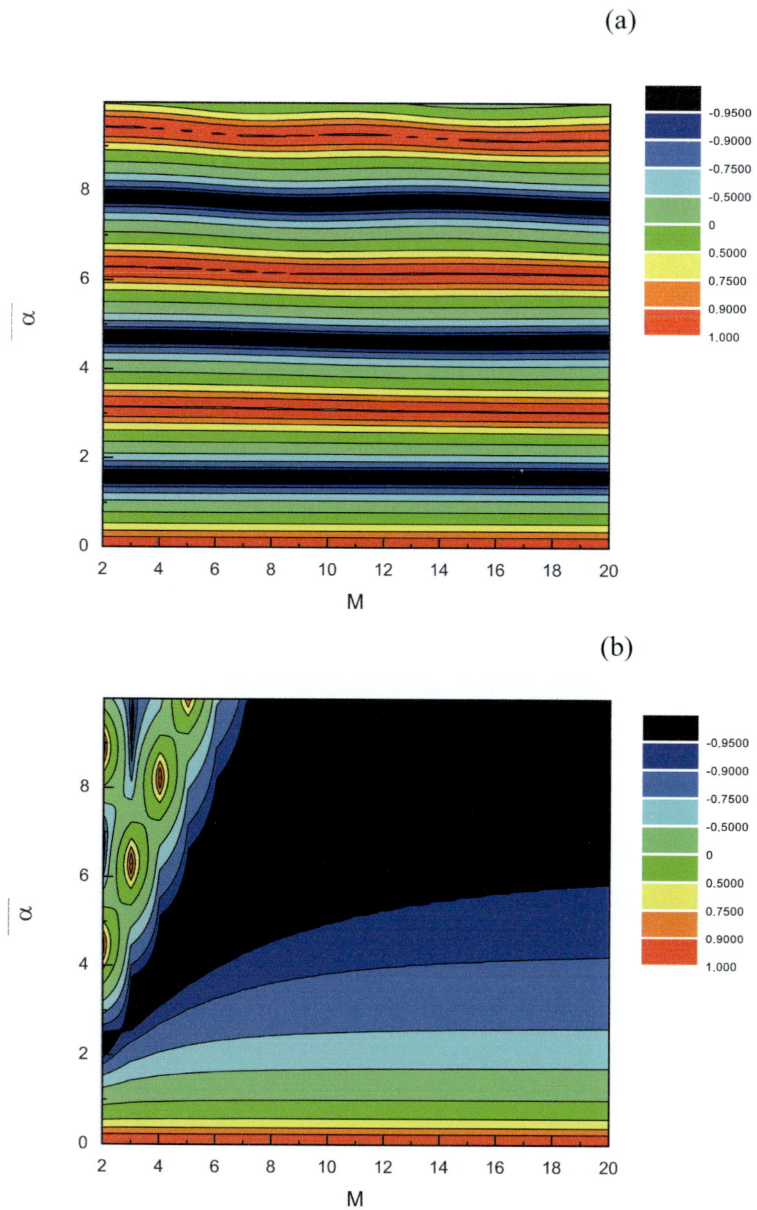

Figure 17.4 Contour map of P as a function of the Rashba strength $\overline{\alpha}$ and half of the edges' number of the inscribed polygon M: (a) in the AB elliptical ring, (b) in the circular ring.

17.3 Summary

We have studied spin transport, especially spin flip in single quantum ring with RSOI to find suitable geometry and other conditions for spin-inverter. The elliptical ring and the circular ring are studied as typical AB rings. We found that if these two kinds of ring work as spin-inverters, they have much different character. The elliptical ring can realize spin flip with a relatively small RSOI strength α, but the spin flip degree P will oscillate quickly and the spin-inverter can only work in a small range of α. The circular ring can not realize spin flip perfectly, but P in circular ring can be very close to -1 with a relatively large α, and it is insensitive to α at this time. In addition, we have investigated the influence of the defect of the AB ring's geometry on the spin transport. Results shows that in an elliptical ring spin transport is insensitive to the defect of the ring, but in a circular ring, the defect should be small to give P closed to that of the ideal circular ring. These results show that if RSOI strength is relatively small and can be controlled well, the elliptical ring is a good choice for a spin-inverter, and we just need a approximate ellipse. Another conclusion is that if large RSOI strength is realized, a circular ring can work as a spin-inverter with a high stability.

Acknowledgments

This work was supported by The National Natural Science Foundation of China No. 11504016.

References

1. E. I. Rashba. 1960 *Sov. Phys. Solid. State* **2**, 1109.
2. T. P. Pareek. 2004 *Phys. Rev. Lett.* **92**, 076601.
3. P. Foldi, O. Kalman, M. G. Benedict and F. M. Peeters. 2006 *Phys. Rev. B* **73**, 155325.

4. M. Busl and G. Platero. 2010 *Phys. Rev. B* **82**, 205304.
5. J. L. Cardoso and P. Pereyra. 2008 *EPL* **83** 38001.
6. T. Koga, J. Nitta and M. Veenhuizen. 2004 *Phys. Rev. B* **70**, 161302(R).
7. T. Koga, Y. Sekine and J. Nitta. 2006 *Phys. Rev. B* **74**, 041302(R).
8. H. Simchi H, M. Esmaeilzadeh and H. Mazidabadi. 2013 *Physica E* **54**, 220.
9. A. Dyrdał, M. Inglot, V. K. Dugaev and J. Barnas. 2013 *Phys. Rev. B* **87**, 245309.
10. Y. Ban and E. Y. Sherman. 2013 *J. Appl. Phys.* **113**, 043716.
11. F. Fallah and M. Esmaeilzadeh. 2011 *AIP Adv.* **1**, 032113.
12. J. Splettstoesser, M. Governale, and U. Zulicke. 2003 *Phys. Rev. B* **68**, 165341.
13. J. S. Sheng and K. Chang. 2006 *Phys. Rev. B* **74**, 235315.
14. B. Molnar, P. Vasilopoulos and F. M. Peeters. 2005 *Phys. Rev. B* **72**, 075330.
15. M. Nita, D. C. Marinescu, A. Manolescu, B. Ostahie and V. Gudmundsson. 2012 *Physica E* **46**, 12.
16. L. Eslami, M. Esmaeilzadeh and E. Namvar. 2012 *Phys. Lett. A*, **376**, 2141.
17. V. Moldoveanu and B. Tanatar. 2010 *Phys. Rev. B* **81**, 035326.
18. A. S. Naeimi A S, L. Eslami, M. Esmaeilzadeh M and M. R. Abolhassani. 2013 *J. Appl. Phys.* **113**, 014303.
19. F. E. Meijer, A. F. Morpurgo and T. M. Klapwijk. 2002 Phys. Rev. B **66**, 033107.
20. H. Mathur and A. D. Stone. 1992 *Phys. Rev. Lett.* **68**, 2964.
21. B. Molnar, F. M. Peeters and P. Vasilopoulos. 2004 *Phys. Rev. B* **69**, 155335.
22. D.-Y. Liu, J.-B. Xia and Y.-C. Chang. 2009 *J. Appl. Phys.* **106**, 093705.
23. D. Y. Liu, J.-B. Xia and Y.-C. Chang 2010 *Sci. China. Ser G* **53**, 16.
24. S. Griffith. 1953 *Trans. Faraday Soc.* **49**, 345.
25. J. B. Xia. 1992 *Phys. Rev. B* **45**, 3593.
26. S. Datta and B. Das. 1990 *Appl. Phys. Lett.* **58**, 665.
27. J. Nitta, T. Akazaki, H. Takayanagi, and T. Enoki. 1997 *Phys. Rev. Lett* **78**, 1335.
28. D. Grundler. 2000 *Phys. Rev. Lett.* **84**, 6074.
29. H. Murakawa, M. S. Bahramy, M. Tokunaga, Y. Kohama, C. Bell, Y. Kaneko, N. Nagaosa, H. Y. Hwang and Y. Tokura. 2013 *Science* **342**, 1490.

Index

For Product Safety Concerns and Information please contact our EU representative GPSR@taylorandfrancis.com Taylor & Francis Verlag GmbH, Kaufingerstraße 24, 80331 München, Germany